Introduction to Analysis of the Infinite
Book I

Euler

Introduction to Analysis of the Infinite

Book I

Translated by John D. Blanton

Springer-Verlag
New York Berlin Heidelberg
London Paris Tokyo

Translator

John D. Blanton
Department of Mathematics
St. John Fisher College
Rochester, NY 14618
USA

Mathematical Subject Classification: 01-XX

Library of Congress Cataloging-in-Publication Data
Euler, Leonard, 1707–1783.
Introduction to analysis of the infinite.
Translation of: Introductio in analysin infinitorum.
1. Series, Infinite—Early works to 1800.
2. Products, Infinite—Early works to 1800.
3. Fractions, Continued—Early works to 1800. I. Title.
QA295.E8413 1988 515′.243 88-18475

Printed on acid-free paper

Camera-ready copy supplied by the translator.
Printed and bound by R.R. Donnelley & Sons, Harrisonburg, Virginia.
Printed in the United States of America.

9 8 7 6 5 4 3 2 1

ISBN 0-387-96824-5 Springer-Verlag New York Berlin Heidelberg
ISBN 3-540-96824-5 Springer-Verlag Berlin Heidelberg New York

PREFACE

Often I have considered the fact that most of the difficulties which block the progress of students trying to learn analysis stem from this: that although they understand little of ordinary algebra, still they attempt this more subtle art. From this it follows not only that they remain on the fringes, but in addition they entertain strange ideas about the concept of the infinite, which they must try to use. Although analysis does not require an exhaustive knowledge of algebra, even of all the algebraic techniques so far discovered, still there are topics whose consideration prepares a student for a deeper understanding. However, in the ordinary treatise on the elements of algebra, these topics are either completely omitted or are treated carelessly. For this reason, I am certain that the material I have gathered in this book is quite sufficient to remedy that defect.

I have striven to develop more adequately and clearly than is the usual case those things which are absolutely required for analysis. Moreover, I have also unraveled quite a few knotty problems so that the reader gradually and almost imperceptibly becomes acquainted with the idea of the infinite.

There are also many questions which are answered in this work by means of ordinary algebra, although they are usually discussed with the aid of analysis. In this way the interrelationship between the two methods becomes clear.

I have divided this work into two books; in the first of these I have confined myself to those matters concerning pure analysis. In the second book I have explained those things which must be known from geometry, since analysis is ordinarily developed in such a way that its application to geometry is shown. In both parts, however, I have omitted the elementary matters and developed only those things which, in other places, are either completely omitted or only cursorily treated or, finally, follow from new arguments.

Thus, in the first book, since all of analysis is concerned with variable quantities and functions of such variables, I have given a full treatment to functions. I have also treated the transformation of functions and functions as the sum of infinite series. In addition I have developed functions in infinite series.

Many kinds of functions whose characteristic qualities are discovered by higher analysis are classified. First I have distinguished between algebraic and transcendental functions: the former are formed from the ordinary algebraic operations on variable quantities; the latter arise from other procedures or from the infinite repetition of algebraic operations.

The primary subdivision of algebraic functions is into that of non-irrational and irrational. I have shown how the former can not only be simplified, but also factored, and this is very useful in integral calculus. It has been shown to what extent irrational functions can be brought to non-irrational form by means of suitable substitutions. Both types can be developed in infinite series, but this method is usually applied with the greatest usefulness to transcendental functions. It is clear that the theory of infinite series has greatly extended higher analysis. Several chapters have been included in which I have examined the properties and summation of many infinite series; some of these are arranged in such a way that it can be seen that they could hardly be investigated without the aid of analysis. Series of this type are those whose summations are expressed either through logarithms or circular arcs. However, since these are transcendental quantities which can be defined by quadratures of the hyperbola and the circle, for the most part they are usually treated in analysis. After that I shall have progressed from powers of quantities to exponential quantities, which are simply powers whose exponents are variables. From the inverse of these I have arrived at the most natural and fruitful concept of logarithms. Whence not only are very ample uses of these immediately obtained, but also from them it is possible to obtain all those infinite series by which ordinarily these quantities are represented. Then there is produced a method of reasonably simple construction of tables of logarithms. In a like manner I have turned my attention to circular arcs. This type of quantity, although quite different from logarithms, nevertheless, there is such a close mutual relationship that when the latter is viewed as a complex quantity, it is converted into the former. Just as logarithms have their own particular algorithm, which has most useful applications in all of analysis, I have derived algorithms for the trigonometric quantities, so that these calculations can be made as easily as for the logarithmic and algebraic quantities. The extent of the usefulness of this for the solution of very difficult problems becomes clear in several chapters of this book. Indeed, very many other examples from analysis could be offered were they not sufficiently known already, and in fact more are being found almost daily. But this investigation brings the greatest help to the resolution of rational functions into real factors. Since this is so important for integral calculus, I have given this diligent attention. I have investigated those infinite series which arise from the development of this type of function, and are known as recurrent series. For these I have given both summations and general terms and also other important properties. Since the resolution into factors has led to these series, so in turn, I have pondered to what extent the product of several factors, and even infinite products, can be expressed in a series. This business opened the way to knowledge of a myriad of series. Since a series can be expressed as an infinite product, I have found rather convenient numerical expressions with the aid of which the logarithms of sines, cosines, and tangents can easily be computed. Furthermore, from this same source we can derive the solutions of many problems which are concerned with the partition of numbers. Questions of this sort would seem to defeat analysis without this help. Such a diversity of material might easily have grown into several volumes, but I have, as far as possible, expressed everything so succinctly that everywhere the foundation is very clearly explained. The further development is left to the industry of the readers. In this way they will have an opportunity to try their own ingenuity and further develop analysis itself. Nor do I hesitate to proclaim that within this book there are contained many things which are clearly new, but also some sources have been uncovered from which many significant further discoveries can be drawn.

I have used the same arrangement in the second volume, where I have treated those topics which are commonly called higher geometry. Before I discuss conic sections, which in other treatments almost always come first, I have proposed a theory of curves with enough generality that it can advantageously be applied to an examination of the nature of any curve whatsoever. I use only an equation by which the nature of every curve is expressed, and I show how to derive from this both the shape and its primary characteristics. It has seemed to me that this is most especially advantageous in the case of conic sections. Until this time they have ordinarily been treated only from the geometric viewpoint, or if by analysis, in an awkward and unnatural way. I have first explained their general properties from the general equation for a second order curve. Then I have subdivided them into genera and species, considering whether they have branches going to infinity or the whole curve is included in a bounded region. In the former case something else has to be considered, that is, how many branches extend to infinity and of what nature each of these is; namely, whether or not the branch has a straight line asymptote. In this way I have obtained the customary three types of conic section. The first is the ellipse, totally contained in a bounded region; the second is the hyperbola, which has four infinite branches asymptotic to two straight lines; the third is the parabola with two infinite branches without asymptotes. In a like manner I have described third order curves, which I have divided into sixteen kinds, after considering their general properties. Indeed, to these kinds I have reduced the seventy-two of Newton's classification. I have described this method with such clarity that for curves of any higher order whatsoever, a classification may easily be made. I have made a test of the method in the case of fourth order curves. Having explained whatever concerns the order of a curve, I returned to uncovering the general properties of all curves. Thus I have explained a method for defining tangents to curves, their normals, and curvature, which is ordinarily measured by the radius of the osculating circle. Although all of these nowadays are ordinarily accomplished by means of differential calculus, nevertheless, I have here presented them using only ordinary algebra, in order that the transition from finite analysis to analysis of the infinite might be rendered easier. I have also investigated points of inflection of curves, cusps, double points, and multiple points. The definitions of these kinds of points follow without difficulty from the equations. At the same time I readily admit that these matters can be much more easily worked out by differential calculus. I have touched on the controversy over a second order cusp, where both arcs which come together in the cusp curve in the same direction. It seems to me that I have settled this question in such a way that there can remain no doubt. Finally, I have adjoined several chapters in which I have explained how to find curves with certain stated properties. I give the solution to a number of problems concerned with circles. Since there are some topic from geometry which seem to offer strong support for learning analysis, I have added an appendix in which I have presented, using calculus, the theory of solids and the surfaces of solids. I have shown, insofar as it is possible to do so through equations in three variables, the nature of these surfaces. Then having divided surfaces into orders, as was done in the case of curves, according to the power of the variables in the equation, I have shown that the only surface of the first order is the plane. I have divided surfaces of the second order, by considering their parts which extend to infinity, into six types. In a similar way a division can be made for surfaces of other orders. I have also considered the intersections of two surfaces. These, insofar as they can be understood through equations, I have shown generally to be curves which do not lie in a single plane.

For the rest, since many things here will be met which have already been treated by others I ought to ask pardon, since I shall not have given credit in every place to all who have toiled herein. I have endeavored to develop everything as briefly as possible. If the history of each problem had been discussed, that would have increased the size of this work beyond reasonable bounds. Many of these problems, whose solutions can be found elsewhere, in this work have solutions which arise from different arguments. For this reason I would seem in no little part to be exonerated. I do hope that both these things, but also especially those which are entirely new, will be acceptable to most of those who enjoy this work.

TRANSLATOR'S INTRODUCTION

In October, 1979, Professor Andre Weil spoke at the University of Rochester on the Life and Works of Leonard Euler. One of his remarks was to the effect that he was trying to convince the mathematical community that our students of mathematics would profit much more from a study of Euler's *Introductio in Analysin Infinitorum,* rather than of the available modern textbooks. I found that this work had been translated into French, German, and Russian; I could find no trace of a translation of the complete work into English. This was the impetus for the present translation.

It is clear from Euler's preface that the work is strictly pre-calculus; there are several places within the text where he remarks that certain topics are better left for a treatment by differential or integral calculus.

I would add a remark on the significance of the plural *Infinitorum* in the title. A literal translation of the title might be *An Introduction to the Analysis of the Infinities,* which seems a bit awkward. I take the plural to refer to the three main topics: infinite series, infinite products, and continued fractions.

In the translation I have taken some liberties with regard to notation and terminology for the sake of ease in reading. When this work was originally published in 1748, Euler had not yet adopted the notation i for $\sqrt{-1}$, but I have made this substitution. With regard to terminology, Euler called any function "rational" provided the variable was not involved in an irrationality. In order to conform to the modern understanding of rational function I have translated

Euler's "rational" by the rather awkward "non- irrational ".

On the other hand, Euler's reference to "infinitely large" and "infinitely small" numbers has been preserved in the translation, since any attempt which avoided this would no longer be a work of Euler's. I have also preserved in translation Euler's terminology for "bifid" and "trifid" functions, due to the lack of a better alternative, as well as the fact that these are genuine English words.

I would like to express thanks to St. John Fisher College for granting me a sabbatical leave in order to complete this work. I would also express gratitude for the encouragement of the late Walter Kaufmann-Buehler, mathematics editor of Springer-Verlag. Finally thanks are due for the patience of my children Jack, Paul, Drew, and Anne and for the love of my wife Claire, who typed and retyped the whole manuscript.

One final remark on this work of Euler's is due to Professor Alberto Dou, S.J., of the University of Barcelona, who has translated a number of the works of Euler into Spanish. In a recent conversation with him about this translation, he said that of Euler's works, "This is the very best."

CONTENTS

INTRODUCTION
TO
ANALYSIS
OF THE
INFINITE

BOOK I

Containing

an explanation of functions of variable quantities; the resolution of functions into factors and their development in infinite series; together with the theory of logarithms, circular arcs, and their sines and tangents, also many other things which are no little aid in the study of analysis.

CHAPTER I

On Functions in General

1. *A constant quantity is a determined quantity which always keeps the same value.*

Quantities of this type are numbers of any sort which keep the same constant value, once they have been assigned; if it is required to represent constant quantities by a symbol, the initial letters of the alphabet are used a, b, c, etc. In algebra, where only fixed quantities are considered, these first letters of the alphabet usually denote known quantities, while the final letters represent unknown quantities, but in analysis this distinction is not so much used, since here it is more a question of considering the former as constants and the latter as variables.

2. *A variable quantity is one which is not determined or is universal, which can take on any value.*

Since all determined values can be expressed as numbers, a variable quantity takes on all possible numbers (all numbers of all types). Just as from the ideas of individuals the ideas of species and genus are formed, so a variable quantity is a genus in which are contained all determined quantities. Variable quantities of this kind are usually represented by the final letters of the alphabet z, y, x, etc.

3. *A variable quantity is determined when some definite value is assigned to it.*

Hence a variable quantity can be determined in infinitely many ways, since absolutely all numbers can be substituted for it. Nor is the symbol of the variable quantity exhausted until all definite numbers have been assigned to it. Thus a variable quantity encompasses within itself absolutely all numbers, both positive and negative, integers and rationals, irrationals and transcendentals. Even zero and complex numbers are not excluded from the signification of a variable quantity.

4. *A function of a variable quantity is an analytic expression composed in any way whatsoever of the variable quantity and numbers or constant quantities.*

Hence every analytic expression, in which all component quantities except the variable z are constants, will be a function of that z; thus $a + 3z$; $az - 4z^2$; $az + b\sqrt{a^2 - z^2}$; c^z; etc. are functions of z.

5. *Hence a function itself of a variable quantity will be a variable quantity.*

Since it is permitted to substitute all determined values for the variable quantity, the function takes on innumerable determined values; nor is any determined value excluded from those which the function can take, since the variable quantity includes complex values. Thus, although the function $\sqrt{9 - z^2}$, with real numbers substituted for z, never attains a value greater than 3, nevertheless, by giving z complex values, for instance $5i$, there is no determined value which cannot be obtained from the formula $\sqrt{9 - z^2}$. There do occur, however, some apparent functions which retain the same value no matter in what way the variable quantity is changed, for example z^0; 1^z; $\frac{a^2 - az}{a - z}$ which assume the appearance of functions, but really are constant quantities.

6. *The principal distinction between functions, as to the method of combining the variable quantity and the constant quantities is here set down.*

Indeed, it depends on the operations by which the quantities can be arranged and mixed together. These operations are addition, subtraction, multiplication, division, raising to a power, and extraction of roots. Also the solution of equations have to be considered. Besides there operations, which are usually called algebraic, there are many others which are transcendental, such as exponentials, logarithms, and others which integral calculus supplies in abundance.

In the meantime certain kinds of functions can be noted, such as multiples $2z$, $3z$, $\frac{3}{5}z$, az, etc. and powers of z itself as z^2, z^3, $z^{\frac{1}{2}}$, z^{-1}, etc. which, as they arise from a single operation, so, expressions which come from any type of operation are distinguished by the name of function.

7. *Functions are divided into algebraic and transcendental. The former are those made up from only algebraic operations, the latter are those which involve transcendental operations.*

Thus multiples and powers of z are algebraic functions; also absolutely all expressions which are formed by the algebraic operations previously recalled, such as $\dfrac{a + bz^n - c\sqrt{2z - z^2}}{a^2z - 3bz^3}$. Indeed frequently algebraic functions cannot be expressed explicitly. For example, consider the function Z of z defined by the equation, $Z^5 = az^2Z^3 - bz^4Z^2 + cz^3Z - 1$. Even if this equation cannot be solved, still it remains true that Z is equal to some expression composed of the variable z and constants, and for this reason Z shall be a function of z. Something else about transcendental functions should be noted, and this is the fact that the function will be transcendental only if the transcendental operation not only enters in, but actually affects the variable quantity. If the transcendental operations pertain only to the constants, the function is to be considered algebraic. For instance, if c denotes the circumference of a circle with radius equal to 1, c will be a transcendental quantity, nevertheless, these expressions: $c + z$, cz^2, $4z^c$, etc. are algebraic functions of z. The doubt raised by some as to whether such expressions as z^c are correctly classified as algebraic is of little importance. Indeed some people prefer to call powers of z, in which the exponents are irrational such as $z^{\sqrt{2}}$, intercendental functions rather than algebraic.

8. *Algebraic functions are subdivided into non-irrational and irrational functions: the former are such that the variable quantity is in no way involved with irrationality; the latter are those in which the variable quantity is affected by radical signs.*

Thus in non-irrational functions there are no other operations besides addition, subtraction, multiplication, division, and raising to powers in which the exponents are integers: such, for example, as $a + z$; $a - z$; az; $\dfrac{a^2 + z^2}{a + z}$;

$az^3 - bz^5$; etc. are non-irrational functions of z. Such expressions as $\sqrt{z}$; $a + \sqrt{a^2 - z^2}$; $(a - 2z + z^2)^{\frac{1}{3}}$; $\dfrac{a^2 - z\sqrt{a^2 + z^2}}{a + z}$ are irrational functions of z.

It is convenient to distinguish these into explicit and implicit irrational functions.

The explicit functions are those which are expressed with radical signs, as in the given examples. The implicit are those irrational functions which arise from the solution of equations. Thus Z is an implicit irrational function of z if it is defined by an equation such as $Z^7 = az$ or $Z^2 = bz^5$. Indeed, an explicit value of Z may not be expressed even with radical signs, since common algebra has not yet developed to such a degree of perfection.

9. *Non-irrational functions are subdivided into polynomial and rational functions.*

In a polynomial function the variable z has no negative exponents whatsoever, nor does it contain fractional expressions in which the variable z enters into the denominators. Rational functions are understood to be such that the denominators contain z or those in which negative exponents of z occur. This then is the general formula for polynomial functions:

$$a + bz + cz^2 + dz^3 + ez^4 + fz^5 + \cdots$$

etc. It is impossible to think up a polynomial function which is not included in this expression. All rational functions, however, since the sum of several fractions can be expressed as a single fraction, are contained in the following formula:

$$\frac{a + bz + cz^2 + dz^3 + ez^4 + fz^5 + \cdots}{\alpha + \beta z + \gamma z^2 + \delta z^3 + \epsilon z^4 + \zeta z^5 + \cdots}$$

where it should be noted that the constant quantities a, b, c, d, etc. and α, β, γ, δ, etc., whether they be positive or negative, integers or fractions, rational or irrational, or even transcendental, do not change the nature of the functions.

10. *Finally, we must make a distinction between single-valued and multiple-valued functions.*

A single-valued function is one for which, no matter what value is assigned to the variable z, a single value of the function is determined. On the other hand, a multiple-valued function is one such that, for some value substituted for the variable z, the function determines several values. Hence, all non-irrational functions, whether polynomial or rational, are single-valued functions, since expressions of this kind, whatever value be given to the variable z, produce a single-value. However, irrational functions are all multiple-valued, because the radical signs are ambiguous and give paired values. There are also among the transcendental functions, both single-valued and multiple-valued functions; indeed, there are infinite-valued functions. Among these are the arcsine of z, since there are infinitely many circular arcs with the same sine. We shall use letters P, Q, R, S, T, etc. for the individual single-valued functions of z.

11. *A two-valued function of z is one which gives a pair of values for each determined value of z.*

Functions of this kind display a square root, such as $\sqrt{2z+z^2}$. Whatever value is assigned to z, the expression $\sqrt{2z+z^2}$ has a twofold signification, either positive or negative. In general, Z will be a two-valued function of z if it is determined by a quadratic equation $Z^2 - PZ + Q = 0$, provided P and Q are single-valued functions of z. In this case, $Z = \frac{1}{2}P \pm \sqrt{(1/4)P^2 - Q}$, from which it is obvious that for each definite value of z, there correspond a pair of determined values of Z. However, it should be noted that the values of the function Z are both either real or both are complex. Further, as is clear from the nature of the equation, the sum of both values of Z is always equal to P and the product is equal to Q.

12. *A three-valued function of z is one which gives three determined values for any value of z.*

Functions of this kind originate from the solution of cubic equations; if P, Q, and R are single-valued functions, and $Z^3 - PZ^2 + QZ - R = 0$, then Z is a three-valued function of z, since for an arbitrary definite value of z, three values are obtained. These three values of Z, corresponding to each value of z, are either all real or one is real and the remaining two are complex. Further, it is clear that the sum of the three values is always equal to P, that the sum of the products taken in pairs is equal to Q, and that the product of all three is equal to R.

13. *A four-valued function of z is one which gives four determined values for any value of z.*

Functions of this kind come from the solution of biquadratic equations; if P, Q, R, and S stand for single-valued functions of z, and $Z^4 - PZ^3 + QZ^2 - RZ + S = 0$, then Z is a four-valued function of z, since to each value of z there correspond four values of Z. Hence these four values are either all real, or two are real and two complex, or all four are complex. Further, the sum of these four values is always equal to P; the sum of the products of all pairs of values is equal to Q; the sum of the products taken three at a time is equal to R, and the product of all four is equal to S. In an analogous manner the nature of five-valued functions and the subsequent multiple-valued functions can be seen.

14. *Thus Z is a multiple-valued function of z which for each value of z, exhibits n values of Z where n is a positive integer. If Z is defined by this equation* $Z^n - PZ^{n-1} + QZ^{n-2} - RZ^{n-3} + SZ^{n-4} - \cdots = 0.$

Here it should especially be noted that n must be an integer, and in order that the degree of multiple-valuedness of Z as a function of z be decided, the equation by which Z is defined must always be reduced to rationality: when this is done, the largest power of Z will indicate the number of values corresponding to each value of z. Further, it should be kept in mind that the letters P, Q, R, S, etc. should denote single-valued function of z. If any of them is already a multiple-valued function, then the function Z will have many more values, corresponding to each value of z, than the exponent of Z would indicate. It is always true that if some of the values are complex, then there will be an even number of them. From this we know that if n is an odd number, there will always be at least one real value of Z. On the other hand, it is possible, if n is even, that no value of Z is real.

15. *If Z is a multiple-valued function of z such that it always exhibits a single real value, then Z imitates a single-valued function of z, and frequently can take the place of a single-valued function.*

Functions of this kind are $P^{\frac{1}{3}}$, $P^{\frac{1}{5}}$, $P^{\frac{1}{7}}$, etc. which indeed give only one real value, the others all being complex, provided P is a single-valued function of z. For this reason, an expression of the form $P^{\frac{m}{n}}$, whenever n is odd, can be counted as a single-valued function, whether m is odd or even. However, if n is even then $P^{\frac{m}{n}}$ will have either no real value or two; for this reason, expressions of the form $P^{\frac{m}{n}}$, with n even, can be considered to be two- valued functions, provided the fraction $\frac{m}{n}$ cannot be reduced to lower terms.

16. *If y is any kind of function of z, then likewise, z will be a function of y.*

Since y is a function of z, whether single-valued or multiple-valued there is given an equation by which y is defined through z and constant quantities. From the same equation z can be defined through y and constants. Since y is a variable quantity, z will be equal to an expression composed of y and constants, and for this reason it will be a function of y. From this also it is clear to what extent z will be a multiple-valued function of y. Thus if y is defined through z by the following equation $y^3 = ayz - bz^2$, y is a three- valued function of z, while z is a two-valued function of y.

17. *If y and x are both functions of z, then y is also a function of x, and x in turn is also a function of y.*

Since y is a function of z, z is also a function of y; in like manner, z is also a function of x. Hence a function of y is equal to a function of x. From this equation both y through x and, vice versa, x can be defined through y. Now it is clear that y is a function of x and x is a function y. Frequently these functions are not explicit due to defects in algebraic techniques; nevertheless, if the reciprocity of the functions is considered attentively, almost all equations can be solved. For the rest, through the traditional algebraic methods, from two given equations, of which one contains y and z, the other x and z, through the elimination of z, one equation is formed which expresses the relationship between x and y.

18. *Next we should note some special kinds of functions: thus an even function of z is one which gives the same value whether $+ k$ or $- k$ is assigned to z.*

An even function of z is z^2; whether we assign $z = k$ or $z = -k$, the expression z^2 gives the same value, namely $z^2 = +k^2$. In like manner, other even functions of z are these powers of z: z^4, z^6, z^8, and in general, all powers z^m, provided m is an even integer, whether positive or negative. Indeed, since $z^{\frac{m}{n}}$ imitates a single-valued function of z, if n is odd, it is clear that $z^{\frac{m}{n}}$ is an even function of z if m is even and n is odd. For the same reason any expression composed in any way from such powers, gives an even function; thus Z is an even function of z if $Z = a + bz^2 + cz^4 + dz^6 + \cdots$, also if $Z = \dfrac{a + bz^2 + cz^4 + dz^6 + \cdots}{\alpha + \beta z^2 + \gamma z^4 + \delta z^6 + \cdots}$. In like manner, by introducing fractional exponents of z, Z is an even function of z if $Z = a + bz^{\frac{2}{3}} + cz^{\frac{4}{7}} + \cdots$, or if

$$Z = a + bz^{-\frac{2}{3}} + cz^{-\frac{4}{3}} + dz^{-\frac{2}{5}} + \cdots \text{, or if}$$

$$Z = \frac{a + bz^{\frac{2}{7}} + cz^{-\frac{4}{5}} + dz^{\frac{8}{3}} + \cdots}{\alpha + \beta z^{\frac{2}{3}} + \gamma z^{-\frac{2}{5}} + \delta z^{\frac{4}{7}} + \cdots}.$$ Expressions of this kind, since all are single-valued functions of z, can be called single-valued even functions of z.

19. *A multiple-valued even function of z is one which, even if for each value of z it gives several determined values, it still gives the same value whether $z = k$ or $z = -k$.*

Let Z be a multiple-valued even function of z. Indeed, the nature of a multiple-valued function is expressed through an equation in Z and z in which Z has an exponent equal to the number of values involved. It is clear that Z will be a multiple-valued even function if, in the equation expressing the nature of Z, the variable z everywhere has an even exponent. Thus if $Z^2 = azZ^4 + bz^2$, then Z is a two-valued function of z; but if $Z^3 - az^2Z^2 + bz^4Z - cz^8 = 0$ then Z is a three-valued even function of z. In general, if P, Q, R, S, etc. denote single-valued even functions of z, then Z will be a two-valued even function of z if $Z^2 - PZ + Q = 0$. But Z will be a three-valued even function of z if $Z^3 - PZ^2 + QZ - R = 0$, and so forth.

20. *Hence an even function of z, whether it is single-valued or multiple-valued, will be an expression consisting of constants and the variable z in which everywhere the exponent of z is even.*

Thus, functions of this kind, besides the single-valued functions, examples of which were given above, are expressions such as $a + \sqrt{b^2 - z^2}$; $az^2 + (a^6z^4 - bz^2)^{\frac{1}{3}}$; and $az^{\frac{2}{3}} + \left(z^2 + \sqrt{a^4 - z^4}\right)^{\frac{1}{3}}$; etc.

Hence it is clear that even functions can thus be defined as functions of z^2.

For if $y = z^2$ and Z is any function of y, then replacing z^2 for y, Z then becomes a function of z in which z everywhere has an even exponent. We must exclude those cases in which $\sqrt{y}$ occurs in the expression of Z, and likewise any other forms in which upon substituting z^2 for y, radical signs are eliminated.

Although $y + \sqrt{ay}$ is a function of y, still when we make the substitution $y = z^2$, the same expression is not an even function of z, since $y + \sqrt{ay}$ becomes $z^2 + z\sqrt{a}$. Provided we make these exceptions, this final definitions of even functions is good, and quite suitable for the formulation of such functions

21. *An odd function of z is a function whose value changes sign when - z is substituted for z.*

Thus, all odd powers of z are odd functions of z, such as z^1, z^3, z^5, z^7, etc. Likewise, z^{-1}, z^{-3}, z^{-5}, etc. Then also $z^{\frac{m}{n}}$ is an odd function if m and n are both odd integers. In general, every expression composed of terms of this kind will be an odd function. The following are of this type; $az + bz^3$, $az + az^{-1}$, $z^{\frac{1}{3}} + az^{\frac{3}{5}} + bz^{-\frac{5}{3}}$, etc. However, the discovery of functions of this kind and their nature is more easily seen from a consideration of even functions.

22. *If an even function of z is multiplied by z or by any odd function of z, the product is an odd function of z.*

Let P be an even function of z, so that the value of P remains the same when $-z$ is substituted for z. Then, if in the product Pz, $-z$ is substituted for z, the product becomes $-Pz$, so that Pz is an odd function of z. Now let P be an even function of z and let Q be an odd function of z, and from the definitions it is clear that if $-z$ is substituted for z, the value of P remains the same while the value of Q becomes its negative, $-Q$. Hence the product PQ, when $-z$ is substituted for z, becomes $-PQ$, that is , its negative. Thus PQ is an odd function of z. Also, since $a + \sqrt{a^2 + z^2}$ is an even function and z^3 is an odd function of z, the product $az^3 + z^3\sqrt{a^2 + z^2}$ is an odd function of z. In like manner $z\left(\dfrac{a + bz^2}{\alpha + \beta z^2}\right) = \dfrac{az + bz^3}{\alpha + \beta z^2}$ is an odd function of z. From what has been discussed, it can be seen that if of the two functions P and Q, one is odd and the other even and one is divided by the other, the quotient is an odd function. Thus $\dfrac{P}{Q}$ and $\dfrac{Q}{P}$ are both odd functions of z.

23. *If an odd function is either multiplied or divided by another odd function, the resulting function is even.*

Let Q and S be odd functions of z, so that if z is replaced by $-z$, Q becomes $-Q$ and S becomes $-S$. It is clear that both the product QS and the quotient $\dfrac{Q}{S}$ retain the same value even when z is replaced by $-z$. Hence both are even functions of z. It is also obvious that the square of any odd function is even, its cube is odd, its fourth power is even, and so forth.

24. *If y is an odd function of z, then z is also an odd function of y.*

Since y is an odd function of z, if z is replaced by $-z$, y becomes $-y$. Now if z is defined as a function of y, it follows that when y is replaced by $-y$, z must necessarily become $-z$. Hence z is an odd function of y. Thus, if $y = z^3$, so that y is an odd function of z, from the given equation we have $z = y^{\frac{1}{3}}$ and z is an odd function of y. Also if $y = az + bz^3$, y is an odd function of z and in turn from the equation $bz^3 + az = y$, the value of z expressed in terms of y is an odd function of y.

25. *If the nature of the function y is defined by an equation of the type just discussed, in which for each term the sum of the exponents of y and z are either all odd or all even, then y is an odd function of z.*

If in this kind of equation, in place of z, we write $-z$, and at the same time $-y$ in place of y, all terms of the equation remain the same or become negative. In either case the equation remains the same. From this it is clear that $-y$ is determined by $-z$ just as $+y$ is determined by $+z$. For this reason, if $-z$ is substituted for z, the value of y becomes $-y$, or y is an odd function of z. Thus if $y^2 = ayz + bz^2 + c$ or if $y^3 + ay^2z = byz^2 + cy + dz$, from either equation, y is an odd function of z.

26. *If Z is a function of z and Y a function of y such that Y is defined through y and constants, in the same way as Z is defined through z and constants, then the functions Y and Z are said to be similar functions of y and z respectively.*

For example if $Z = a + bz + cz^2$ and $Y = a + by + cy^2$, then Z and Y are similar functions of y and z. From this it follows that if Y and Z are similar functions of y and z, then if y is written in place of z, the function Z

becomes the function Y. This similarity is usually expressed by saying that Y is such a function of y as Z is of z. We use such expressions whether or not the variables z and y depend on each other; thus such a function of y as $ay + by^3$, is similar to the function $a(y + n) + b(y + n)^3$ of $y + n$, since for example $z = y + n$. Further, the function $\frac{a + bz + cz^2}{\alpha + \beta z + \gamma z^2}$ of z is similar to the function $\frac{az^2 + bz + c}{\alpha z^2 + \beta z + \gamma}$ of $\frac{1}{z}$, which is seen by letting $y = \frac{1}{z}$, and so from these examples the idea of similar functions, which is fruitfully used throughout all of higher analysis becomes quite clear. This should be enough general information about the nature of functions of one variable, since a rather full explanation will be given in the following application.

CHAPTER II

On the Transformation of Functions.

27. *The form of a function is changed, either by introducing a different variable, or if the same variable is kept, the transformation consists in expressing the same function in a different way.*

This is clear from algebra, where the same quantity can be expressed in various ways. Transformations of this kind are exemplified by writing $(1 - z)(2 - z)$ for $2 - 3z + z^2$, or $(a + z)^3$ instead of $a^3 + 3a^2z + 3az^2 + z^3$, or $\frac{a}{a - z} + \frac{a}{a + z}$ instead of $\frac{2a^2}{a^2 - z^2}$, or $\sqrt{1 + z^2} + z$ for $\frac{1}{\sqrt{1 + z^2} - z}$. All of these respective expressions differ in form but are truly equivalent.

Frequently it happens that one of these several expressions is more apt for a particular task than the others, and for this reason the most convenient form should be chosen.

The other way in which a function is transformed, namely, the introduction of a different variable y for the variable z, where y has a given relationship to z, is said to be by substitution. In this way a function may be expressed more succinctly or more conveniently. For example, if the function of z is $a^4 - 4a^3z + 6a^2z^2 - 4az^3 + z^4$, and if for $a - z$ we put y, we obtain this much simpler function of y, y^4. If we have the irrational function $\sqrt{a^2 + z^2}$ of

z and let $z = \frac{a^2 - y^2}{2y}$, this function becomes a rational function $\frac{a^2 + y^2}{2y}$ of y. We shall treat this second way of transforming functions in the next chapter, while the first way, without substitution, we shall explain in this chapter.

28. *It is frequently convenient to factor a polynomial function and thus express it as a product.*

When a polynomial function is factored in this way, its nature is more easily seen; it is immediately clear for what values of z the function is equal to zero. Thus, this function of z, $6 - 7z + z^3$ is transformed into the product $(1 - z)(2 - z)(3 + z)$, and from this it is obvious for which three values the given function vanishes; namely, if $z = 1$, or $z = 2$, or $z = -3$, but these properties of the function are not so easily understood from the form $6 - 7z + z^3$. Factors of this kind, in which the variable z occurs without a higher power, are called linear factors to distinguish them from composite factors in which z is squared or cubed or to some other higher power. In general, then, $f + gz$ is the form of a linear factor, $f + gz + hz^2$ the form of a quadratic factor, $f + gz + hz^2 + jz^3$ the form of a cubic factor, and so forth. It is clear that a quadratic factor is the product of two linear factors, that a cubic factor is the product of three linear factors, and so forth. Thus, a polynomial function of z in which the highest power of z is n will contain n linear factors; from this we also know the number of factors if some of the factors are quadratic, cubic, etc.

29. *The linear factors of any polynomial function Z of z are found if the function Z is set equal to zero and the roots of this equation are investigated: all of the roots give all of the linear factors of the function Z.*

If from the equation $Z = 0$, there is some root $z = f$, then $z - f$ is a divisor and hence a factor of the function Z. Thus by investigating all of the roots of the equation $Z = 0$, namely, $z = f$, $z = g$, $z = h$, etc. The function Z is resolved into its simple factors and transformed into the product $Z = (z - f)(z - g)(z - h)$ etc. Here it is to be noted that if the highest power of z in Z has a coefficient different from $+1$, then the product $(z - f)(z - g)$ etc. must be multiplied by that coefficient. Thus if $Z = Az^n + Bz^{n-1} + Cz^{n-2} + \cdots$, then $Z = A(z - f)(z - g)(z - h)\cdots$. Also, if $Z = A + Bz + Cz^2 + Dz^3 + Ez^4 + \cdots$ and the roots of the equation $Z = 0$ are found to be $f, g, h, j, \cdots$, then $Z = A\left(1 - \frac{z}{f}\right)\left(1 - \frac{z}{g}\right)\left(1 - \frac{z}{h}\right)\cdots$. On the other hand, from these forms it is known if the function Z has a factor $z - f$ or $1 - \frac{z}{f}$ since if in place of z we put f, the value of the function is zero. For if $z = f$, one factor $z - f$ or $1 - \frac{z}{f}$ of the function Z, hence the function Z itself must vanish.

30. *Linear factors are either real or complex; if the function Z has complex factors, the number of such factors is always even.*

Since linear factors come from roots of the equation $Z = 0$, real roots give real factors and complex roots give complex factors. In every equation the number of complex roots is even so that the function Z has either no complex factor, or it has two or four or six etc. If the function Z has only two complex factors, the product of these two will be real, which gives a quadratic real factor, since if we let P be the product of all of the real factors, then the product of the two complex factors will be $\frac{Z}{P}$, which is real. In the same way, if the function Z has four, or six, or eight complex factors, their product is always real, since it is equal to the quotient which arises when Z is divided by the product of all the real factors.

31. *If Q is the real product of four complex linear factors, then this product Q can also be represented as the product of two real quadratic factors.*

Now Q has the form $z^4 + Az^3 + Bz^2 + Cz + D$ and if we suppose the Q cannot be represented as the product of two real quadratic factors, then we show that is can be represented as the product of two complex quadratic factors having the following forms: $z^2 - 2(p + qi)z + r + si$ and $z^2 - 2(p - qi)z + r - si$. No other form is possible, since the product is real, namely, $z^4 + Az^3 + Bz^2 + Cz + D$. From these complex quadratic factors we derive the following four complex linear factors:

$$\begin{array}{ll} I. & z - (p + qi) + \sqrt{p^2 + 2pqi - q^2 - r - si} \\ II. & z - (p + qi) - \sqrt{p^2 + 2pqi - q^2 - r - si} \\ III. & z - (p - qi) + \sqrt{p^2 - 2pqi - q^2 - r + si} \\ IV. & z - (p - qi) - \sqrt{p^2 - 2pqi - q^2 - r + si}. \end{array}$$

For the sake of brevity we let $t = p^2 - q^2 - r$ and $u = 2pq - s$. When the first and third of these factors are multiplied, the product is equal to

$$z^2 - \left(2p - \left(2t + 2\sqrt{t^2 + u^2}\right)^{\frac{1}{2}}\right)z + p^2 + q^2 - p\left(2t + 2\sqrt{t^2 + u^2}\right)^{\frac{1}{2}}$$

$$+ \sqrt{t^2 + u^2} + q\left(- 2t + 2\sqrt{t^2 + u^2}\right)^{\frac{1}{2}}$$ which is real. In like manner the product of the second and fourth factors is the real

$$z^2 - \left[2p + \left(2t + 2\sqrt{t^2 + u^2}\right)^{\frac{1}{2}}\right]z + p^2 + q^2 + p\left(2t + 2\sqrt{t^2 + u^2}\right)^{\frac{1}{2}}$$

$+ \sqrt{t^2 + u^2} + q\left(-2t + 2\sqrt{t^2 + u^2}\right)^{\frac{1}{2}}$. Thus the proposed product Q, which we supposed could not be expressed as two real factors, can be expressed as the product of two real quadratic factors.

32. *Whatever number of complex linear factors a polynomial function Z of z may have, they can always be paired in such a way that the product of such pairs is real.*

Since the number of complex roots is always even, let it be $2n$; it is also clear that the product of all of these factors is real. Now if there are only two complex roots, certainly the product is real; however, if there are four complex factors, then, as we have seen the product of all four can be expressed as the product of two real quadratic factors of the form $fz^2 + gz + h$. Although the same method of proof is not valid for higher powers, nevertheless, there is no doubt that the same property holds for any number of complex factors; hence, instead of $2n$ complex linear factors, n real quadratic factors can be introduced. Thus, every polynomial function of z can be expressed as the product of real factors, either linear or quadratic. Granted that this has not been proved with complete rigor, still the truth of the statement will be corroborated in what follows, where functions with the form $a + bz^n$, $a + bz^n + cz^{2n}$, $a + bz^n + cz^{2n} + dz^{3n}$, etc. will actually be resolved into such real quadratic factors.

33. *If the polynomial function Z takes the value A when $z = a$ and takes the value B when $z = b$, then there is a value of z between a and b for which the*

function Z takes any value between A and B.

Since Z is a single valued function of z, for whatever real value is assigned to z, there is a real value for the function Z. So in the first case, where $z = a$, Z has the value A, and in the second case, where $z = b$, Z has the value B; but the function cannot pass from A to B without taking on all of the intermediate values. Now if the equation $Z - A = 0$ has a real root and also if to $Z - B = 0$ there corresponds a real root, then the equation $Z - C = 0$ also has a real root provided that C lies between A and B. Thus if the expressions $Z - A$ and $Z - B$ each has a real linear factor, then any expression $Z - C$ also has real linear factor whenever C lies between A and B.

34. *If in a polynomial function Z the greatest exponent of z is the odd number $2n + 1$, then the function Z has at least one real linear factor.*

For instance, if $Z = z^{2n+1} + \alpha z^{2n} + \beta z^{2n-1} + \gamma z^{2n-2} + \cdots$, and if z is set equal to ∞, since the values of all of the other terms vanish before that of the first term, Z becomes equal to $\infty^{2n+1} = \infty$. It follows that $Z - \infty$ has a real linear factor, namely, $z - \infty$. On the other hand, if we let $z = -\infty$, then $Z = (-\infty)^{2n+1} = -\infty$, so that $Z + \infty$ has the real linear factor $z + \infty$. Since both $Z - \infty$ and $Z + \infty$ have real linear factors, it follows that also $Z + C$ will have a real linear factor provided that C is between $+\infty$ and $-\infty$, that is if C is any real number, whether positive, negative, or zero. For this reason, when $C = 0$, the function Z has a real linear factor $z - c$; the number c lies between $+\infty$ and $-\infty$ and hence will be either positive, negative or zero.

35. *A polynomial function Z in which the greatest exponent of z is odd will have either one real linear factor or three, or five, or seven etc.*

Since it has been shown that the function Z certainly has at least one real linear factor $z - c$; we suppose that there is another factor $z - d$. When Z, in which the greatest exponent of z is $2n + 1$, is divided by $(z - c)(z - d)$, the quotient has z with the exponent $2n - 1$, which is also odd. It follows that Z has another real linear factor. Hence, if Z has more than one real linear factor, it will have three, or (since the same argument is valid) five or seven, etc. Therefore, the number of real linear factors is odd and, since the total number of linear factors is $2n + 1$, the number of complex factors is even.

36. *A polynomial function Z in which the greatest exponent of z is the even number $2n$, has either two real linear factors, or four, or six, or etc.*

We suppose that the function Z has an odd number, $2m + 1$, of real linear factors. If Z is divided by the product of all of these factors, in the quotient the greatest exponent of z is $2n - 2m - 1$, which is odd. Hence, the function Z has another real linear factor, so that the number of real linear factors is at least $2m + 2$, which is even. It follows that every polynomial function has an even number of complex linear factors, which fact we had already established.

37. *If in a polynomial function Z the greatest exponent of z is an even number and the constant term has a negative sign, then the function Z has at least two real linear factors.*

Now the function Z with which we are concerned has the following form: $z^{2n} \pm az^{2n-1} \pm bz^{2n-2} \pm \cdots \pm nz - A$. If we let $z = \infty$, then, as we have seen above, $Z = \infty$, while if we let $z = 0$, then $Z = -A$. Hence $Z - \infty$ has a real factor $z - \infty$ and $Z + A$ has a factor $z - 0$; from this it follows that, since 0 lies between $-\infty$ and A, $Z + 0$ has a real linear factor $z - c$ where c lies between 0 and ∞. Also, since $Z = \infty$ when $z = -\infty$, it follows that $Z - \infty$ has a factor $z + \infty$ and $Z + A$ has a factor $z + 0$, so that $Z + 0$ has a real linear factor $z + d$ where d lies between 0 and ∞; thus the proposition is proved. From this argument it is clear that if Z is such a function as has been here described, then the equation $Z = 0$ must have at least two real roots, one of which is positive and the other is negative. Thus the equation $z^4 + \alpha z^3 + \beta z^2 + \gamma z - a^2 = 0$ has two real roots, one positive and the other negative.

38. *If in a rational function the greatest power of the variable z in the numerator is greater than or equal to the greatest power of z in the denominator, then this function can be expressed in two parts, the first of which is a polynomial function, and the second part is a rational function in which the greatest power of z in the numerator is less than the greatest power of z in the denominator.*

If the greatest power of z in the denominator is less than that in the numerator, then the numerator is divided by the denominator in the usual way until the power of z in the quotient would be negative; at this point the division process is interrupted so that the quotient will consist of a polynomial part and a rational part in which the numerator has a lower degree than the denominator. This quotient is equal to the function in question. Thus if the proposed rational function is $\frac{1 + z^4}{1 + z^2}$, it is determined by division, to obtain

$\dfrac{1 + z^4}{1 + z^2} = z^2 - 1 + \dfrac{2}{1 + z^2}$. This kind of rational function in which the degree of the numerator is greater than or equal to the degree of the denominator can be called, as is done in arithmetic, improper fractions or improper rational functions, as opposed to proper rational functions in which the degree of the numerator is less than the degree of the denominator. Hence, an improper rational function can be expressed as the sum of a polynomial function and a proper rational function; this is accomplished by the ordinary division process.

39. *If the denominator of a rational function has two relatively prime factors, then this rational function can be expressed as the sum of two fractions whose denominators are equal to the two factors.*

Although this resolution applies to both improper and proper rational functions, we will for the most part confine our attention to the proper case. In this case, when the denominator is expressed as two relatively prime factors, the proper rational function is resolved into two other proper rational functions with denominators equal to the respective factors. This resolution is unique, provided the original function was proper. This fact can be seen more clearly by an example than by an argument. Let us consider the following rational function: $\dfrac{1 - 2z + 3z^2 - 4z^3}{1 + 4z^4}$, whose denominator is equal to the product $(1 + 2z + 2z^2)(1 - 2z + 2z^2)$. The given function is to be expressed as the sum of two fractions of which the first has $1 + 2z + 2z^2$ and the second has $1 - 2z + 2z^2$ for denominators. In order to find these fractions, which are proper, we let the numerators be $a + bz$ and $c + dz$ respectively. By hypothesis

$\frac{1 - 2z + 3z^2 - 4z^3}{1 + 4z^4} = \frac{a + bz}{1 + 2z + 2z^2} + \frac{c + dz}{1 - 2z + 2z^2}$. When we add the two fractions we find that the denominator is equal to $1 + 4z^4$ and the numerator is equal to $a - 2az + 2az^2 + bz - 2bz^2 + 2bz^3 + c + 2cz + 2cz^2 + dz + 2dz^2 + 2dz^3$. Since the denominator of the sum is equal to the denominator of the original fraction, the two numerators must be equal. Since the number of unknown values, namely, a, b, c, d is the same as the number of terms in the numerator to be equated, and this is unique, we obtain the following four equations:

$$\text{I.} \quad a + c = 1 \qquad \text{III.} \; 2a - 2b + 2c + 2d = 3$$
$$\text{II.} \quad -a + b + 2c + d = -2 \qquad \text{IV.} \; 2b + 2d = -4.$$

Equations II and III yield $a - c = 0$ and $d - b = \frac{1}{2}$ which together with $a + c = 1$ and $b + d = -2$, give us $a = \frac{1}{2}$, $c = \frac{1}{2}$, $b = -\frac{5}{4}$, $d = -\frac{3}{4}$.

Therefore, the proposed fraction $\frac{1 - 2z + 3z^2 - 4z^3}{1 + 4z^4}$ is transformed into these two: $\frac{\frac{1}{2} - \frac{5}{4}z}{1 + 2z + 2z^2} + \frac{\frac{1}{2} - \frac{3}{4}z}{1 - 2z + z^2}$. In like manner it is easily seen that this resolution can always be accomplished, since we always introduce the same number of unknown values as there are terms in the given numerator. It is common knowledge that this resolution will be successful only if the factors of the denominator are relatively prime.

40. *A rational function* $\frac{M}{N}$ *can be resolved into as many simple fractions of the form* $\frac{A}{p - qz}$ *as there are different linear factors in the denominator* N.

Let this rational function $\frac{M}{N}$ represent any proper fraction, so that M and N are polynomial function of z in which the degree of M is less than that of N. If the denominator N is resolved into its linear factors, and they are all different, then the expression $\frac{M}{N}$ can be resolved into as many fractions as there are linear factors in the denominator N, since each factor becomes the denominator of a partial fraction. Thus, if $p - qz$ is a factor of N, then it will be the denominator of a certain partial fraction, and since the degree of the numerator of this fraction must be less than the degree of the denominator $p - qz$ the numerator must be a constant. It follows that from each linear factor $p - qz$ of the denominator N there arises a simple fraction $\frac{A}{p - qz}$ and the sum of all of these fractions is equal to the given fraction $\frac{M}{N}$.

EXAMPLE

Let the given rational function be $\frac{1 + z^2}{z - z^3}$. Since the linear factors of the denominator are z, $1 - z$, and $1 + z$, this function is resolved into these three simple fractions $\frac{A}{z} + \frac{B}{1 - z} + \frac{C}{1 + z} = \frac{1 + z^2}{z - z^3}$, where the constant numerators A, B, and C have to be defined. When the three fractions are expressed with a common denominator, which is $z - z^3$, and the numerator of the sum is equated to $1 + z^2$, there arises the equation : $A + Bz + Cz - Az^2 + Bz^2 - Cz^2 = 1 + z^2 = 1 + 0z + z^2$. This gives rise to as many equations as there are unknown quantities, A, B, C, that is

$$\begin{aligned} &I. \quad A = 1 \\ &II. \quad B + C = 0 \\ &III. \quad -A + B - C = 1. \end{aligned}$$

It follows that $B - C = 2$, so that $A = 1$, $B = 1$, and $C = -1$. Thus the given function $\frac{1 + z^2}{z - z^3}$ is resolved into the expression $\frac{1}{z} + \frac{1}{1 - z} - \frac{1}{1 + z}$.

In like manner it can be understood that whatever linear factors the denominator N may have, provided they be different, the rational function $\frac{M}{N}$ can always be resolved into the same number of simple fractions. On the other hand, if some of the factors are equal, then the resolution must be accomplished in a different way, to be explained later.

41. *Since each linear factor of the denominator N gives rise to a simple fraction in the resolution of the given function* $\frac{M}{N}$, *it is shown how, from the knowledge of a linear factor of the denominator N, the corresponding simple fraction can be found.*

Let $p - qz$ be a linear factor of N, so that $N = (p - qz)S$, where S is a polynomial function of z. Let $\frac{A}{p - qz}$ be the fraction which corresponds to the factor $p - qz$, and let the fraction corresponding to the other factor S be equal to $\frac{P}{S}$, so that, according to section 39, $\frac{M}{N} = \frac{A}{p - qz} + \frac{P}{S} = \frac{M}{(p - qz)S}$; hence, $\frac{P}{S} = \frac{M - AS}{(p - qz)S}$. Since these two fractions are equal $M - AS$ must be divisible by $p - qz$ and the quotient of $M - AS$ by $p - qz$ is equal to the polynomial function P. Since $p - qz$ is a divisor of $M - AS$, $M - AS$ vanishes when we let $z = \frac{p}{q}$. Therefore, if we substitute the constant $\frac{p}{q}$ for z in M and

S, then $M - AS = 0$ so that $A = \frac{M}{S}$. In this way we have found the numerator A of the required fraction $\frac{A}{p - qz}$. If for each of the linear factors of the denominator N, provided they are all different, the corresponding simple fractions are found, the sum of all of these simple fractions will be equal to the given function $\frac{M}{N}$.

EXAMPLE

Thus, in the preceding example $\frac{1 + z^2}{z - z^3}$, where $M = 1 + z^2$ and $N = z - z^3$, if we take z as the linear factor, then $S = 1 - z^2$ and for the simple fraction $\frac{A}{Z}$, the numerator $A = \frac{1 + z^2}{1 - z^2} = 1$ when $z = 0$, which value z has when the linear factor is set equal to zero. In like manner, if we take $1 - z$ for the factor of the denominator, then $S = z + z^2$ and $A = \frac{1 + z^2}{z + z^2}$, so that $A = 1$ when $1 - z = 0$ and from the factor $1 - z$ there arises the fraction $\frac{1}{1 - z}$. The third factor $1 + z$, since $S = z - z^2$ and $A = \frac{1 + z^2}{z - z^2}$ gives $A = -1$ when $1 + z = 0$ or $z = -1$ and we have the simple fraction $\frac{-1}{1 + z}$. From this rule we see that $\frac{1 + z^2}{z - z^3} = \frac{1}{z} + \frac{1}{1 - z} - \frac{1}{1 + z}$, as in the previous example.

42. *A rational function with the form* $\frac{P}{(p - qz)^n}$, *where the degree of the numerator* P *is less than the degree of the denominator* $(p - qz)^n$, *can be transformed into the sum of partial fractions of the following form:*

$$\frac{A}{(p - qz)^n} + \frac{B}{(p - qz)^{n-1}} + \frac{C}{(p - qz)^{n-2}} + \cdots + \frac{K}{p - qz},$$ *where all of the numerators are constants.*

Since the degree of P is less than n, P has the form $a + bz + cz^2 + dz^3 + \cdots + kz^{n-1}$ with n terms, which must equal the numerator of the sum of all of the partial fractions after they have been expressed with the common denominator $(p - qz)^n$; this numerator has the form $A + B(p - qz) + C(p - qz)^2 + D(p - qz)^3 + \cdots + K(p - qz)^{n-1}$. The degree of this expression is also $n - 1$ and has n unknown quantities $A, B, C, \cdots, K$, which number n is the same as the number of terms in P. For this reason, the quantities A, B, C, etc. can be so defined that the rational function $\frac{P}{(p - qz)^n}$ is equal to

$$\frac{A}{(p - qz)^n} + \frac{B}{(p - qz)^{n-1}} + \frac{C}{(p - qz)^{n-2}} + \frac{D}{(p - qz)^{n-1}} + \cdots + \frac{K}{p - qz}.$$

An easy method for determining these constant numerators will soon be shown.

43. *If the denominator N of the rational function $\frac{M}{N}$ has a factor $(p - qz)^2$, the partial fractions arising from this factor are found in the following way.*

We have already shown how the partial fractions corresponding to linear factors which differ from all other factors are found; now we suppose that there are two such equal factors which when combined give a factor $(p - qz)^2$ of the denominator N. From this factor, according to the preceding section, there arise the two partial fractions $\frac{A}{(p - qz)^2} + \frac{B}{p - qz}$. Let $N = (p - qz)^2 S$, then

$\frac{M}{N} = \frac{M}{(p - qz)^2 S} = \frac{A}{(p - qz)^2} + \frac{B}{p - qz} + \frac{P}{S}$, where $\frac{P}{S}$ stands for the sum of all of the simple fractions which arise from the factor S of the denominator. It follows that $\frac{P}{S} = \frac{M - AS - B(p - qz)S}{(p - qz)^2 S}$, and $P = \frac{M - AS - B(p - qz)S}{(p - qz)^2}$, which is a polynomial function. Hence, $M - AS - B(p - qz)S$ is divisible by $(p - qz)^2$, so that it is certainly divisible by $p - qz$. The whole expression $M - AS - B(p - qz)S$ vanishes when $p - qz = 0$ or $z = \frac{p}{q}$; if we let $z = \frac{p}{q}$, then $M - AS = 0$, so that $A = \frac{M}{S}$; that is, the fraction $\frac{M}{S}$, with z everywhere replaced by $\frac{p}{q}$, gives the value of the constant A. After A has been calculated, since $M - AS - B(p - qz)S$ is divisible by $(p - qz)^2$, or $\frac{M - AS}{p - qz} - BS$ is divisible by $p - qz$. If we put $z = \frac{p}{q}$ everywhere, then $\frac{M - AS}{p - qz} = BS$, so that $B = \frac{M - AS}{(p - qz)S} = \frac{1}{p - qz}\left(\frac{M}{S} - A\right)$, where we note that since $M - AS$ is divisible by $p - qz$, the division must first be carried out before substituting $\frac{p}{q}$ for z. That is, let $\frac{M - AS}{p - qz} = T$, then $B = \frac{T}{S}$ when we let $z = \frac{p}{q}$. Now that we have calculated the constant numerators A and B, the partial fractions arising from the factor $(p - qz)^2$ in the denominator N are $\frac{A}{(p - qz)^2} + \frac{B}{p - qz}$.

EXAMPLE I

If the rational function is $\dfrac{1 - z^2}{z^2(1 + z^2)}$, then because of the quadratic factor z^2, $S = 1 + z^2$ and $M = 1 - z^2$. If the partial fractions arising from z^2 are $\dfrac{A}{z^2} + \dfrac{B}{z}$, then $A = \dfrac{M}{S} = \dfrac{1 - z^2}{1 + z^2}$ which is equal to 1 when $z = 0$. Then $M - AS = -2z^2$, which when divided by the linear factor z gives $T = -2z$, so that $B = \dfrac{T}{S} = \dfrac{-2z}{1 + z^2}$. When $z = 0$, $B = 0$, so that from the factor z^2 in the denominator there arises only one partial fraction $\dfrac{1}{z^2}$.

EXAMPLE II

Let the given rational function be $\dfrac{z^3}{(1 - z)^2(1 + z^4)}$. Since $(1 - z)^2$ is a quadratic factor of the denominator, we have the partial fractions $\dfrac{A}{(1 - z)^2} + \dfrac{B}{1 - z}$. Hence, $M = z^3$ and $S = 1 + z^4$ while $A = \dfrac{M}{S} = \dfrac{z^3}{1 + z^4}$. If we let $1 - z = 0$, or $z = 1$, then $A = \dfrac{1}{2}$. So that $M - AS = z^3 - \dfrac{1}{2} - \dfrac{1}{2}z^4 = -\dfrac{1}{2} + z^3 - \dfrac{1}{2}z^4$, which quantity divided by $1 - z$ gives $T = -\dfrac{1}{2} - \dfrac{1}{2}z - \dfrac{1}{2}z^2 + \dfrac{1}{2}z^3$. Thus $B = \dfrac{T}{S} = \dfrac{-1 - z - z^2 + z^3}{2 + 2z^4}$, which gives $B = -\dfrac{1}{2}$ when $z = 1$. The partial fractions are $\dfrac{1}{2(1 - z)^2} - \dfrac{1}{2(1 - z)}$.

44. *If in the rational function* $\dfrac{M}{N}$ *the denominator* N *has a factor* $(p - qz)^3$, *then the following partial fractions arise from this factor:*

$$\frac{A}{(p - qz)^3} + \frac{B}{(p - qz)^2} + \frac{C}{p - qz}.$$

Let $N = (p - qz)^3 S$ and let $\frac{P}{S}$ be the rational function which arises from the factor S. Then $P = \frac{M - AS - B(p - qz)S - C(p - qz)^2 S}{(p - qz)^3}$, which is a polynomial. Hence, the numerator, $M - AS - B(p - qz)S - C(p - qz)^2 S$ is certainly divisible by $p - qz$ and so it vanishes when $p - qz = 0$ or $z = \frac{p}{q}$. Thus $M - AS = 0$ or $A = \frac{M}{S}$ when $z = \frac{p}{q}$. First we calculate A in this way. Since $M - AS$ is divisible by $p - qz$, we let $T = \frac{M - AS}{p - qz}$, so that $T - BS - C(p - qz)S$ is still divisible by $(p - qz)^2$ and hence is equal to zero when $p - qz = 0$. It follows that $B = \frac{T}{S}$ when $z = \frac{p}{q}$. Now we let $V = \frac{T - BS}{p - qz}$, and $V - CS = 0$ when $p - qz = 0$. It follows that $C = \frac{V}{S}$ when $z = \frac{p}{q}$. Thus we have found the numerators A, B, and C, and the partial fractions arising from the factor $(p - qz)^3$ of the numerator N will be

$$\frac{A}{(p - qz)^3} + \frac{B}{(p - qz)^2} + \frac{C}{p - qz}.$$

EXAMPLE

Let the given rational function be $\frac{z^2}{(1 - z)^3(1 + z^2)}$. From the cubic factor of the denominator $(1 - z)^3$ there arise the following partial fractions: $\frac{A}{(1 - z)^3} + \frac{B}{(1 - z)^2} + \frac{C}{1 - z}$. Then $M = z^2$ and $S = 1 + z^2$, so that $A = \frac{z^2}{1 + z^2}$ when $1 - z = 0$ or $z = 1$. It follows that $A = \frac{1}{2}$. Since

$T = \dfrac{M - AS}{1 - z}$, we have $T = \dfrac{\frac{1}{2}z^2 - \frac{1}{2}}{1 - z} = -\dfrac{1}{2} - \dfrac{1}{2z}$. Since $B = \dfrac{-\frac{1}{2} - \frac{1}{2}z}{1 + z^2}$ when $z = 1$, we obtain $B = -\dfrac{1}{2}$. Finally, since $V = \dfrac{T - BS}{1 - z} = \dfrac{T + \frac{1}{2}S}{1 - z}$, we have $V = \dfrac{-\frac{1}{2}z + \frac{1}{2}z^2}{1 - z} = -\dfrac{1}{2}z$. It follows that $C = \dfrac{V}{S} = \dfrac{-\frac{1}{2}z}{1 + z^2}$ when $z = 1$, so that $C = -\dfrac{1}{4}$. Now we have the partial fractions which arise from the factor $(1 - z)^3$ of the numerator:

$$\frac{1}{2(1 - z)^3} - \frac{1}{2(1 - z)^2} - \frac{1}{4(1 - z)}.$$

45. *If the denominator N of the rational function $\dfrac{M}{N}$ has a factor $(p - qz)^n$, then the partial fractions*

$$\frac{A}{(p - qz)^n} + \frac{B}{(p - qz)^{n-1}} + \frac{C}{(p - qz)^{n-2}} + \cdots + \frac{K}{p - qz}$$

which arise therefrom are calculated in the following way.

Let the denominator $N = (p - qz)^n Z$, then we follow the same kind of argument as before. First, we find that $A = \dfrac{M}{Z}$ when $z = \dfrac{p}{q}$, and we let $P = \dfrac{M - AZ}{p - qz}$. Secondly, $B = \dfrac{P}{Z}$ when $z = \dfrac{p}{q}$, and we let $Q = \dfrac{P - BZ}{p - qz}$. Thirdly, $C = \dfrac{Q}{Z}$ when $z = \dfrac{p}{q}$, and we let $R = \dfrac{Q - CZ}{p - qz}$. Fourthly, $D = \dfrac{R}{Z}$ when $z = \dfrac{p}{q}$, and we let $S = \dfrac{R - DZ}{p - qz}$. Fifthly, $E = \dfrac{S}{Z}$ when $z = \dfrac{p}{q}$, etc. In this way we calculate the individual constant numerators A, B, C, D, etc., and so we find all of the partial fractions which arise from the factor $(p - qz)^n$ of the denominator N.

EXAMPLE

Let the given rational function be $\frac{1 + z^2}{z^5(1 + z^3)}$. From the factor z^5 of the denominator there arise the partial fractions $\frac{A}{z^5} + \frac{B}{z^4} + \frac{C}{z^3} + \frac{D}{z^2} + \frac{E}{z}$. In order to find the constant numerators, we let $M = 1 + z^2$ and $Z = 1 + z^3$ with $\frac{p}{q} = 0$. Now we begin the calculations. First $A = \frac{M}{Z} = \frac{1 + z^2}{1 + z^3}$ when $z = 0$, so that $A = 1$. We let $P = \frac{M - AZ}{z} = \frac{z^2 - z^3}{z} = z - z^2$, then $B = \frac{P}{Z} = \frac{z - z^2}{1 + z^3}$ when $z = 0$, so that $B = 0$. We let $Q = \frac{P - BZ}{z} = \frac{z - z^2}{z} = 1 - z$, then $C = \frac{Q}{Z} = \frac{1 - z}{1 + z^3}$ when $z = 0$, so that $C = 1$. We let $R = \frac{Q - CZ}{z} = \frac{-z - z^3}{z} = -1 - z^2$, then $D = \frac{R}{Z} = \frac{-1 - z^2}{1 + z^3}$ when $z = 0$, so that $D = -1$. We let $S = \frac{R - DZ}{z} = \frac{-z^2 + z^3}{z} = -z + z^2$, then $E = \frac{S}{Z} = \frac{-z + z^2}{1 + z^3}$ when $z = 0$, so that $E = 0$. Thus the required partial fractions are $\frac{1}{z^5} + \frac{0}{z^4} + \frac{1}{z^3} - \frac{1}{z^2} + \frac{0}{z}$.

46. *Given any rational function whatsoever* $\frac{M}{N}$, *it can be resolved into parts and transformed into its simplest form in the following way.*

First one obtains all of the linear factors, whether real or complex. Of these factors, those which are not repeated are treated individually, and from each of them a partial fraction is obtained from section 41. If a linear factor occurs two or more times, then these are taken together, and from their product, which will be of the form $(p - qz)^n$, we obtain the corresponding partial fractions from section 45. In this way, since for each of the linear factors partial fractions have been found, the sum of all of these partial fractions will equal the given function $\frac{M}{N}$ unless it is improper. If it is improper, then the polynomial part must be found and then added to the computed partial fractions in order to obtain the function $\frac{M}{N}$ expressed in its simplest form. This is the form whether the polynomial part is extracted before or after the partial fractions are obtained, since the same partial fraction arises from an individual factor of the denominator N whether the numerator M itself is used or M increased or diminished by some multiple of N. This fact is easily seen by anyone who considers the given rules.

EXAMPLE

We seek the simplest expression for the function $\frac{1}{z^3(1 - z)^2(1 + z)}$. First we take the single factor of the denominator $1 + z$, which gives $\frac{p}{q} = -1$ while $M = 1$ and $Z = z^3 - 2z^4 + z^5$. In order to find the fraction $\frac{A}{1 + z}$, we let $A = \frac{1}{z^3 - 2z^4 + z^5}$ when $z = -1$. Hence $A = -\frac{1}{4}$ and from the factor $1 + z$ there arises the partial fraction $\frac{-1}{4(1 + z)}$. Now take the quadratic factor $(1 - z)^2$, which gives $\frac{p}{q} = 1$, $M = 1$, and $Z = z^3 + z^4$. We let the partial fractions arising from this be $\frac{A}{(1 - z)^2} + \frac{B}{1 - z}$, then $A = \frac{1}{z^3 + z^4}$ when

$z = 1$, so that $A = \frac{1}{2}$. We let

$$P = \frac{M - \frac{1}{2}Z}{1 - z} = \frac{1 - \frac{1}{2}z^3 - \frac{1}{2}z^4}{1 - z} = 1 + z + z^2 + \frac{1}{2}z^3$$ so that

$B = \frac{P}{Z} = \frac{1 + z + z^2 + \frac{1}{2}z^3}{z^3 + z^4}$ when $z = 1$, hence, $B = \frac{7}{4}$ and the desired partial fractions are $\frac{1}{2(1 - z)^2} + \frac{7}{4(1 - z)}$. Finally, the cubic factor z^3 gives $\frac{p}{q} = 0$, $M = 1$ and $Z = 1 - z - z^2 + z^3$. We let the corresponding partial fractions be $\frac{A}{z^3} + \frac{B}{z^2} + Cz$. Then $A = \frac{M}{Z} = \frac{1}{1 - z - z^2 + z^3}$ when $z = 0$, hence $A = 1$. We let $P = \frac{M - Z}{z} = 1 + z - z^2$ so that $B = \frac{P}{Z}$ when $z = 0$, hence $B = 1$. We let $Q = \frac{P - Z}{z} = 2 - z^2$ so that $C = \frac{Q}{Z}$ when $z = 0$, hence $C = 2$. Thus the given function $\frac{1}{z^3(1 - z)^2(1 + z)}$ is expressed in the form

$\frac{1}{z^3} + \frac{1}{z^2} + \frac{2}{z} + \frac{1}{2(1 - z)^2} + \frac{7}{4(1 - z)} - \frac{1}{4(1 + z)}$. There is no polynomial part since the given function is not improper.

CHAPTER III

On the Transformation of Functions by Substitution.

46. *If y is some function of z, and z is defined by a new variable x, then y can also be defined by x.*

Since previously y was a function of z, now a new variable quantity x is introduced by which both the y and the z can be defined. Thus, if $y = \frac{1 - z^2}{1 + z^2}$, we let $z = \frac{1 - x}{1 + x}$. When this value is substituted for z, $y = \frac{2x}{1 + x^2}$. If any definite value is assigned to x, values of z and y are then determined so that the value of y corresponding to z is found at the same time as z. If we let $x = \frac{1}{2}$, then $z = \frac{1}{3}$ and $y = \frac{4}{5}$; we find that $y = \frac{4}{5}$ also, if in the expression $\frac{1 - z^2}{1 + z^2}$ for y we let $z = \frac{1}{3}$. A new variable may be introduced for either of two reasons: if the expression of y by z contains a radical, it may be removed, or if the relationship between y and z is given by an equation of higher order so that y cannot be expressed explicitly in terms of z, then a new variable x is introduced in terms of which both y and z can conveniently be defined. The main uses of substitutions should be sufficiently clear from these remarks, but they may become even clearer from the following.

47. *If $y = \sqrt{a + bz}$, then a new variable x, by which both z and y can be*

expressed without radicals, is found in the following manner.

Since both z and y are to be functions of x, whose expressions are not to contain radicals, it is clear that this can be done if we let $\sqrt{a + bz} = bx$. First we have $y = bx$ and $a + bz = b^2x^2$, so that $z = bx^2 - \frac{a}{b}$. Thus we have expressed both y and z by functions of x without radicals, and since $y = \sqrt{a + bz}$, z becomes $bx^2 - \frac{a}{b}$, so that $y = bx$.

48. *If* $y = (a + bz)^{\frac{m}{n}}$, *then a new variable* x *by means of which* y *can be expressed without radicals, is found as follows.*

Let $y = x^m$ so that $(a + bz)^{\frac{m}{n}} = x^m$ and hence $(a + bz)^{\frac{1}{n}} = x$. It follows that $a + bz = x^n$ and $z = \frac{x^n - a}{b}$. Thus both y and z are defined by x without radicals by means of the substitution $z = \frac{x^n - a}{b}$ which gives $y = x^m$. Although neither y can be expressed in terms of z, nor z in terms of y without radicals, nevertheless each becomes a function without radicals of the new variable x which is introduced by a substitution quite apt for this purpose.

49. *If* $y = \left(\frac{a + bz}{f + gz}\right)^{\frac{m}{n}}$, *then a new variable quantity* x *is required to express both* y *and* z *without radicals.*

It is clear that if we let $y = x^m$, then we have the required substitution, since then $\left(\frac{a + bz}{f + gz}\right)^{\frac{m}{n}} = x^m$, and so $\frac{a + bz}{f + gz} = x^n$. From this equation we have $z = \frac{a - fx^n}{gx^n - b}$, and when this substitution is made we obtain $y = x^m$.

From this we can also see that if $\left(\dfrac{\alpha + \beta y}{\gamma + \delta y}\right)^n = \left(\dfrac{a + bz}{f + gz}\right)^m$, then both y and z can be expressed in terms of x without radicals, if each expression is set equal to x^{mn}. It will result then that $y = \dfrac{\alpha - \gamma x^m}{\delta x^m - \beta}$ and $z = \dfrac{a - fx^n}{gx^n - b}$, and in this case there is no difficulty.

50. *If* $y = \sqrt{(a + bz)(c + dz)}$, *then a suitable substitution can be found, by which* y *and* z *can be expressed without radicals, as follows.*

Let $\sqrt{(a + bz)(c + dz)} = (a + bz)x$, then it is easy to see that an expression for z without a radical results, because the value of z is determined by an equation which is linear in z. That is, since $c + dz = (a + bz)x^2$, it follows that

$$z = \frac{c - ax^2}{bx^2 - d}.$$

Because

$$a + bz = \frac{a(bx^2 - d)}{bx^2 - d} + \frac{b(c - ax^2)}{bx^2 - d} = \frac{bc - ad}{bx^2 - d}$$

and $y = \sqrt{(a + bz)(c + dz)} = (a + bz)x$ we have $y = \dfrac{(bc - ad)x}{bx^2 - d}$. It follows that the function $y = \sqrt{(a + bz)(c + dz)}$ containing a radical is expressed, by the substitution $z = \dfrac{c - ax^2}{bx^2 - d}$, as a function without a radical, namely $y = \dfrac{(bc - ad)x}{bx^2 - d}$. For example, if $y = \sqrt{a^2 - z^2} = \sqrt{(a + z)(a - z)}$, since $b = 1$, $c = a$, and $d = -1$, we let $z = \dfrac{a - ax^2}{1 + x^2}$ and then $y = \dfrac{2ax}{1 + x^2}$. Hence whenever there are two linear real factors under a radical sign the radical can be removed by this method. If, however, the two linear factors under the radical are complex, then the following method may be used.

51. *Let* $y = \sqrt{p + qz + rz^2}$. *A suitable substitution for* z *is to be found, so that* y *may be expressed without a radical.*

This can happen in many different ways, depending on whether p and q are positive or negative quantities. Suppose first that p is positive, then let $p = a^2$. Even if p is not a perfect square, the irrationality of a will not affect the present argument.

I. If $y = \sqrt{a^2 + bz + cz^2}$, we let $\sqrt{a^2 + bz + cz^2} = a + xz$. Then, after squaring both sides of the equation, $b + cz = 2ax + x^2 z$, so that $z = \dfrac{b - 2ax}{x^2 - c}$. It follows that

$y = a + xz = \dfrac{a(x^2 - c)}{x^2 - c} + \dfrac{x(b - 2ax)}{x^2 - c} = \dfrac{bx - ax^2 - ac}{x^2 - c}$ and thus z and y are both functions of x without radicals.

II. If $y = \sqrt{a^2 z^2 + bz + c}$ and we let $\sqrt{a^2 z^2 + bz + c} = az + x$, then $bz + c = 2axz + x^2$ and $z = \dfrac{x^2 - c}{b - 2ax}$. Thus

$$y = az + x = \frac{-ac + bx - ax^2}{b - 2ax}.$$

III. If p and r are both negative, then, unless $q^2 > 4pr$, the value of y is always complex. If indeed $q^2 > 4pr$, the expression $p + qz + rz^2$ can be expressed as the product of two linear factors, which reduces to the case of the preceding section. Frequently, however, it is more convenient in this case if $y = \sqrt{a^2 + (b + cz)(d + ez)}$, we let $y = a + (b + cz)x$, so that $d + ez = 2ax + bx^2 + cx^2 z$. It follows that $z = \dfrac{d - 2ax - bx^2}{cx^2 - e}$ and $y = \dfrac{-ae + (cd - be)x - acx^2}{cx^2 - e}$. Sometimes it may be more convenient to express y as $\sqrt{a^2z^2 + (b + cz)(d + ez)}$ and let $y = az + (b + cz)x$, so that $d + ez = 2axz + bx^2 + cx^2z$. It follows that $z = \dfrac{bx^2 - d}{e - 2ax - cx^2}$ and

$$y = \frac{-ad + (be - cd)x - abx^2}{e - 2ax - cx^2}.$$

EXAMPLE

If we have the following function of z containing a radical $y = \sqrt{-1 + 3z - z^2}$, then it can be expressed as $y = \sqrt{1 - 2 + 3z - z^2} = \sqrt{1 - (1 - z)(2 - z)}$. We let $y = 1 - (1 - z)x$, so that $-2 + z = -2x + x^2 - x^2z$. Then $z = \dfrac{2 - 2x + x^2}{1 + x^2}$ and since $1 - z = \dfrac{-1 + 2x}{1 + x^2}$, we have $y = 1 - (1 - z)x = \dfrac{1 + x - x^2}{1 + x^2}$. Now these are the cases which indeterminate algebra or the Diophantine method supplies; other cases, which are not discussed in this treatise, cannot be reduced to a form without radicals by a substitution without radicals. For this reason it is time to proceed to another use for substitution.

52. *If y as a function of z is expressed as $ay^\alpha + bz^\beta + cy^\gamma z^\delta = 0$, we seek a new variable x by means of which y and z can be expressed explicitly.*

Since we do not have a general solution of the given equation $ay^\alpha + bz^\beta + cy^\gamma z^\delta = 0$, neither y can be expressed as a function of z, nor z as a function of y. The following is offered as a remedy for this inconvenience. Let $y = x^m z^n$ so that $ax^{\alpha m} z^{\alpha n} + bz^\beta + cx^{\gamma m} z^{\gamma n + \delta} = 0$. Now the exponent n is to be determined so that from the equation the value of z can be defined. There are three ways this can be done.

I. If we let $\alpha n = \beta$, then $n = \beta/\alpha$ and having divided the equation by $z^{\alpha n} = z^{\beta}$, we have $ax^{\alpha m} + b + cx^{\gamma m} z^{\gamma n - \beta + \delta} = 0$. It follows that

$$z = \left(\frac{-ax^{\alpha m} - b}{cx^{\gamma m}}\right)^{\frac{1}{\gamma n - \beta + \delta}} \quad \text{or} \quad z = \left(\frac{-ax^{\alpha m} - b}{cx^{\gamma m}}\right)^{\frac{\alpha}{\gamma\beta - \alpha\beta + \alpha\delta}} \quad \text{and}$$

$$y = x^m \left(\frac{-ax^{\alpha m} - b}{cx^{\gamma m}}\right)^{\frac{\beta}{\beta\gamma - \alpha\beta + \alpha\delta}}$$

II. If we let $\beta = \gamma n + \delta$ or $n = \dfrac{\beta - \delta}{\gamma}$, then after having divided the equation by z^{β} we have $ax^{\alpha m} z^{\alpha n - \beta} + b + cx^{\gamma m} = 0$. It follows that

$$z = \left(\frac{-b - cx^{\gamma m}}{ax^{\alpha m}}\right)^{\frac{1}{\alpha n - \beta}} = \left(\frac{-b - cx^{\gamma m}}{ax^{\alpha m}}\right)^{\frac{\gamma}{\alpha\beta - \alpha\delta - \beta\gamma}} \quad \text{and also}$$

$$y = x^m \left(\frac{-b - cx^{\gamma m}}{ax^{\alpha m}}\right)^{\frac{\beta - \delta}{\alpha\beta - \alpha\delta - \beta\gamma}}$$

III. If we let $\alpha n = \gamma n + \delta$, or $n = \delta/(\alpha - \gamma)$ then after dividing the equation by $z^{\alpha n}$, we have $ax^{\alpha m} + bz^{\beta - \alpha n} + cx^{\gamma m} = 0$. It follows that

$$z = \left(\frac{-ax^{\alpha m} - cx^{\gamma m}}{b}\right)^{\frac{1}{\beta - \alpha n}} = \left(\frac{-ax^{\alpha m} - cx^{\gamma n}}{b}\right)^{\frac{\alpha - \gamma}{\alpha\beta - \beta\gamma - \alpha\delta}}, \quad \text{and also}$$

$$y = x^m \left(\frac{-ax^{\alpha m} - cx^{\gamma m}}{b}\right)^{\frac{\delta}{\alpha\beta - \beta\gamma - \alpha\delta}}.$$

By these three methods we obtain functions of x which are equal to z and y. Furthermore, any integer except zero can be substituted for m and thus the formulas can be reduced to the most convenient form.

EXAMPLE

Let the function y be determined by the equation $y^3 + z^3 - cyz = 0$. We seek functions of x which are equal to y and z. In this case $a = -1$, $b = -1$, $\alpha = 3$, $\beta = 3$, $\gamma = 1$, and $\delta = 1$. Then by the first method when $m = 1$, we have $z = \left(\frac{x^3+1}{cx}\right)^{-1}$ and $y = x\left(\frac{x^3+1}{cx}\right)^{-1}$ or $z = \frac{cx}{1+x^3}$ and $y = \frac{cx^2}{1+x^3}$ and each of these is a rational function. The second method gives the following: $z = \left(\frac{cx-1}{x^3}\right)^{\frac{1}{3}}$ and $y = x\left(\frac{cx-1}{x^3}\right)^{\frac{2}{3}}$ or $z = \frac{1}{x}(cx-1)^{\frac{1}{3}}$ and $y = \frac{1}{x}(cx-1)^{\frac{2}{3}}$. The third method gives $z = (cx - x^3)^{\frac{2}{3}}$ and $y = x(cx - x^3)^{\frac{1}{3}}$.

53. *These equations, by which y and z as functions of the new variable x are determined, can be found* a posteriori.

We think it may be useful, once we have found the solution, to consider the following. Suppose $z = \left(\frac{ax^{\alpha} + bx^{\beta} + cx^{\gamma} + \cdots}{A + Bx^{\mu} + Cx^{\nu} + \cdots}\right)^{\frac{p}{r}}$ and $y = x\left(\frac{ax^{\alpha} + bx^{\beta} + cx^{\gamma} + \cdots}{A + Bx^{\mu} + Cx^{\nu} + \cdots}\right)^{\frac{q}{r}}$, while $y^p = x^p z^q$, so that we have $x = yz^{\frac{-q}{p}}$. Since $z^{\frac{r}{p}} = \frac{ax^{\alpha} + bx^{\beta} + cx^{\gamma} + \cdots}{A + Bx^{\mu} + Cx^{\nu} + \cdots}$, if instead of x we substitute $yz^{\frac{-q}{p}}$, then we have the following result: $z^{\frac{r}{p}}$ is equal to a quotient whose numerator is $ay^{\alpha}z^{\frac{-\alpha q}{p}} + by^{\beta}z^{\frac{-\beta q}{p}} + cy^{\gamma}z^{\frac{-\gamma q}{p}} + \cdots$ and whose denominator is $A + By^{\mu}z^{\frac{-\mu q}{p}} + Cy^{\nu}z^{\frac{-\nu q}{p}} + \cdots$. This can also be written as

$$Az^{\frac{r}{p}} + By^{\mu}z^{\frac{r - \mu q}{p}} + Cy^{\nu}z^{\frac{r - \nu q}{p}} + \cdots$$

$= ay^{\alpha}z^{\frac{-\alpha q}{p}} + by^{\beta}z^{\frac{-\beta q}{p}} + cy^{\gamma}z^{\frac{-\gamma q}{p}} + \cdots$. When this equation is multiplied by $z^{\frac{\alpha q}{p}}$ we obtain

$$Az^{\frac{\alpha q + r}{p}} + By^{\mu}z^{\frac{\alpha q - \mu q + r}{p}} + Cy^{\nu}z^{\frac{\alpha q - \nu q + r}{p}} + \cdots$$

$= ay^{\alpha} + by^{\beta}z^{\frac{\alpha q - \beta q}{p}} + cy^{\gamma}z^{\frac{\alpha q - \gamma q}{p}} + \cdots$. If we let $\frac{\alpha q + r}{p} = m$, $\frac{\alpha q - \beta q}{p} = n$, when $P = \alpha - \beta$, then $q = n$, and $r = \alpha m - \beta m - \alpha n$ so that there arises the equation

$$Az^{m} + By^{\mu}z^{\frac{m - \mu n}{\alpha - \beta}} + Cy^{\nu}z^{\frac{m - \nu n}{\alpha - \beta}} + \cdots$$

$= ay^{\alpha} + by^{\beta}z^{n} + cy^{\gamma}z^{\frac{(\alpha - \gamma)n}{(\alpha - \beta)}} + \cdots$ which is solved to give

$$z = \left(\frac{ax^{\alpha} + bx^{\beta} + cx^{\gamma} + \cdots}{A + Bx^{\mu} + Cx^{\nu} + \cdots}\right)^{\frac{\alpha - \beta}{\alpha m - \beta m - \alpha n}}$$ and

$$y = x\left(\frac{ax^{\alpha} + bx^{\beta} + cx^{\gamma} + \cdots}{A + Bx^{\mu} + Cx^{\nu} + \cdots}\right)^{\frac{n}{\alpha m - \beta m - \alpha n}}.$$ But if we let $\frac{\alpha q + r}{p} = m$ and $\frac{\alpha q - \mu q + r}{p} = n$, then $m - n = \frac{\mu q}{p}$, $\frac{q}{p} = \frac{m - n}{\mu}$, and $\frac{r}{p} = m - \frac{\alpha m - \alpha n}{\mu}$. When $p = \mu$, then $q = m - n$ and $r = \mu m - \alpha m + \alpha n$, so that the following equations result:

$$\alpha q - \nu q + r = \alpha m - \alpha n - \nu m + \nu n + \mu m - \alpha m + \alpha n$$
$$= \nu n - \nu m + \mu m.$$

$$Az^{m} + By^{\mu}z^{n} + Cy^{\nu}z^{\frac{\mu m - \nu(m - n)}{\mu}} + \cdots$$

$= ay^{\alpha} + by^{\beta}z^{\frac{(\alpha - \beta)(m - n)}{\mu}} + \cdots$, which is solved to give

$$z = \left(\frac{ax^{\alpha} + bx^{\beta} + cx^{\gamma} + \cdots}{A + Bx^{\mu} + Cx^{\nu} + \cdots} \right)^{\frac{\mu}{\mu m - \alpha m + \alpha n}} \text{ and}$$

$$y = x \left(\frac{ax^{\alpha} + bx^{\beta} + cx^{\gamma} + \cdots}{A + Bx^{\mu} + Cx^{\nu} + \cdots} \right)^{\frac{m - n}{\mu m - \alpha m + \alpha n}}.$$

54. *If y depends on z in such a way that $ay^2 + byz + cz^2 + dy + ez = 0$, then both y and z can be expressed as rational functions of a new variable x in the following way.*

Let $y = xz$, then having divided by z we obtain $ax^2z + bxz + cz + dx + e = 0$ from which we see that $z = \frac{-dx - e}{ax^2 + bx + c}$ and $y = \frac{-dx^2 - ex}{ax^2 + bx + c}$, but if the given equation between y and z is $ay^2 + byz + cz^2 + dy + ez + f = 0$, then by adding to or subtracting certain constants from the two variables, this equation also can be satisfied by rational functions of a new variable x.

55. *If y depends on z in such a way that $ay^3 + by^2z + cyz^2 + dz^3 + ey^2 + fyz + gz^2 = 0$, then both y and z can be expressed as rational functions of a new variable x in the following way.*

Let $y = xz$ and when the substitution is made, the whole equation can be divided by z^2. The following equation then results: $ax^3z + bx^2z + cxz + dz + ex^2 + fx + g = 0$. From this we have $z = \frac{-ex^2 - fx - g}{ax^3 + bx^2 + cx + d}$ so that $y = \frac{-ex^3 - fx^2 - gx}{ax^3 + bx^2 + cx + d}$. From these cases it is easily understood how equations of higher degree, by means of which y is defined as a function of z should be treated in order to find solutions of the kind we seek. Some of these cases are covered by the formulas given above in section 53, but since general formulas for cases which frequently occur are not so easily found, we have considered some of these separately.

56. *If y depends on z in such a way that $ay^2 + byz + cz^2 = d$, then both y and z can be expressed by a new variable x in the following way.*

Let $y = xz$ so that $(ax^2 + bx + c)z^2 = d$, $z = \sqrt{d/(ax^2 + bx + c)}$ and $y = x\sqrt{d/(ax^2 + bx + c)}$. In a similar way, if $ay^3 + by^2z + cyz^2 + dz^3 = ey + fz$, then we let $y = xz$. After making the substitution and dividing the whole equation by z, the following equation results: $(ax^3 + bx^2 + cx + d)z^2 = ex + f$. Thus $z = \sqrt{(ex + f)/(ax^3 + bx^2 + cx + d)}$ and $y = x\sqrt{(ex + f)/(ax^3 + bx^2 + cx + d)}$. These cases and also some other similar solutions are contained in the following section.

57. *If y depends on z in such a way that $ay^m + by^{m-1}z + cy^{m-2}z^2 + dy^{m-3}z^3 + \cdots = \alpha y^n + \beta y^{n-1}z + \gamma y^{n-2}z^2 + \delta y^{n-3}z^3 + \cdots$, then both z and y can conveniently be expressed by a new variable x.*

Let $y = xz$ and after the substitution is made, provided m is greater than n, we divide the whole equation by z^n to obtain $(ax^m + bx^{m-1} + cx^{m-2} + \cdots)z^{m-n} = \alpha x^n + \beta x^{n-1} + \gamma x^{n-2} + \cdots$. From this it follows that

$$z = \left(\frac{\alpha x^n + \beta x^{n-1} + \gamma x^{n-2} + \delta x^{n-3} + \cdots}{ax^m + bx^{m-1} + cx^{m-2} + dx^{m-3} + \cdots}\right)^{\frac{1}{m-n}} \quad \text{and}$$

$$y = x\left(\frac{\alpha x^n + \beta x^{n-1} + \gamma x^{n-2} + \delta x^{n-3} + \cdots}{ax^m + bx^{m-1} + cx^{m-2} + dx^{m-3} + \cdots}\right)^{\frac{1}{m-n}}.$$ Now this solution holds if in the equation which expressed the relationship between y and z, there are in each term only two possible total degrees of y and z. In the case treated here the total degree of y and z is either m or n.

58. *If in the equation giving the relationship between y and z there are three different combined powers, such that, the largest total degree differs from the middle total degree by the same amount as the middle total degree differs from the smallest, then y and z can be expressed in terms of the new variable x by solving a quadratic equation.*

If we let $y = xz$ and then divide by the least power of z, then z can be expressed in terms of x by means of the quadratic formula. This will become clear from the following examples.

EXAMPLE I

Let $ay^3 + by^2z + cyz^2 + dz^3 = 2ey^2 + 2fyz + 2gz^2 + hy + jz$. When we let $y = xz$ and then divide by z we obtain $(ax^3 + bx^2 + cx + d)z^2 = 2(ex^2 + fx + g)z + hx + j$. From this quadratic equation in z we find that

$$z = \frac{ex^2 + fx + g \pm \sqrt{(ex^2 + fx + g)^2 + (ax^3 + bx^2 + cx + d)(hx + j)}}{ax^3 + bx^2 + cx + d}$$

and from this we find $y = xz$.

EXAMPLE II

Let $y^5 = 2az^3 + by + cz$. When we let $y = xz$ and divide by z we have $x^5z^4 = 2az^2 + bx + c$. From this we obtain $z^2 = \dfrac{a \pm \sqrt{a^2 + bx^6 + cx^5}}{x^5}$ and so $z = \dfrac{\sqrt{a \pm \sqrt{a^2 + bx^6 + cx^5}}}{x^2\sqrt{x}}$ and $y = \dfrac{\sqrt{a \pm \sqrt{a^2 + bx^6 + cx^5}}}{x\sqrt{x}}$.

EXAMPLE III

Let $y^{10} = 2ayz^6 + byz^3 + cz^4$ in which the total degrees are 10, 7, and 4. We let $y = xz$ then divide by z^4 which results in $x^{10}z^6 = 2axz^3 + bx + c$ or $z^6 = \dfrac{2axz^3 + bx + c}{x^{10}}$. Hence $z^3 = \dfrac{ax \pm x\sqrt{a^2 + bx^9 + cx^8}}{x^{10}}$ so that

$z = \dfrac{\left(a \pm \sqrt{a^2 + bx^9 + cx^8}\right)^{\frac{1}{3}}}{x^3}$ and $y = \dfrac{\left(a \pm \sqrt{a^2 + bx^9 + cx^8}\right)^{\frac{1}{3}}}{x^2}$. From these examples the use of this kind of substitution is quite clear.

CHAPTER IV

On the Development of Functions in Infinite Series.

59. Since both rational functions and irrational functions of z are not of the form of polynomials $A + Bz + Cz^2 + Dz^3 + \cdots$, where the number of terms is finite, we are accustomed to seek expressions of this type with an infinite number of terms which give the value of the rational or irrational function. Even the nature of transcendental functions seems to be better understood when it is expressed in this form, even though it is an infinite expression. Since the nature of polynomial functions is very well understood, if other functions can be expressed by different powers of z in such a way that they are put in the form $A + Bz + Cz^2 + Dz^3 + \cdots$, then they seem to be in the best form for the mind to grasp their nature, even though the number of terms is infinite. It is clear that no function which is not a polynomial in z can be expressed as $A + Bz + Cz^2 + \cdots$, in which the number of terms is finite, since in that case it would be a polynomial function by definition. If there is any doubt that a function can be thus expressed with an infinite series, this doubt should be removed by the following discussion. In order that the following explanation be rather general, besides positive integral powers of z we will allow the exponents to be any real number. Thus there is no doubt that any function of z can be given the form $Az^\alpha + Bz^\beta + Cz^\gamma + Dz^\delta + \cdots$, where the exponents

α, β, γ, δ, etc. are any real numbers.

60. *By a continued division procedure the rational function* $\frac{a}{\alpha + \beta z}$ *can be expressed as the infinite series*

$\frac{a}{\alpha} - \frac{a\beta z}{\alpha^2} + \frac{a\beta^2 z^2}{\alpha^3} - \frac{a\beta^3 z^3}{\alpha^4} + \frac{a\beta^4 z^4}{\alpha^5} - \cdots$. *Since the quotient of any two successive terms is* $-\frac{\alpha}{\beta z}$, *this is called a geometric series.*

This series can also be found by setting $\frac{a}{\alpha + \beta z} = A + Bz + Cz^2 + Dz^3 + Ez^4 + \cdots$, and then find the coefficients $A, B, C, D, \cdots$, which give equality. Since $a = (\alpha + \beta z)(A + Bz + Cz^2 + Dz^3 + \cdots)$, after the indicated multiplication is performed, $a = \alpha A + \alpha Bz + \alpha Cz^2 + \alpha Dz^3 + \alpha Ez^4 + \cdots + \beta Az + \beta Bz^2 + \beta Cz^3 + \beta Dz^4 + \cdots$. Hence $a = \alpha A$ so that $A = \frac{a}{\alpha}$. The coefficients of each power of z must have a zero sum so that we have the equations $\alpha B + \beta A = 0, \alpha C + \beta B = 0$, $\alpha D + \beta C = 0, \alpha E + \beta D = 0$. Once any coefficient is known, the following coefficient is easily found as follows. If the coefficient P is known and Q is the next coefficient, then $\alpha Q + \beta P = 0$ or $Q = \frac{-\beta P}{\alpha}$. Since the first term A is found to be equal to $\frac{a}{\alpha}$, from this the letters B, C, D, etc. are defined in the same way as they would arise from the division procedure. For the rest, by inspection it is clear that in the infinite series for $\frac{a}{\alpha + \beta z}$, the coefficient of z^n will be $\frac{\pm a\beta^n}{\alpha^{n+1}}$, where the positive sign occurs when n is even and the negative sign occurs when n is odd, or the coefficients can be expressed as $\frac{a}{\alpha}(-\beta/\alpha)^n$.

61. *In a similar way by means of a continued division procedure the rational function* $\frac{a+bz}{\alpha+\beta z+\gamma z^2}$ *can be converted to an infinite series.*

Since division becomes tedious and there is no easy way to show the nature of the resulting infinite series, it is more convenient to suppose that there is such an infinite series, and then find it in the way just shown. Hence we let $\frac{a+bz}{\alpha+\beta z+\gamma z^2} = A + Bz + Cz^2 + Dz^3 + Ez^4 + \cdots$. When both sides are multiplied by $\alpha + \beta z + \gamma z^2$, we obtain

$$a + bz = \alpha A + \alpha Bz + \alpha Cz^2 + \alpha Dz^3 + \alpha Ez^4 + \cdots$$

$$+ \beta Az + \beta Bz^2 + \beta Cz^3 + \beta Dz^4 + \cdots \qquad + \gamma Az^2 + \gamma Bz^3 + \gamma Cz^4 + \cdots .$$

From this we see that $\alpha A = a$ and $\alpha B + \beta A = b$, so that $A = \frac{a}{\alpha}$ and $B = \frac{b}{\alpha} - \frac{a\beta}{\alpha^2}$. The remaining coefficients are determined from the following equations: $\alpha C + \beta B + \gamma A = 0$, $\alpha D + \beta C + \gamma B = 0$, $\alpha E + \beta D + \gamma C = 0$, $\alpha F + \beta E + \gamma D = 0$, etc. Thus if two consecutive coefficients are know, then the next coefficient can be found. Thus if P and Q are two known consecutive coefficients and R is the next coefficient after P and Q, then $\alpha R + \beta Q + \gamma P = 0$ or $R = \frac{-\beta Q - \gamma P}{\alpha}$. Since the first two coefficients A and B have already been found the following C, D, E, F, etc. and all succeeding coefficients can be found from those already found. Thus we find the infinite series $A + Bz + Cz^2 + Dz^3 + \cdots$ equal to the given rational function $\frac{a+bz}{\alpha+\beta z+\gamma z^2}$.

EXAMPLE

If the given rational function is $\frac{1 + 2z}{1 - z - z^2}$ and this is set equal to the series $A + Bz + Cz^2 + Dz^3 + \cdots$, since

$a = 1$, $b = 2$, $\alpha = 1$, $\beta = -1$, and $\gamma = -1$, we have $A = 1$, $B = 3$. Then $C = B + A$, $D = C + B$, $E = D + C$, $F = E + D$, etc., so that any coefficient is the sum of the two immediately preceding it. If P and Q are known successive coefficients and R is the next coefficient, then $R = P + Q$. Since the first two coefficients A and B are known, the given rational function $\frac{1 + 2z}{1 - z - z^2}$ is transformed into the infinite series $1 + 3z + 4z^2 + 7z^3 + 11z^4 + 18z^5 + \cdots$, which can be continued as long as desired with no trouble.

62. From this discussion the nature of infinite series which express rational functions should be sufficiently clear. There is the law by which any coefficient is determined by a certain number of its predecessors. That is, if the denominator of the proposed function is $\alpha + \beta z$ and the infinite series is set equal to

$$A + Bz + Cz^2 + \cdots + Pz^n + Qz^{n+1} + Rz^{n+2} + Sz^{n+3} + \cdots,$$

any coefficient Q is determined only by its predecessor P since $\alpha Q + \beta P = 0$. If the denominator is the trinomial $\alpha + \beta z + \gamma z^2$, any coefficient of the series R is determined by the two predecessors Q and P, since $\alpha R + \beta Q + \gamma P = 0$. In like fashion, if the denominator is a quadrinomial $\alpha + \beta z + \gamma z^2 + \delta z^3$, then any coefficient S is determined by the three predecessors R, Q, and P, since $\alpha S + \beta R + \gamma Q + \delta P = 0$. A similar argument covers the cases with denominators of higher degree. In all of these series each term is determined from certain preceding terms according to some fixed law which is immediately clear from the denominator of the rational function which gives rise to the series. This kind of series if called recurrent by the celebrated DeMoivre, who has examined their nature very carefully. The name comes from the fact that if we wish to investigate subsequent terms we have to "run back", to previous terms.

63. For the formation of these series it is necessary that the constant α in the denominator be non-zero, since if $\alpha = 0$, the first term $A = \frac{a}{\alpha}$ and indeed all other terms would be infinite. With the exception of this case, which will be discussed later we can transform into an infinite series any rational function of the form $\frac{a + bz + cz^2 + dz^3 + \cdots}{1 - \alpha z - \beta z^2 - \gamma z^3 - \delta z^4 - \cdots}$, where the first term of the denominator is equal to 1. But any rational function can be reduced to this form unless the constant term is zero. The remaining terms in the denominator are written with negative signs in order that the resulting infinite series may have all positive signs. Thus, if the recurrent series is written as $A + Bz + Cz^2 + Dz^3 + Ez^4 + \cdots$, then the coefficients are determined to be $A = a$, $B = \alpha A + b$, $C = \alpha B + \beta A + c$, $D = \alpha C + \beta B + \gamma A + d$, $E = \alpha D + \beta C + \gamma B + \delta A + e$. Hence each coefficient is a weighted sum of the preceding coefficients in addition to a number from the numerator. Unless the numerator has an infinite number of terms this added number soon is absent and then succeeding coefficients are determined by a fixed law from certain of the preceding coefficients. In order that the law of formation remain unviolated, the rational function should be a proper fraction. If the fraction happens to be

improper, then the polynomial part will either add to or subtract from the corresponding terms and in these the fixed law will not be followed. For example, the improper rational function $\frac{1 + 2z - z^3}{1 - z - z^2}$ gives the series $1 + 3z + 4z^2 + 6z^3 + 10z^4 + 16z^5 + 26z^6 + 42z^7 + \cdots$, where by the law, each coefficient should be the sum of the two preceding coefficients, however, the fourth term $6z^3$ is an exception.

64. A recurrent series deserves special consideration if the denominator of the fraction which gives rise to it happens to be a power. Thus, if the rational function $\frac{a + bz}{(1 - \alpha z)^2}$ is expressed as a series, the result is

$$\begin{aligned} &a + 2\alpha az + 3\alpha^2 az^2 + 4\alpha^3 az^3 + 5\alpha^4 az^4 + \cdots \\ &+ bz + 2\alpha bz^2 + 3\alpha^2 bz^3 + 4\alpha^3 bz^4 + \cdots \end{aligned}$$

in which the coefficient of z^n will be $(n + 1)\alpha^n a + n\alpha^{n-1}b$. But this is a recurrent series, since each term is determined by the two preceding terms. The law of formation is clearly seen when the denominator is expanded to $1 - 2\alpha z + \alpha^2 z^2$. If we let $\alpha = 1$ and $z = 1$, the series becomes a general arithmetic progression $a + (2a + b) + (3a + 2b) + (4a + 3b) + \cdots$ whose terms have a constant difference. Thus every arithmetic progression is a recurrent series, for if $A + B + C + D + E + F + \cdots$ is an arithmetic progression then $C = 2B - A$, $D = 2C - B$, $E = 2D - C$, etc.

65. The function $\frac{a + bz + cz^2}{(1 - \alpha z)^3}$, since $\frac{1}{(1 - \alpha z)^3} = (1 - \alpha z)^{-3}$ $= 1 + 3\alpha z + 6\alpha^2 z^2 + 10\alpha^3 z^3 + 15\alpha^4 z^4 + \cdots$, is transformed into this infinite series:

$$a + 3\alpha az + 6\alpha^2 az^2 + 10\alpha^3 az^3 + 15\alpha^4 az^4 + \cdots$$
$$+ bz + 3\alpha bz^2 + 6\alpha^2 bz^3 + 10\alpha^3 bz^4 + \cdots$$
$$+ cz^2 + 3\alpha cz^3 + 6\alpha^2 cz^4 + \cdots$$

in which z^n has the coefficient

$\frac{(n+1)(n+2)}{1\cdot 2}\alpha^n a + \frac{(n)(n+1)}{1\cdot 2}\alpha^{n-1} b + \frac{(n-1)(n)}{1\cdot 2}\alpha^{n-2} c$. If in this series we let $\alpha = 1$ and $z = 1$, the series becomes a general progression of the second order, whose second differences are constant. If we let $A + B + C + D + E + \cdots$ be this kind of progression, it will be a recurrent series in which a term depends on the three preceding terms in such a way that $D = 3C - 3B + A$, $E = 3D - 3C + B$, $F = 3E - 3D + C$, etc. Since the second differences of an arithmetic progression are equal, namely equal to zero, these properties apply also to arithmetic progressions.

66. In a similar way the function $\frac{a + bz + cz^2 + dz^3}{(1 - \alpha z)^4}$ gives an infinite series in which the coefficient of z^n has the form

$$\frac{(n+1)(n+2)(n+3)}{1\cdot 2\cdot 3}\alpha^n a + \frac{n(n+1)(n+2)}{1\cdot 2\cdot 3}\alpha^{n-1} b$$
$$+ \frac{(n-1)n(n+1)}{1\cdot 2\cdot 3}\alpha^{n-2} c + \frac{(n-2)(n-1)n}{1\cdot 2\cdot 3}\alpha^{n-3} d.$$

If we let $\alpha = 1$ and $z = 1$, then this series represents all algebraic progression of the third order, in which the third differences are all constant. Every progression of this order, which has the form

$A + B + C + D + E + F + \cdots$ will be recurrent from the denominator $1 - 4z + 6z^2 - 4z^3 + z^4$. Hence

$E = 4D - 6C + 4B - A$, $F = 4E - 6D + 4C - B$, etc. This property belongs also to every progression of a lower order.

67. In this way it can be shown that every algebraic progression of any order which finally leads to constant differences is a recurrent series, whose law is given by the denominator $(1 - z)^n$, where $n - 1$ is the order of the progression. Since $a^m + (a + b)^m + (a + 2b)^m + (a + 3b)^m + \cdots$ is a progression of the mth order, it is a recurrent series and we have

$$\begin{aligned} 0 = a^m &- \frac{n}{1}(a + b)^m + \frac{n(n - 1)}{1 \cdot 2}(a + 2b)^m \\ &- \frac{n(n - 1)(n - 2)}{1 \cdot 2 \cdot 3}(a + 3b)^m + \cdots \pm (a + nb)^m, \end{aligned}$$

where the upper sign appears if n is odd and the lower sign if n is even. But this equation is always true if $n = m + 1$. From this it can be understood how far the applications of recurrent series extend.

68. If the denominator happens to be a power of a multinomial rather than a binomial, the nature of the series can be explained in a different way. Suppose the function is $\frac{1}{(1 - \alpha z - \beta z^2 - \gamma z^3 - \delta z^4 - \cdots)^{m+1}}$, then the infinite series representation is

$$\begin{aligned} 1 &+ \frac{(m + 1)}{1}\alpha z + \frac{(m + 1)(m + 2)}{1 \cdot 2}\alpha^2 z^2 \\ &+ \frac{(m + 1)(m + 2)(m + 3)}{1 \cdot 2 \cdot 3}\alpha^3 z^3 + \cdots \\ &+ \frac{(m + 1)}{1}\beta z^2 + \frac{(m + 1)(m + 2)}{1 \cdot 2}2\alpha\beta z^3 + \cdots \\ &+ \frac{(m + 1)}{1}\gamma z^3 + \cdots . \end{aligned}$$

In order to examine this series more carefully, let us write it with general letters as follows:

$$\begin{aligned} 1 &+ Az + Bz^2 + Cz^3 + \cdots + Kz^{n-3} \\ &+ Lz^{n-2} + Mz^{n-1} + Nz^n + \cdots . \end{aligned}$$

Any coefficient N is determined by the same number of preceding coefficients as there are letters α, β, γ, δ, etc. as follows:

$N = \frac{m + n}{n}\alpha M + \frac{2m + n}{n}\beta L + \frac{3m + n}{n}\gamma K + \frac{4m + n}{n}\delta I + \cdots$. This law is not constant since it depends on the power of z, but in other respects it is similar to the constant law of a progression which is determined by the denominator in a recurrent series. Note that this non-constant law applies only if the numerator is 1 or some constant. If the numerator also contains some powers of z, then the law becomes much more complicated, as will be more easily seen after the study of some of the principles of differential calculus.

69. Up until this time we have supposed that the constant term in the denominator is not equal to zero and in that place we have let it be equal to 1. Now let us see what series arises, if in the denominator we allow the constant term to vanish. In this case the rational function has the form $\frac{a + bz + cz^2 + \cdots}{z(1 - \alpha z - \beta z^2 - \cdots)}$. If we neglect the factor z, the rest of the function $\frac{a + bz + cz^2 + \cdots}{1 - \alpha z - \beta z^2 - \cdots}$ can be written in a recurrent series $A + Bz + Cz^2 + Dz^3 + \cdots$. Then it is clear that

$$\frac{a + bz + cz^2 + \cdots}{z(1 - \alpha z - \beta z^2 - \cdots)} = \frac{A}{z} + B + Cz + Dz^2 + Ez^3 + \cdots .$$

Similarly $\frac{a + bz + cz^2 + \cdots}{z^2(1 - \alpha z - \beta z^2 - \cdots)} = \frac{A}{z^2} + \frac{B}{z} + C + Dz + Ez^2 + \cdots$

and in general

$$\frac{a + bz + cz^2 + \cdots}{z^m(1 - \alpha z - \beta z^2 - \cdots)} = \frac{A}{z^m} + \frac{B}{z^{m-1}}$$

$+ \frac{C}{z^{m-2}} + \frac{D}{z^{m-3}} + \cdots$, whatever the exponent m may be.

70. Since by substituting for z another variable x in a rational function, and indeed this can be done in many different ways, it follows that the same rational function can be expressed in a recurrent series in infinitely many different ways. For instance, let $y = \frac{1 + z}{1 - z - z^2}$, so that we have the recurrent series $y = 1 + 2z + 3z^2 + 5z^3 + 8z^4 + \cdots$. If we let $z = \frac{1}{x}$, then $y = \frac{x^2 + x}{x^2 - x - 1} = \frac{-x(1 + x)}{1 + x - x^2}$. But

$\frac{1 + x}{1 + x - x^2} = 1 + 0x + x^2 - x^3 + 2x^4 - 3x^5 + 5x^6 - \cdots$ so that $y = -x + 0x^2 - x^3 + x^4 - 2x^5 + 3x^6 - 5x^7 + \cdots$. However, if we let $z = \frac{1 - x}{1 + x}$, then $y = \frac{-2 - 2x}{1 - 4x - x^2}$ and

$y = -2 - 10x - 42x^2 - 178x^3 - 754x^4 - \cdots$ and so innumerable recurrent series of this kind can be found to represent y.

71. Irrational functions can be transformed into series by the following universal theorem:

$$(P + Q)^{\frac{m}{n}} = P^{\frac{m}{n}} + \frac{m}{n}P^{\frac{m-n}{n}}Q + \frac{m(m - n)}{n \cdot 2n}P^{\frac{m-2n}{n}}Q^2$$

$$+ \frac{m(m - n)(m - 2n)}{n \cdot 2n \cdot 3n}P^{\frac{m-3n}{n}}Q^3 + \cdots,$$ where there are an infinite number of terms unless $\frac{m}{n}$ is a positive integer. Thus when we choose fixed integers for m and n,

$$(P + Q)^{\frac{1}{2}} = P^{\frac{1}{2}} + \frac{1}{2}P^{\frac{-1}{2}}Q - \frac{1 \cdot 1}{2 \cdot 4}P^{\frac{-3}{2}}Q^2 + \frac{1 \cdot 1 \cdot 3}{2 \cdot 4 \cdot 6}P^{\frac{-5}{2}}Q^3 - \cdots$$ or

$$(P + Q)^{\frac{-1}{2}} = P^{\frac{-1}{2}} - \frac{1}{2}P^{\frac{-3}{2}}Q + \frac{1 \cdot 3}{2 \cdot 4}P^{\frac{-5}{2}}Q^2 - \frac{1 \cdot 3 \cdot 5}{2 \cdot 4 \cdot 6}P^{\frac{-7}{2}}Q^3 + \cdots$$ or

$$(P+Q)^{\frac{1}{3}} = P^{\frac{1}{3}} + \frac{1}{3}P^{\frac{-2}{3}}Q - \frac{1\cdot 2}{3\cdot 6}P^{\frac{-5}{3}}Q^2 + \frac{1\cdot 2\cdot 5}{3\cdot 6\cdot 9}P^{\frac{-8}{3}}Q^3 - \cdots \quad \text{or}$$

$$(P+Q)^{\frac{-1}{3}} = P^{\frac{-1}{3}} - \frac{1}{3}P^{\frac{-4}{3}}Q + \frac{1\cdot 4}{3\cdot 6}P^{\frac{-7}{3}}Q^2 - \frac{1\cdot 4\cdot 7}{3\cdot 6\cdot 9}P^{\frac{-10}{3}}Q^3 + \cdots \quad \text{or}$$

$$(P+Q)^{\frac{2}{3}} = P^{\frac{2}{3}} + \frac{2}{3}P^{\frac{-1}{3}}Q - \frac{2\cdot 1}{3\cdot 6}P^{\frac{-4}{3}}Q^2 + \frac{2\cdot 1\cdot 4}{3\cdot 6\cdot 9}P^{\frac{-7}{4}}Q^3 - \cdots .$$

72. In a series of this type any term can be determined from the form of the preceding term. If in a series which arises from $(P+Q)^{\frac{m}{n}}$, a certain term has the form $MP^{\frac{m-kn}{n}}Q^k$, then the next term has the form $\frac{m-kn}{(k+1)n}MP^{\frac{m-(k+1)n}{n}}Q^{k+1}$. It should be noted that in succeeding terms the exponent of P decreases by 1 and the exponent of Q increases by 1. In certain cases it may be be more convenient to express the general form $(P+Q)^{\frac{m}{n}}$ as $P^{\frac{m}{n}}(1+Q/P)^{\frac{m}{n}}$. Then when $(1+Q/P)^{\frac{m}{n}}$ is expressed in an infinite series the result may be multiplied by $P^{\frac{m}{n}}$ to give the series in its first form. On the other hand, if m designates not only integers, but also fractions, then n can always equal 1. Further, if instead of $\frac{Q}{P}$, which is some function of z, we put Z, then we have

$$(1+Z)^m = 1 + \frac{m}{1}Z + \frac{m(m-1)}{1\cdot 2}Z^2 + \frac{m(m-1)(m-2)}{1\cdot 2\cdot 3}Z^3 + \cdots .$$

In what follows, it will be convenient to have noted the following

$$(1+Z)^{m-1} = 1 + \frac{(m-1)}{1}Z + \frac{(m-1)(m-2)}{1\cdot 2}Z^2 + \frac{(m-1)(m-2)(m-3)}{1\cdot 2\cdot 3}Z^3 + \cdots .$$

73. If $Z = \alpha z$, then

$$(1 + \alpha z)^{m-1} = 1 + \frac{m-1}{1}\alpha z + \frac{(m-1)(m-2)}{1\cdot 2}\alpha^2 z^2$$

$+ \dfrac{(m-1)(m-2)(m-3)}{1\cdot 2\cdot 3}\alpha^3 z^3 + \cdots$. If we write this series in the general form $1 + Az + Bz^2 + Cz^3 + \cdots + Mz^{n-1} + Nz^n + \cdots$, then any coefficient N is determined by the preceding term M as follows: $N = \dfrac{m-n}{n}\alpha M$. Thus, if $n = 1$, then $N = A = \dfrac{m-1}{1}\alpha$, while if $n = 2$, since $M = A = \dfrac{m-1}{1}\alpha$, then

$N = B = \dfrac{m-2}{2}\alpha m = \dfrac{(m-1)(m-2)}{1\cdot 2}\alpha^2$. In like manner

$C = \dfrac{m-3}{3}\alpha B = \dfrac{(m-1)(m-2)(m-3)}{1\cdot 2\cdot 3}\alpha^3$ as was seen in the original statement of the series.

74. Let $Z = \alpha z + \beta z^2$, then

$$(1 + \alpha z + \beta z^2)^{m-1} = 1 + \frac{m-1}{1}(\alpha z + \beta z^2)$$

$+ \dfrac{(m-1)(m-2)}{1\cdot 2}(\alpha z + \beta z^2)^2 + \cdots$. Now if the terms are arranged according to powers of z then

$$(1 + \alpha z + \beta z^2)^{m-1} = 1 + \frac{m-1}{1}\alpha z + \frac{(m-1)(m-2)}{1\cdot 2}\alpha^2 z^2$$

$$+ \frac{(m-1)(m-2)(m-3)}{1\cdot 2\cdot 3}\alpha^3 z^3$$

$+ \cdots + \dfrac{m-1}{1}\beta z^2 + \dfrac{(m-1)(m-2)}{1\cdot 2}2\alpha\beta z^3 + \cdots$. If we write the general form for this series:

$1 + Az + Bz^2 + Cz^3 + \cdots + Lz^{n-2} + Mz^{n-1} + Nz^n + \cdots$, then any coefficient can be determined from the two preceding coefficients as follows: $N = \frac{m-n}{n}\alpha M + \frac{2m-n}{n}\beta L$. Thus all terms are defined by the first one, which of course is 1. That is

$$A = \frac{m-1}{1}\alpha, \qquad B = \frac{m-1}{2}\alpha A + \frac{2m-2}{2}\beta,$$

$$C = \frac{m-3}{3}\alpha B + \frac{2m-3}{3}\beta A, \qquad D = \frac{m-4}{4}\alpha C + \frac{2m-4}{4}\beta B.$$

75. If $Z = \alpha z + \beta z^2 + \gamma z^3$, then

$(1 + \alpha z + \beta z^2 + \gamma z^3)^{m-1} = 1 + \frac{m-1}{1}(\alpha z + \beta z^2 + \gamma z^3)$

$+ \frac{(m-1)(m-2)}{1\cdot 2}(\alpha z + \beta z^2 + \gamma z^3)^2 + \cdots$. If we order all the terms according to increasing powers of z we obtain the series $1 + \frac{m-1}{1}\alpha z + \frac{(m-1)(m-2)}{1\cdot 2}\alpha^2 z^2 + \frac{(m-1)(m-2)(m-3)}{1\cdot 2\cdot 3}\alpha^3 z^3$

$+ \cdots + \frac{m-1}{1}\beta z^2 + \frac{(m-1)(m-2)}{1\cdot 2}2\alpha\beta z^3 + \cdots + \frac{m-1}{1}\gamma z^3$

$+ \cdots$. In order that the law of the progression may more easily become clear, we write the series as $1 + Az + Bz^2 + Cz^3 + \cdots + Kz^{n-3}$

$+ Lz^{n-2} + Mz^{n-1} + Nz^n + \cdots$. In this series the coefficient of each term is determined by the coefficient of the three preceding terms as follows: $N = \frac{m-n}{n}\alpha M + \frac{2m-n}{n}\beta L + \frac{3m-n}{n}\gamma K$. Since the first term is 1, and it has no preceding terms, $A = \frac{m-1}{1}\alpha$, $B = \frac{m-2}{2}\alpha A + \frac{2m-2}{2}\beta$,

$$C = \frac{m-3}{3}\alpha B + \frac{2m-3}{3}\beta A + \frac{3m-3}{3}\gamma,$$

$$D = \frac{m-4}{4}\alpha C + \frac{2m-4}{4}\beta B + \frac{3m-4}{4}\gamma A,$$

$$E = \frac{m-5}{5}\alpha D + \frac{2m-5}{5}\beta C + \frac{3m-5}{5}\gamma B.$$

76. In general if

$(1 + \alpha z + \beta z^2 + \gamma z^3 + \delta z^4 + \cdots)^{m-1} = 1 + Az + Bz^2$

$+ Cz^3 + Dz^4 + Ez^5 + \cdots$. Then all of the terms are so defined by preceding terms, that $A = \frac{m-1}{1}\alpha$, $B = \frac{m-2}{2}\alpha A + \frac{2m-2}{2}\beta$,

$$C = \frac{m-3}{3}\alpha B + \frac{2m-3}{3}\beta A + \frac{3m-3}{3}\gamma,$$

$$D = \frac{m-4}{4}\alpha C + \frac{2m-4}{4}\beta B + \frac{3m-4}{4}\gamma A + \frac{4m-4}{4}\delta,$$

$$E = \frac{m-5}{5}\alpha D + \frac{2m-5}{5}\beta C + \frac{3m-5}{5}\gamma B + \frac{4m-5}{5}\delta A + \frac{5m-5}{5}\epsilon,$$

etc. where each term is determined by as many predecessors as there are letters α, β, γ, δ, etc. in the function of z whose power is converted into the series. Furthermore, the nature of the law agrees with that which was discussed above in section 68, where a similar form $(1 - \alpha z - \beta z^2 - \gamma z^3 - \cdots)^{-m-1}$ was developed in an infinite series. Indeed, if instead of m we write $-m$ and the letters α, β, γ, δ, etc. are made negative, the resulting series is exactly the same. In this place we will not prove *a priori* the law of this progression, since it can be done so much more easily with the aid of some principles of differential calculus. In the meantime let it suffice that the truth has been made reasonable by the application to examples of so many different kinds.

CHAPTER V

Concerning Functions of Two or More Variables.

77. Although we have heretofore considered many variable quantities, still all of them were so constituted that they were all functions of one variable. If one of them was determined, then all of them were determined. Now we shall consider variable quantities which are independent of each other. In this case, although a definite value is given to one, the others remain indeterminate and variable. Variable quantities of this sort, such as x, y, z, can take on any definite values. If these values are compared to one another, they are completely different, since when any definite value is substituted for z, then x and y remain undetermined as before the substitution. We are concerned here with dependent and independent variables. In the former case, if one is determined, then the others are determined; in the latter however, the determination of one leaves the others completely unrestricted.

78. *A function of two or more variable quantities x, y, z is an expression composed in any way from these.*

The expression $x^3 + xyz + az^3$ is a function of three variables x, y, z. This function, if one variable, say z, is determined in that a constant number is substituted for it, then it still is a variable quantity, that is, a function of x and y. If in addition, besides z, y is also determined, then it is still a function of x.

A function of this kind, which is of several variables, is not determined until each of the variable quantities is determined. Since any variable quantity can be determined in an infinite number of ways, a function of two variables, when one of the variables is determined in any of an infinite number of ways, still admits of an infinite number of determinations. Thus it admits an infinity of infinite determinations. Further, in a function of three variables, the determinations will be greater by infinity; it grows in this manner with the number of variables.

79. *Functions of several variables, just as functions of a single variable, are primarily divided into algebraic and transcendental.*

The former are those in which the rule of composition involves only algebraic operations; the latter are those in which at least one transcendental operation enters. In this case we note that the transcendental operation can involve all of the variables, some of the variables, or only one. This expression, $z^2 + y \log z$, since it contains the expression $\log z$, is a transcendental function of y and z. We can consider it less transcendent though, since if the variable z is determined, it becomes an algebraic function of y. For the present it is not necessary to pursue this subdivision any further.

80. *Algebraic functions are subdivided into irrational and non-irrational functions while the latter is further subdivided into polynomials and rational functions.*

The rationale for these distinctions should be clear from the first chapter. A non-irrational function is completely free of any irrationality affecting the variables of the function. If the non-irrational function has involved division with a variable in the divisor, then it is a rational function; if it is not so encumbered, then it is a polynomial. The general form of a polynomial in two variables y and z is

$$\alpha + \beta y + \gamma z + \delta y^2 + \epsilon yz + \zeta z^2 \\ + \eta y^3 + \theta y^2 z + \iota yz^2 + \kappa z^3 + \cdots .$$

If P and Q both designate polynomials, whether of two or several variables, then P/Q is the general form for a rational function. An irrational function can be either explicit or implicit. The former, which involves a radical sign, has been sufficiently developed. The latter is given by an equation which cannnot be resolved. The function V of y and z is implicity irrational, if it is given by $V^5 = (ayz + z^3)V^2 + (y^4 + z^4)V + y^5 + 2ayz^3 + z^5$.

81. *We must note that functions of several variables may be multiple valued just as those of a single variable.*

The non-irrational functions will be single valued since when each of the variables is determined, the function takes only one value. Let P, Q, R, S, etc. designate single valued functions of the variables x, y, z, then V will be a two-valued function of the same variables if $V^2 - PV + Q = 0$. This is the case since, whatever values are given to x, y, and z, the function V always has two determined values. Likewise, V will be a three-valued function if $V^3 - PV^2 + QV - R = 0$; it will be a four-valued function if $V^4 - PV^3 + QV^2 - RV + S = 0$. In a similar way other multiple-valued functions can be defined.

82. Just as when a function of one variable z is set equal to zero, the variable z is determined to one or several values, so when a function of two variables

y and z is set equal to zero, each of the variables is determined by the other. In this way each is a function of the other, while previously the two were mutually independent. In the same way, if a function of three variables x, y, z is set equal to zero, then each of the variables is determined by the other two, and so becomes a function of the other two variables. The same thing happens if the function is set equal to some constant other than zero, or even set equal to some other function. Indeed, from each equation, whatever variables may be involved, one variable is determined by, and so becomes a function of, the remaining variables. Two different equations in the same variables will define a pair of the variables as functions of the other variables. In general the number of equations will determine the number of functions defined.

83. *A distinction quite worthy of notice in functions of several variables is that between homogeneous and heterogeneous.*

A homogeneous function is one in which each term has the same degree. A heterogeneous function is one in which different degrees occur. We think of each variable as having one degree, a quadratic in one variable or the product of two different variables is of degree two. The product of three variables, whether they are the same or different, is of degree three, and so forth. Constants are not given a degree. In the following formulas: αy and βz have one degree; αy^2, βyz, γz^2 are second degree; αy^3, $\beta y^2 z$, γyz^2, δz^3 are third degree; αy^4, $\beta y^3 z$, $\gamma y^2 z^2$, δyz^3, ϵz^4 are fourth degree, and so forth.

84. First we apply this distinction to polynomials, and we consider only two variables since the concept is clear when there are more variables.

A polynomial is homogeneous when each term has the same degree.

It is very convenient to classify functions of this kind by the degree of the terms which constitute the function. The general form of a first degree polynomial is $\alpha y + \beta z$; $\alpha y^2 + \beta yz + \gamma z^2$ is the general form of a second degree function; $\alpha y^3 + \beta y^2 z + \gamma yz^2 + \delta z^3$ is third degree;
$\alpha y^4 + \beta y^3 z + \gamma y^2 z^2 + \delta yz^3 + \epsilon z^4$ is fourth degree, and so forth. In analogy with this terminology we will say that a constant function is zero degree.

85. *A rational function is homogeneous if its numerator and denominator are both homogeneous.*

An example of a homogeneous rational function in y and z is $\frac{ay^2 + bz^2}{\alpha y + \beta z}$. The degree of the function is given by subtracting the degree of the denominator from the degree of the numerator. The given example is a first degree function. The expression $\frac{y^5 + z^5}{y^2 + z^2}$ is a third degree function. If the numerator and denominator have the same degree, then the function will be zero degree. For example, $\frac{y^3 + z^3}{y^2 z}$, $\frac{y}{z}$, $\frac{\alpha z^2}{y^2}$, $\frac{\beta y^3}{z^3}$ are all zero degree. If the degree of the denominator is greater than the degree of the numerator, then the degree of the rational function will be negative. A function of degree -1 is $\frac{y}{z^2}$, while $\frac{y + z}{y^4 + z^4}$ is of degree -3, and $\frac{1}{y^5 + ayz^4}$ is of degree -5, since the numerator is zero degree. For the rest, many homogeneous functions are easily recognized when each part with the same degree are either added or subtracted. The expression $\alpha y + \frac{\beta z^2}{y} + \frac{\gamma y^4 - \delta z^4}{y^2 z + yz^2}$ has degree 1, while

$$\alpha + \frac{\beta y}{z} + \frac{\gamma z^2}{y^2} + \frac{y^2 + z^2}{y^2 - z^2}$$ is of degree zero.

86. The nature of a homogeneous function can be extended also to irrational functions. If P is a homogeneous function of degree n, then $\sqrt{P}$ has degree $\frac{1}{2}n$, and $P^{\frac{1}{3}}$ has degree $\frac{1}{3}n$. In general $P^{\frac{\mu}{\nu}}$ is a function of degree $\frac{\mu}{\nu}n$. Thus $\sqrt{y^2 + z^2}$ is a first degree function, $(y^9 + z^9)^{\frac{1}{3}}$ is a third degree function, $(yz + z^2)^{\frac{3}{4}}$ is of degree $\frac{3}{2}$, and $\frac{y^2 + z^2}{\sqrt{y^4 + z^4}}$ is of zero degree. Putting this all together, we see that the expression

$$\frac{1}{y} + \frac{y\sqrt{y^2 + z^2}}{z^3} - \frac{y}{(y^6 - z^6)^{\frac{1}{3}}} + \frac{y\sqrt{z}}{z^2\sqrt{y} + \sqrt{y^5 + z^5}}$$ is a function of degree -1.

87. Whether an implicit irrational function is homogeneous or not should be clear from this. Let V be such an implicit function and $V^3 + PV^2 + QV + R = 0$, where P, Q, and R are functions of y and z. First it should be clear that V cannot be homogeneous unless P, Q, and R are all homogeneous. Furthermore, if we let V be a function of degree n then V^2 is of degree $2n$ and V^3 is of degree $3n$. Since each term must have the same degree, it is necessary that P have degree n, Q have degree $2n$, and R have degree $3n$. If now P, Q, and R are homogeneous functions of respective degrees n, $2n$, $3n$, then we conclude that V is of degree n. Now let $V^5 + (y^4 + z^4)V^3 + ay^8V - z^{10} = 0$, then V is a homogeneous second degree function of y and z.

88. *If V is a homogeneous function of degree n in y and z, and we make the substitution $y = uz$, then V can be expressed as the product of z^n and some function of u.*

By this substitution $y = uz$ in each term we introduce only that number of degrees in z which before were in y. Since in each term the joint degrees of y and z was n, now the single variable z has degree n. It follows that each term now contain z^n. In this way the function V becomes divisible and the quotient is a function of the single variable u. This becomes most clear in the case of polynomials. If $V = \alpha y^3 + \beta y^2 z + \gamma y z^2 + \delta z^3$, when we set $y = uz$, then $V = z^3(\alpha u^3 + \beta u^2 + \gamma u + \delta)$. But then the same is clear in the case of a rational function. Let $V = \frac{\alpha y + \beta z}{y^2 + z^2}$, which is of degree -1. When we let $y = uz$, $V = z^{-1}\frac{\alpha u + \beta}{u^2 + 1}$. Even irrational functions follow the same rule. If $V = \frac{y + \sqrt{y^2 + z^2}}{z\sqrt{y^3 + z^3}}$, which is of degree $-\frac{3}{2}$, when we let $y = uz$, we have $V = z^{-\frac{3}{2}}\frac{u + \sqrt{u^2 + 1}}{\sqrt{u^3 + 1}}$. In this way a homogeneous function of two variables is reduced to a function of one variable. The power of z, since it is a factor, does not contaminate the function of u.

89. *A homogeneous function V of two variables y and z of degree zero is transformed into a function of one variable u when we let $y = uz$.*

Since the degree is zero, the power of z which multiplies the function of u, is $z^0 = 1$. For this reason the variable z does not appear in the expression. Thus if $V = \frac{y + z}{y - z}$ and we let $y = uz$, then $V = \frac{u + 1}{u - 1}$. In the irrational case, if $V = \frac{y - \sqrt{y^2 - z^2}}{z}$ and we let $y = uz$, $V = u - \sqrt{u^2 - 1}$.

90. *A homogeneous polynomial in two variables y and z can be expressed as the product of the same number of factors of the form $\alpha y + \beta z$ as there are degrees in the function.*

Since the function is homogeneous, when we let $y = uz$, it becomes the product of z^n and some polynomial in u. But this polynomial can be expressed as the product of linear factors of the form $\alpha u + \beta$. When each of the linear factors is multiplied by z, it takes the form $\alpha uz + \beta z = \alpha y + \beta z$, since $uz = y$. There will be the same number of these factors as there are z's in z^n. Note that these linear factors may be real or complex; that is, the coefficients α and β can be real or complex.

From this it follows that a second degree function $ay^2 + byz + cz^2$ has two linear factors of the form $\alpha y + \beta z$. The function $ay^3 + by^2z + cyz^2 + dz^3$ has three linear factors of the form $\alpha y + \beta z$; and so forth for homogeneous polynomials of higher degree.

91. Insofar as $\alpha y + \beta z$ is the general form for a homogeneous polynomial of degree 1, $(\alpha y + \beta z)(\gamma y + \delta z)$ is the general form of a homogeneous polynomial of degree 2, and every homogeneous polynomial of degree 3 can be expressed as $(\alpha y + \beta z)(\gamma y + \delta z)(\epsilon y + \zeta z)$, thus all homogeneous polynomials can be expressed as the product of factors of the form $\alpha y + \beta z$ where the number of factors is the degree of the polynomial. These factors may be found by the solution of equations in the same way as we previously found the linear

factors of a polynomial in one variable. We cannnot extend this property of a function of two variables to functions of three or more variables. The reason is that the general form $ay^2 + byz + cyx + dxy + ex^2 + fz^2$ does not generally reduce to a product of the form $(\alpha y + \beta z + \gamma x)(\delta y + \epsilon z + \zeta x)$. Even less so can the function of higher degree be put into a product of this form.

92. From all that has been said about homogeneous functions, we may understand what a heterogeneous function is. That is, in each of the terms there is not always the same degree. However, heterogeneous functions can be classified according to the number of different degrees which occur in them. Thus a bifid function is one in which two different degrees occur; it is the sum of two homogeneous functions with different degrees. For example, $y^5 + 2y^3z^2 + y^2 + z^2$ is a bifid function since part of it is of degree five and the other part of degree two. A trifid function is one in which three different degrees occur, that is, one which is made up of three homogeneous functions; for example, $y^6 + y^2z^2 + z^4 + y - z$. Besides these, there are rational and irrational functions which are so mixed up that they cannot be resolved into homogeneous functions; for example $\dfrac{y^3 + ayz}{by + z^2}$ and $\dfrac{a + \sqrt{y^2 + z^2}}{y^2 - bz}$.

93. Sometimes a heterogeneous function can be reduced to a homogeneous function by means of a suitable substitution for one or another of the variables. It is not very easy to state under what conditions this may be possible. Let it suffice to give a few examples of this procedure. If the given function is $y^5 + z^2y + y^3z + \dfrac{z^3}{y}$, after a bit of attention it becomes clear that the substi-

tution $z = x^2$ will result in $y^5 + x^4y + y^3x^2 + \frac{x^6}{y}$, which is a homogeneous function of degree 5 in x and y. Further, the function $y + y^2x + y^3x^2 + y^5x^4 + \frac{a}{x}$ can be reduced to homogeneity by the substitution $x = \frac{1}{z}$. This results in the degree 1 function $y + \frac{y^2}{z} + \frac{y^3}{z^2} + \frac{y^5}{z^4} + az$. There are much more difficult cases which can be reduced to homogeneity, but with substitutions which are not so simple.

94. Finally we should note the classification of polynomials according to order. The order is defined by the greatest degree of any term in the function. An example of a second order polynomial is $x^2 + y^2 + z^2 + ay - a^2$ since there are terms of degree 2.
Also $y^4 + yz^3 - ay^2z + abyz - a^2y^2 + b^4$ is a fourth order polynomial. This classification is usually made in the discussion of curves, and for this reason it merits notice.

95. There remains the classification of polynomials into reducible and irreducible. A reducible function is one which can be expressed as the product of two or more non-irrational functions. For example, $y^4 - z^4 + 2az^3 - 2byz^2 - a^2z^2 + 2abzy - b^2y^2$ is the product of the two functions $(y^2 + z^2 - az + by)(y^2 - z^2 + az - by)$. We have seen that every homogeneous polynomial in two variables is a reducible function, since it is the product of as many factors of the form $(\alpha y + \beta z)$ as there are degrees in the polynomial. A polynomial is irreducible if it cannot be expressed as the product of non-irrational factors. It is easily seen that there are no non-irrational factors

of $y^2 + z^2 - a^2$. From a consideration of divisors it appears whether a given function is reducible or irreducible.

CHAPTER VI

On Exponentials and Logarithms.

96. Although the concept of a transcendental function depends on integral calculus, still, before we come to that, there are certain kinds of functions which are more obvious, which can be conveniently developed, and which open the door to further investigations. First of all we will consider exponentials, or powers in which the exponent itself is a variable. It is clear that quantities of this kind are not algebraic functions, since in those the exponents must be constant. There are different kinds of exponentials, according to whether only the exponent is a variable or both the base and the exponent are variables. The first case is exemplified by a^z , while the second by y^z . Indeed the exponent itself may be an exponential as in the following: a^{a^z} , a^{y^z} , y^{a^z} , x^{y^z} . We will not consider these different forms to be different genera, since their nature will be sufficiently clear if we develop only a^z.

97. Let the exponential to be considered be a^z where a is a constant and the exponent z is a variable. Since the exponent z stands for all determined numbers, it is clear at least that all positive integers can be substituted for z to give determined values a^1, a^2, a^3, a^4, a^5, a^6, etc. If for z we substitute the negative integers -1, -2, -3, etc., we obtain $\frac{1}{a}$, $\frac{1}{a^2}$, $\frac{1}{a^3}$, $\frac{1}{a^4}$, etc. If $z = 0$, then we have $a^0 = 1$. If we substitute a fraction for z, for instance $\frac{1}{2}$, $\frac{1}{3}$, $\frac{2}{3}$, $\frac{1}{4}$, $\frac{3}{4}$,

etc. we obtain the values $\sqrt{a}$, $a^{\frac{1}{3}}$, $a^{\frac{2}{3}}$, $a^{\frac{1}{4}}$, $a^{\frac{3}{4}}$, etc. These symbols can have two or more values, since the extraction of roots gives several values. However we will consider only their primary value, that is the real positive values since a^z is to be thought of as a single valued function. For this reason $a^{\frac{5}{2}}$ lies between a^2 and a^3, and so it is a quantity of the same genus. Although $a^{\frac{5}{2}}$ is equal to both $-a^2\sqrt{a}$ and $a^2\sqrt{a}$, we consider only the second. In like manner we let z take irrational values, even though it is more difficult to understand this concept. However, we consider only real values for z. Thus $a^{\sqrt{7}}$ has a value which lies between a^2 and a^3.

98. The values of the exponential a^z depend primarily on the magnitude of the constant a. If $a = 1$, then we always have $a^z = 1$, no matter what value is given to z. If $a > 1$, then a^z will have a greater value if the value of z is greater than it was originally and as z goes to infinity, so also a^z increases to infinity. If $z = 0$, then $a^z = 1$; if $z < 0$, then the values of a^z become less than 1 and as z goes to $-\infty$, a^z goes to 0. On the other hand if $a < 1$ but still positive, then the values of a^z decrease when z increases above 0. The exponential increases as z increases in the negative direction. Since when $a < 1$, we have $\frac{1}{a} > 1$, and if we let $\frac{1}{a} = b$, then $a^z = b^{-z}$. For this reason we can examine the case when $a < 1$ from the case when $a > 1$.

99. If $a = 0$, we take a huge jump in the values of a^z. As long as the value of z remains positive, or greater than zero, then we always have $a^z = 0$. If $z = 0$, then $a^0 = 1$. However if z is a negative number, then a^z takes on an

infinitely large value; for example, if $z = -3$, then $a^z = 0^{-3} = \frac{1}{0^3} = \frac{1}{0}$, which is infinite. Much greater jumps occur if the constant a takes on a negative value, for instance -2 In this case, when z takes on integral values, a^z takes positive and negative values alternately, as can be seen from the sequence $a^{-4}, a^{-3}, a^{-2}, a^{-1}, a^0, a^1, a^2, a^3, a^4$, etc.

$+\frac{1}{16}, -\frac{1}{8}, +\frac{1}{4}, -\frac{1}{2}, +1, -2, +4, -8, +16$. Furthermore if the exponent z takes fractional values, then $a^z = (-2)^z$ sometimes has real values and sometimes complex values. For instance $a^{1/2} = \sqrt{-2}$ which is a pure imaginary, while $a^{\frac{1}{3}} = (-2)^{\frac{1}{3}} = -2^{\frac{1}{3}}$ which is real. If the exponent z is given an irrational value, then a^z may give real or complex values, but this cannnot be predicted.

100. After having considered the inconveniences associated with a negative value for a, we decide that a will be a positive number, indeed greater than 1, since from this case it is easy to investigate the case when a lies between 0 and 1. If we let $y = a^z$, and for z substitute all real numbers, which lie between $-\infty$ and $+\infty$, then y takes all positive real values between 0 and $+\infty$. If z goes to ∞, then y also goes to ∞, if $z = 0$, then $y = 1$ and when z goes to $-\infty$, y goes to 0. On the other hand, for any positive value assigned to y, there is a real value corresponding to z such that $a^z = y$. If a negative value is given to y, there is no corresponding real value for z.

101. If $y = a^z$, then y is a function of z, and the extent to which y depends on z is easily understood from the nature of exponents. Thus whatever

value is given to z, the value of y is determined. For instance $y^2 = a^{2z}$, $y^3 = a^{3z}$, and generally $y^n = a^{nz}$. From this it follows that $\sqrt{y} = a^{\frac{1}{2}z}$, $y^{\frac{1}{3}} = a^{\frac{1}{3}z}$, and $\frac{1}{y} = a^{-z}$, $\frac{1}{y^2} = a^{-2z}$, $\frac{1}{\sqrt{y}} = a^{-\frac{z}{2}}$, and so forth. Furthermore, if $v = a^x$, then $vy = a^{x+z}$ and $\frac{v}{y} = a^{x-z}$. A benefit we derive from these properties is that it is easier to determine the value of z when a value of y is given.

EXAMPLE

If $a = 10$, from arithmetic, which we shall use, the number ten makes it easy to see the values of y when we substitute values for z. We see that $10^1 = 10$, $10^2 = 100$, $10^3 = 1000$, $10^4 = 10000$, and $10^0 = 1$. Likewise $10^{-1} = \frac{1}{10} = 0.1$, $10^{-2} = \frac{1}{100} = 0.01$, $10^{-3} = \frac{1}{1000} = 0.001$. If we let z have fractional values, by means of root extraction, we can state the values of y. Thus $10^{\frac{1}{2}} = \sqrt{10} = 3.162277$, etc.

102. Just as, given a number a, for any value of z, we can find the value of y, so, in turn, given a positive value for y, we would like to give a value for z, such that $a^z = y$. This value of z, insofar as it is viewed as a function of y, it is called the LOGARITHM of y. The discussion about logarithms supposes that there is some fixed constant to be substituted for a, and this number is the *base* for the logarithm. Having assumed this base, we say the logarithm of y is the exponent in the power a^z such that $a^z = y$. It has been customary to designate the logarithm of y by the symbol $\log y$. If $a^z = y$, then $z = \log y$. From this

we understand that the base of the logarithms, although it depends on our choice, still it should be a number greater than 1. Furthermore, it is only of positive numbers that we can represent the logarithm with a real number.

103. Whatever logarithmic base we choose, we always have $\log 1 = 0$, since in the equation $a^z = y$, which corresponds to $z = \log y$, when we let $y = 1$ we have $z = 0$. From this it follows that the logarithm of a number greater than 1 will be positive, depending on the base a. Thus $\log a = 1$, $\log a^2 = 2$, $\log a^3 = 3$, $\log a^4 = 4$, etc. and, after the fact, we know what base has been chosen, that is the number whose logarithm is equal to 1 is the logarithmic base. The logarithm of a positive number less than 1 will be negative. Notice that $\log \frac{1}{a} = -1$, $\log \frac{1}{a^2} = -2$, $\log \frac{1}{a^3} = -3$, etc., but the logarithms of negative numbers will not be real, but complex, as we have already noted.

104. In like manner, if $\log y = z$, then $\log y^2 = 2z$, $\log y^3 = 3z$, etc, and in general $\log y^n = nz$ or $\log y^n = n \log y$, since $z = \log y$. It follows that the logarithm of any power of y is equal to the product of the exponent and the logarithm of y. For example, $\log \sqrt{y} = \frac{1}{2} z = \frac{1}{2} \log y$,

$\log \frac{1}{\sqrt{y}} = \log y^{-\frac{1}{2}} = -\frac{1}{2} \log y$, and so forth. It follows that if we know the logarithms of any number, we can find the logarithms of any power of that number. If we already know the logarithms of two number, for example $\log y = z$ and $\log v = x$, since $y = a^z$ and $v = a^x$, it follows that $\log vy = x + y = \log v + \log y$. Hence, the logarithm of the product of two numbers is equal to the sum of the logarithms of the factors. In like manner, $\log \frac{y}{v} = z - x = \log y - \log v$, that is, the logarithms of a quotient is equal to the logarithm of the numerator diminished by the logarithm of the denominator. These rules can be used to find the logarithms of many numbers from a knowledge of the logarithms of a few.

105. From what we have seen, it follows that the logarithm of a number will not be a rational number unless the given number is a power of the base a. That is, unless the number b is a power of the base a, the logarithm of b cannot be expressed as a rational number. In case b is a power of the base a, then the logarithm of b cannot be an irrational number. If, indeed, $\log b = \sqrt{n}$, then $a^{\sqrt{n}} = b$, but this is impossible if both a and b are rational. It is especially desirable to know the logarithms of rational numbers, since from these it is possible to find the logarithms of fractions and also surds. Since the logarithms of numbers which are not the powers of the base are neither rational nor irrational, it is with justice that they are called transcendental quantities. For this reason, logarithms are said to be transcendental.

106. When they are transcendental, logarithms can be only approximately represented by decimal fractions. The discrepancy is less to the extent that more decimal places are used in the approximation. In the following way we can find an approximation for a logarithm by only extracting square roots. Let $\log y = z$ and $\log v = x$, then $\log \sqrt{vy} = \frac{x + z}{2}$. If the proposed number b lies between a^2 and a^3, whose logarithms are 2 and 3 respectively, we look for the value of $a^{\frac{5}{2}}$

or $a^2\sqrt{a}$ and then b lies either between a^2 and $a^{\frac{5}{2}}$ or between $a^{\frac{5}{2}}$ and a^3. Whichever is the case, we then take the geometric mean of these two and we have closer bounds. We repeat the process and the lengths of the intervals between which b lies decrease. In this way we eventually arrive at the value of b with the desired number of decimal places. Since the logarithms of the bounds have been computed, we finally find the logarithm of b.

EXAMPLE

Let $a = 10$ be the base of the logarithms, which is usually the case in the computed tables. We seek an approximate logarithm of the number 5. Since 5 lies between 1 and 10, whose logarithms are 0 and 1, in the following manner we take successive square roots until we arrive at the number 5 exactly.

$A = 1.000000$; $\log A = 0.0000000$ so that

$B = 10.000000$; $\log B = 1.0000000$; $C = \sqrt{AB}$

$C = 3.162277$; $\log C = 0.5000000$; $D = \sqrt{BC}$

$D = 5.623413$; $\log D = 0.7500000$; $E = \sqrt{CD}$

$E = 4.216964$; $\log E = 0.6250000$; $F = \sqrt{DE}$

$F = 4.869674$; $\log F = 0.6875000$; $G = \sqrt{DF}$

$G = 5.232991$; $\log G = 0.7187500$; $H = \sqrt{FG}$

$H = 5.048065$; $\log H = 0.7031250$; $I = \sqrt{FH}$

$I = 4.958069$; $\log I = 0.6953125$; $K = \sqrt{HI}$

$K = 5.002865$; $\log K = 0.6992187$; $L = \sqrt{IK}$

$L = 4.980416$; $\log L = 0.6972656$; $M = \sqrt{KL}$

$M = 4.991627$; $\log M = 0.6982421$; $N = \sqrt{KM}$

$N = 4.997242$; $\log N = 0.6987304$; $O = \sqrt{KN}$

$O = 5.000052$; $\log O = 0.6989745$; $P = \sqrt{NO}$

$P = 4.998647$; $\log P = 0.6988525$; $Q = \sqrt{OP}$

$Q = 4.999350$; $\log Q = 0.6989135$; $R = \sqrt{OQ}$

$R = 4.999701$; $\log R = 0.6989440$; $S = \sqrt{OR}$

$S = 4.999876$; $\log S = 0.6989592$; $T = \sqrt{OS}$

$T = 4.999963$; $\log T = 0.6989668$; $V = \sqrt{OT}$

$V = 5.000008$; $\log V = 0.6989707$; $W = \sqrt{TV}$

$W = 4.999984$; $\log W = 0.6989687$; $X = \sqrt{WV}$

$X = 4.999997$; $\log X = 0.6989697$; $Y = \sqrt{VX}$

$Y = 5.000003$; $\log Y = 0.6989702$; $Z = \sqrt{XY}$

$Z = 5.000000$; $\log Z = 0.6989700$.

Thus the geometric means finally converge to $z = 5.000000$ and so the logarithm of 5 is 0.698970 when the base is 10. That is approximately $10^{\frac{698970}{1000000}} = 5$. This is the way in which the tables of common logarithms have been computed by Briggs and Vlasc. In the meantime much shorter methods have been found by means of which logarithms can be computed more quickly.

107. There are as many different systems of logarithms as there are different numbers which can be taken as the base a. It follows that there are an infinite number of systems of logarithms. Given two different systems of logarithms, there is a constant which relates the logarithms of the same number. If the base of one system is a and that of the other is b, if also the number n has logarithm

p in the first system and logarithm q in the second, then $a^p = n$ and $b^q = n$. Therefore $a^p = b^q$, so that $a = b^{\frac{q}{p}}$ and the value of $\frac{p}{q}$ is constant, no matter what the value of n may be. If the logarithms of all numbers have been computed in one system, then it is an easy task, by means of this golden rule for logarithms, to find the logarithms in any other system. For example, we have logarithms for the base 10. From these we can find the logarithms with any other base, for instance the base 2. We look for the logarithm of a number n for base 2, which will be q, while the logarithm with base 10 of the same number n will be p. Since for base 10, log 2 = 0.3010300 and for base 2, log 2 = 1, then $\frac{p}{q} = \frac{0.3010300}{1}$ and $q = \frac{p}{0.3010300} = 3.3219277$. If every common logarithm is multiplied by 3.3219277 then we will have produced a table of logarithms for base 2.

108. *It follows that the ratio of the logarithms of two different numbers is constant whatever the system of logarithms.*

Let M and N be two numbers of which in base a the logarithms are respectively m and n so that $M = a^m$ and $N = a^n$, then $a^{mn} = M^n = N^m$ so that $M = N^{\frac{m}{n}}$. In this equation a does not appear and it is clear that the ratio m/n is independent of a. If for some other base b, the logarithms of the same numbers M and N are μ and ν respectively, it follows in the same way that $M = N^{\frac{\mu}{\nu}}$. Since $N^{\frac{m}{n}} = N^{\frac{\mu}{\nu}}$, we have $\frac{m}{n} = \frac{\mu}{\nu}$ or $m/n = \mu/\nu$. We have already seen that in any system of logarithms, the ratio of the logarithms of different powers of the same number, for instance y^m and y^n, is the ratio of the two exponents, m/n.

109. In order to construct a table of logarithms with any base, it is sufficient to have computed the logarithms of only the prime numbers by the method already given or some more convenient method. Since the logarithm of a composite number is equal to the sum of the logarithms of its factors, we can complete our table by simple addition. Thus, if we have the logarithms of 3 and 5 then $\log 15 = \log 3 + \log 5$, $\log 45 = 2 \log 3 + \log 5$. Furthermore, since we have already computed the logarithm of 5 with base 10 to be 0.6989700 and $\log 10 = 1$, it follows that $\log \frac{10}{5} = \log 2 = \log 10 - \log 5$. Therefore, $\log 2 = 1 - 0.6989700 = 0.3010300$. From the logarithms of these two prime numbers, 2 and 5, we can find the logarithms of all composite numbers with factors only 2 and 5. Numbers of this sort are 4, 8, 16, 32, 64, etc., 20, 40, 80, 25, 50, etc.

110. A table of logarithms is of great use in carrying out numerical computations, since from such tables we have not only the logarithms of the given numbers, but for any logarithms we can find a corresponding number. Suppose c, d, e, f, g, and h designate any numbers, besides products, we can find the value of an expression like $\frac{c^2 d \sqrt{e}}{f(gh)^{\frac{1}{3}}}$. The logarithm of this expression will be $2 \log c + \log d + \frac{1}{2} \log e - \log f - \frac{1}{3} \log g - \frac{1}{3} \log h$ and the desired number is that to which this logarithm corresponds. Tables of logarithms are especially useful in finding intricate roots, since without them we would have

only the operations of multiplication and division.

EXAMPLE I

We seek the value of $2^{\frac{7}{12}}$, of which the logarithm is $\frac{7}{12}\log 2$. We multiply the logarithm of 2, which is 0.3010300, by $\frac{7}{12}$, that is, by $\frac{1}{2} + \frac{1}{12}$. But then $\log 2^{\frac{7}{12}} = 0.1756008$, to which logarithm corresponds the number 1.498307, which is an approximate value of $2^{\frac{7}{12}}$.

EXAMPLE II

If the population in a certain region increases annually by one thirtieth and at one time there were 100,000 inhabitants, we would like to know the population after 100 years. For the sake of brevity we let the initial population be n, so that $n = 100{,}000$. After one year the new population will be $(1 + 1/30)n = \frac{31}{30}n$. After two years it will equal $\left(\frac{31}{30}\right)^2 n$. After three years it will be $\left(\frac{31}{30}\right)^3 n$. Finally after one hundred years the population will be $\left(\frac{31}{30}\right)^{100} n = \left(\frac{31}{30}\right)^{100} 100{,}000$. The logarithm of this population is $100 \log\left(\frac{31}{30}\right) + \log 100{,}000$. But

$\log\left(\frac{31}{30}\right) = \log 31 - \log 30 = 0.014240439$, so that

$100 \log\left(\frac{31}{30}\right) = 1.4240439$ which, when increased by $\log 100{,}000 = 5$, gives 6.4240439, the logarithm of the desired population. The corresponding population is 2,654,874. So after one hundred years the population will be more than twenty-six and a half times as large.

EXAMPLE III

Since after the flood all men descended from a population of six, if we suppose that the population after two hundred years was 1,000,000, we would like to find the annual rate of growth. We suppose that each year the increase is $1/x$, so that after two hundred years the population is $\left(\frac{1+x}{x}\right)^{200} 6 = 1{,}000{,}000$. It follows that $\frac{1+x}{x} = \left(\frac{1000000}{6}\right)^{\frac{1}{200}}$, and so $\log\left(\frac{1+x}{x}\right) = \frac{1}{200}\log\frac{1000000}{6} = \frac{1}{200}5.2218487 = 0.0261092$. From this we have $\frac{1+x}{x} = \frac{1061963}{1000000}$, and so $1{,}000{,}000 = 61963x$ and finally x is approximately 16. We have shown that if each year the population increases by $\frac{1}{16}$, the desired result takes place. Now if the same rate holds over an interval of four hundred years, then the population becomes $1000000.\frac{1000000}{6} = 166666666666$. However, the whole earth would never be able to sustain that population.

EXAMPLE IV

If each century the human population doubles, what is the annual rate of growth?

If we suppose that each year the population increases by $\frac{1}{x}$ and the initial population is n, then after one hundred years the population will be

$\left(\frac{1 + x}{x}\right)^{100} n = 2n$. It follows that $\frac{1 + x}{x} = 2^{\frac{1}{100}}$ and $\log\left(\frac{1 + x}{x}\right) = \frac{1}{100} \log 2 = 0.0030103$. From this we have $\frac{1 + x}{x} = \frac{10069555}{10000000}$. Therefore $x = \frac{10000000}{69555}$, so that x is approximately equal to 144. It will suffice if the annual increase is by $\frac{1}{144}$. For this reason it is quite ridiculous for the incredulous to object that in such a short space of time the whole earth could not be populated beginning with a single man.

111. The most important application of logarithms is in the solution of equations in which the unknown quantity is involved in the exponent. For example, if we come upon an equation $a^x = b$, from which we must find the value of x, the solution can be found only with logarithms. Since $a^x = b$, we have $\log a^x = x \log a = \log b$. Therefore $x = \frac{\log b}{\log a}$. We note that any logarithmic system may be used, since in any system the ratio is the same.

EXAMPLE I

If the human population increases annually by $\frac{1}{100}$, we would like to know how long it will take for the population to become ten times as large. We suppose that this will occur after x years and that the initial population is n. Hence after x years the population will be $\left(\frac{101}{100}\right)^x n = 10n$, so that $\left(\frac{101}{100}\right)^x = 10$ and $x \log\left(\frac{101}{100}\right) = \log 10$. From this we have $x = \frac{\log 10}{\log 101 - \log 100} = \frac{10000000}{43214} = 231$. After 231 years the human population will be ten times as large with an annual increase of only $\frac{1}{100}$. It also follows that after 462 years the population will be one hundred times as large, and after 693 years the population will be one thousand times as large.

EXAMPLE II

A certain man borrowed 400,000 florins at the usurious rate of five percent annual interest. Suppose that each year he repays 25,000 florins. The question is, how long will it be before the debt is repaid completely. We let a be the original debt of 400,000 florins and we let b be the sum of 25,000 florins which is repaid annually. After one year the debt will be $\frac{105}{100}a - b$. After the second year the debt will be $\left(\frac{105}{100}\right)^2 a - \left(\frac{105}{100}\right) b - b$. After the third year the debt will be $\left(\frac{105}{100}\right)^3 a - \left(\frac{105}{100}\right)^2 b - \frac{105}{100} b - b$. For the sake of brevity let us set $n = \frac{105}{100}$, and after x years, the debt will be

$n^x - n^{x-1}b - n^{x-2}b - n^{x-3}b - \cdots - b$

$= n^x a - b(1 + n + n^2 + \cdots + n^{x-1})$. From the nature of a geometric progression we know that

$1 + n + n^2 + \cdots + n^{x-1} = \frac{n^x - 1}{n - 1}$. After x years the man owes $n^x a - \frac{n^x b - b}{n - 1}$ florins. When the debt is paid, this should equal zero, so we have the equation $n^x a = \frac{n^x b - b}{n - 1}$, or $(n - 1)n^x a = n^x b - b$. From this it follows that $(b - na + a)n^x = b$, and that $n^x = \frac{b}{b - (n - 1)a}$. Finally we have $x = \dfrac{\log b - \log(b - (n-1)a)}{\log n}$. Since

$a = 400{,}000$, $b = 25{,}000$, $n = \dfrac{105}{100}$, we have

$(n - 1)a = 20{,}000$, $b - (n - 1)a = 5{,}000$ and the number of years after which the debt is completely paid is

$$x = \frac{\log 25000 - \log 5000}{\log \frac{105}{100}} = \frac{\log 5}{\log \frac{21}{20}} = \frac{6989700}{211893}.$$ That is, in a little less than thirty-three years the debt is paid. In fact not only is the debt paid in full, but the creditor owes the debtor

$$\frac{(n^{33} - 1)b}{n - 1} - n^{33}a = \frac{\left(\frac{21}{20}\right)^{33} \cdot 5000 - 25000}{\frac{1}{20}} = 100000\left(\frac{21}{20}\right)^{33} - 500000$$

florins. Since $\log \dfrac{21}{20} = 0.0211892991$, we have $\log\left(\dfrac{21}{20}\right)^{33} = 0.6992469$ and so

$\log 100000\left(\dfrac{21}{20}\right)^{33} = 5.6992469$ which is the logarithm of 500,318.8. Now after 33 years the creditor should return to the debtor 318.8 florins.

112. The common logarithms, that is, those with base 10, have a special usefulness beyond that which logarithms with another base have. This is because our arithmetic is based on the decimal system. Since the logarithms of all numbers are expressed not only with powers of ten but also in decimal fractions, the logarithms of numbers between 1 and 10 lie between 0 and 1, while the logarithms of numbers between 10 and 100 lie between 1 and 2, and so forth. Hence a logarithms is made up of two parts, namely an integral part and a decimal fraction. The integer is called the CHARACTERISTIC, while the decimal part is

called the MANTISSA. The characteristic of a number is one less than the number of digits which express the number. For example, the logarithm of the number 78509 has a characteristic 4, since the number has 5 digits. Thus from the logarithm of a number one knows immediately how many digits the integral part has. For example, to the logarithm 7.5804631 there corresponds an eight digit number.

113. If the mantissae of two logarithms are equal, but if the characteristics are different, then the numbers corresponding to these logarithms have a ratio which is a power of ten. It follows that the same digits represent the two numbers. For example, the numbers whose logarithms are 4.9130187 and 6.9130187 are 81850 and 8185000 respectively. However, the logarithms 3.9130187 corresponds to 8185 and the logarithm 0.9130187 corresponds to 8.185. The mantissa gives only the digits of the number, while from the characteristic it is clear how many digits to the left of the decimal point are to be assigned to the integral part. The remaining digits to the right belong to the decimal fraction. For example, if the logarithm is 2.7603429, then the mantissa provides the digits 5758945, while the characteristic 2 determines the number corresponding to that logarithm to be 575.8945. If the characteristic had been 0, than the number would have been 5.758945. If the characteristic is negative, for instance -1, then the corresponding number is less by one tenth, namely, 0.5758945. The characteristic -2 corresponds to 0.05758945 etc. Instead of writing negative characteristics, such as -1, - 2, -3, etc., we usually write 9,8,7, etc. respectively, and understand that the logarithms are to be diminished by 10. All of this is usually

explained more at length in introductions to tables of logarithms.

EXAMPLE

If the progression 2, 4, 16, 256, $\cdots$ *is formed by letting each term be the square of the preceding term, find the value of the twenty-fifth term.*

If the terms of this progression are expressed by exponents, it is more convenient: 2, 2^2, 2^4, 2^8, etc. Now it becomes clear that the exponents form a geometric progression and that the exponent of the twenty-fifth term will be $2^{24} = 16777216$. It follows that the sought for term is $2^{16777216}$, whose logarithm is 16777216 log 2. Since $\log 2 = 0.301029995663981195$, the logarithm of our number is 5050445.25973367. From the characteristic we see that the number has 5050446 digits. The mantissa, 25973367, in the table of logarithms, gives us at least the initial digits, which are 181858. These digits are followed by 5050440 more digits, a few of which may be found in tables with more places. In fact the eleven initial digits are 18185852986.

CHAPTER VII

Exponentials and Logarithms Expressed through Series.

114. Since $a^0 = 1$, when the exponent on a increases, the power itself increases, provided a is greater than 1. It follows that if the exponent is infinitely small and positive, then the power also exceeds 1 by an infinitely small number. Let ω be an infinitely small number, or a fraction so small that, although not equal to zero, still $a^\omega = 1 + \psi$, where ψ is also an infinitely small number. From the preceding chapter we know that unless ψ were infinitely small, then neither would ω be infinitely small. It follows that $\psi = \omega$, or $\psi > \omega$, or $\psi < \omega$. Which of these is true depends on the value of a, which is not now known, so we let $\psi = k\omega$. Then we have $a^\omega = 1 + k\omega$, and with a as the base for the logarithms, we have $\omega = \log(1 + k\omega)$.

EXAMPLE

In order that it may be clearer how the number k depends on a, let $a = 10$. From the table of common logarithms, we look for the logarithm of a number which exceeds 1 by the smallest possible amount, for instance, $1 + \frac{1}{1000000}$, so that
$$k\omega = \frac{1}{1000000}.$$
Then
$$\log\left(1 + \frac{1}{1000000}\right) = \log\frac{1000001}{1000000} = 0.00000043429 = \omega.$$
Since $k\omega = 0.00000100000$, it follows that $\frac{1}{k} = \frac{43429}{1000000}$ and

$k = \dfrac{100000}{43429} = 2.30258$. We see that k is a finite number which depends on the value of the base a. If a different base had been chosen, then the logarithm of the same number $1 + k\omega$ will differ from the logarithm already given. It follows that a different value of k will result.

115. Since $a^\omega = 1 + k\omega$, we have $a^{j\omega} = (1 + k\omega)^j$, whatever value we assign to j. It follows that

$a^{j\omega} = 1 + \dfrac{j}{1}k\omega + \dfrac{j(j-1)}{1\cdot 2}k^2\omega^2 + \dfrac{j(j-1)(j-2)}{1\cdot 2\cdot 3}k^3\omega^3 + \cdots$. If now we let $j = \dfrac{z}{\omega}$, where z denotes any finite number, since ω is infinitely small, then j is infinitely large. Then we have $\omega = \dfrac{z}{j}$, where ω is represented by a fraction with an infinite denominator, so that ω is infinitely small, as it should be. When we substitute $\dfrac{z}{j}$ for ω then

$a^z = (1 + kz/j)^j = 1 + \dfrac{1}{1}kz + \dfrac{1(j-1)}{1\cdot 2j}k^2z^2 + \dfrac{1(j-1)(j-2)}{1\cdot 2j\cdot 3j}k^3z^3$
$+ \dfrac{1(j-1)(j-2)(j-3)}{1\cdot 2j\cdot 3j\cdot 4j}k^4z^4 + \cdots$. This equation is true provided an infinitely large number is substituted for j, but then k is a finite number depending on a, as we have just seen.

116. Since j is infinitely large, $\dfrac{j-1}{j} = 1$, and the larger the number we substitute for j, the closer the value of the fraction $\dfrac{j-1}{j}$ comes to 1. Therefore, if j is a number larger than any assignable number, then $\dfrac{j-1}{j}$ is equal to 1. For the same reason $\dfrac{j-2}{j} = 1$, $\dfrac{j-3}{j} = 1$, and so forth. It follows that

$\frac{j-1}{2j} = \frac{1}{2}$, $\frac{j-2}{3j} = \frac{1}{3}$, $\frac{j-3}{4j} = \frac{1}{4}$, and so forth. When we substitute these values, we obtain

$a^z = 1 + \frac{kz}{1} + \frac{k^2z^2}{1\cdot 2} + \frac{k^3z^3}{1\cdot 2\cdot 3} + \frac{k^4z^4}{1\cdot 2\cdot 3\cdot 4} + \cdots$. This equation expresses a relationship between the numbers a and k, since when we let $z = 1$, we have $a = 1 + \frac{k}{1} + \frac{k^2}{1\cdot 2} + \frac{k^3}{1\cdot 2\cdot 3} + \frac{k^4}{1\cdot\cdot 2\cdot 3\cdot 4} + \cdots$. When $a = 10$, then k is necessarily approximately equal to 2.30258 as we have already seen.

117. Suppose $b = a^n$, and let a be the base for the logarithms, so that $\log b = n$. Since $b^z = a^{nz}$, we have the infinite series $b^z = 1 + \frac{knz}{1} + \frac{k^2n^2z^2}{1\cdot 2} + \frac{k^3n^3z^3}{1\cdot 2\cdot 3} + \frac{k^4n^4z^4}{1\cdot 2\cdot 3\cdot 4} + \cdots$. Now we substitute $\log b$ for n, so that $b^z = 1 + \frac{kz}{1}\log b + \frac{k^2z^2}{1\cdot 2}(\log b)^2 + \frac{k^3z^3}{1\cdot 2\cdot 3}(\log b)^3 + \frac{k^4z^4}{1\cdot 2\cdot 3\cdot 4}(\log b)^4 + \cdots$. Since we know the value of k from the given value of the base a, the general exponential b^z can be expressed in an infinite series whose terms proceed with the powers of z. Having shown this fact, we now go on to show how logarithms can be expressed by infinite series.

118. Since $a^\omega = 1 + k\omega$, where ω is an infinitely small fraction and the relation between a and k is given by $a = 1 + \frac{k}{1} + \frac{k^2}{1\cdot 2} + \frac{k^3}{1\cdot 2\cdot 3} + \cdots$, if a is taken as the base of the logarithms, then $\omega = \log(1 + k\omega)$ and $j\omega = \log(1 + k\omega)^j$. It is clear that the larger the number chosen for j the more $(1 + k\omega)^j$ will exceed 1. If we let j be an infinite number, the value of the power $(1 + k\omega)^j$ becomes greater than any number greater than 1. Now if we

let $(1 + k\omega)^j = 1 + x$, then $\log(1 + x) = j\omega$. Since $j\omega$ is an finite number, namely the logarithm of $1 + x$, it is clear that j must be an infinitely large number; otherwise, $j\omega$ could not have a finite value.

119. Since we have let $(1 + k\omega)^j = 1 + x$, we have $1 + k\omega = (1 + x)^{\frac{1}{j}}$ and $k\omega = (1 + x)^{\frac{1}{j}} - 1$, so that $j\omega = \frac{j}{k}((1 + x)^{\frac{1}{j}} - 1)$. Since $j\omega = \log(1 + x)$, it follows that $\log(1 + x) = \frac{j}{k}(1 + x)^{\frac{1}{j}} - \frac{j}{k}$ where j is a number infinitely large. But we have

$$(1 + x)^{\frac{1}{j}} = 1 + \frac{1}{jx} - \frac{1(j-1)}{j \cdot 2j}x^2 + \frac{1(j-1)(2j-1)}{j \cdot 2j \cdot 3j}x^3$$
$$- \frac{1(j-1)(2j-1)(3j-1)}{j \cdot 2j \cdot 3j \cdot 4j}x^4 + \cdots .$$

Since j is an infinite number, $\frac{j-1}{2j} = \frac{1}{2}$, $\frac{2j-1}{3j} = \frac{2}{3}$, $\frac{3j-1}{4j} = \frac{3}{4}$, etc. Now it follows that $j(1 + x)^{\frac{1}{j}} = j + \frac{x}{1} - \frac{x^2}{2} + \frac{x^3}{3} - \frac{x^4}{4} + \cdots$. As a result we have $\log(1 + x) = \frac{1}{k}\left(\frac{x}{1} - \frac{x^2}{2} + \frac{x^3}{3} - \frac{x^4}{4} + \cdots\right)$, where a is the base of the logarithm and $a = 1 + \frac{k}{1} + \frac{k^2}{1 \cdot 2} + \frac{k^3}{1 \cdot 2 \cdot 3} + \cdots$.

120. Since we have a series for the logarithm of $1 + x$, we can use this to define the number k when a is the base. If we let $1 + x = a$, since $\log a = 1$, we have

$$1 = \frac{1}{k}\left(\frac{a-1}{1} - \frac{(a-1)^2}{2} + \frac{(a-1)^3}{3} - \frac{(a-1)^4}{4} + \cdots\right).$$

It follows that $k = \frac{a-1}{1} - \frac{(a-1)^2}{2} + \frac{(a-1)^3}{3} - \frac{(a-1)^4}{4} + \cdots$. If we let

$a = 10$, the value of this infinite series must be approximately equal to 2.30258. We have $2.30258 = \frac{9}{1} - \frac{9^2}{2} + \frac{9^3}{3} - \frac{9^4}{4} + \cdots$, but it is difficult to see how this can be since the terms of this series continually grow larger and the sum of several terms does not seem to approach any limit. We will soon have an answer to this paradox.

121. Since $\log(1 + x) = \frac{1}{k}\left(\frac{x}{1} - \frac{x^2}{2} + \frac{x^3}{3} - \cdots\right)$, when we substitute $-x$ for x, we obtain

$\log(1 - x) = -\frac{1}{k}\left(\frac{x}{1} + \frac{x^2}{2} + \frac{x^3}{3} + \frac{x^4}{4} + \cdots\right)$. If we subtract the second series from the first we obtain

$$\log(1 + x) - \log(1 - x) = \log\left(\frac{1 + x}{1 - x}\right)$$

$= \frac{2}{k}\left(\frac{x}{1} + \frac{x^3}{3} + \frac{x^5}{5} + \frac{x^7}{7} + \cdots\right)$. Now if we let $\frac{1 + x}{1 - x} = a$, so that $x = \frac{a - 1}{a + 1}$, and because $\log a = 1$, we have

$k = 2\left(\frac{a - 1}{a + 1} + \frac{(a - 1)^3}{3(a + 1)^3} + \frac{(a - 1)^5}{5(a + 1)^5} + \cdots\right)$. From this equation we can find the value of k when a is given. For example, if $a = 10$, then

$k = 2\left(\frac{9}{11} + \frac{9^3}{3\cdot 11^3} + \frac{9^5}{5\cdot 11^5} + \frac{9^7}{7\cdot 11^7} + \cdots\right)$ and the terms of this series decrease in a reasonable way so that soon a satisfactory approximation for k can be obtained.

122. Since we are free to choose the base a for the system of logarithms, we

now choose a in such a way that $k = 1$. Suppose now that $k = 1$, then the series found above in section 116,

$1 + \frac{1}{1} + \frac{1}{1\cdot 2} + \frac{1}{1\cdot 2\cdot 3} + \frac{1}{1\cdot 2\cdot 3\cdot 4} + \cdots$ is equal to a. If the terms are represented as decimal fractions and summed, we obtain the value for $a = 2.71828182845904523536028 \cdots$. When this base is chosen, the logarithms are called natural or hyperbolic. The latter name is used since the quadrature of a hyperbola can be expressed through these logarithms. For the sake of brevity for this number $2.718281828459 \cdots$ we will use the symbol e, which will denote the base for natural or hyperbolic logarithms, which corresponds to the value $k = 1$, and e represents the sum of the infinite series $1 + \frac{1}{1} + \frac{1}{1\cdot 2} + \frac{1}{1\cdot 2\cdot 3} + \frac{1}{1\cdot 2\cdot 3\cdot 4} + \cdots$.

123. Natural logarithms have the property that the logarithm of $1 + \omega$ is equal to ω, where ω is an infinitely small quantity. From this it follows that $k = 1$ and the natural logarithms of all numbers can be found. Let e stand for the number found above, then

$e^z = 1 + \frac{z}{1} + \frac{z^2}{1\cdot 2} + \frac{z^3}{1\cdot 2\cdot 3} + \frac{z^4}{1\cdot 2\cdot 3\cdot 4} + \cdots$, and the natural logarithms themselves can be found from these series where

$\log(1 + x) = x - \frac{x^2}{2} + \frac{x^3}{3} - \frac{x^4}{4} + \frac{x^5}{5} - \frac{x^6}{6} + \cdots$, and

$\log\left(\frac{1 + x}{1 - x}\right) = \frac{2x}{1} + \frac{2x^3}{3} + \frac{2x^5}{5} + \frac{2x^7}{7} + \frac{2x^9}{9} + \cdots$. This last series is strongly convergent if we substitute an extremely small fraction for x. For instance, if $x = \frac{1}{5}$, then

$\log \frac{6}{4} = \log \frac{3}{2} = \frac{2}{1\cdot 5} + \frac{2}{3\cdot 5^3} + \frac{2}{5\cdot 5^5} + \frac{2}{7\cdot 5^7} + \cdots$. If $x = \frac{1}{7}$, then

$\log \frac{4}{3} = \frac{2}{1\cdot 7} + \frac{2}{3\cdot 7^3} + \frac{2}{5\cdot 7^5} + \frac{2}{7\cdot 7^7} + \cdots$, and if $x = \frac{1}{9}$, then

$\log \frac{5}{4} = \frac{2}{1\cdot 9} + \frac{2}{3\cdot 9^3} + \frac{2}{5\cdot 9^5} + \frac{2}{7\cdot 9^7} + \cdots$. From the logarithms of these fractions we can find the logarithms of integers. From the nature of logarithms we have $\log \frac{3}{2} + \log \frac{4}{3} = \log 2$, and $\log \frac{3}{2} + \log 2 = \log 3$, and $2 \log 2 = \log 4$. Further we have

$\log \frac{5}{4} + \log 4 = \log 5$, $\log 2 + \log 3 = \log 6$, $3 \log 2 = \log 8$, $2 \log 3 = \log 9$, $\log 2 + \log 5 = \log 10$.

EXAMPLE

We can now state the values of the natural logarithms of integers from 1 to 10.

log 1	= 0.00000 00000 00000 00000 00000	log 2	= 0.69314 71805 59945 30941 72321
log 3	= 1.09861 22886 68109 69139 52452	log 4	= 1.38629 43611 19890 61883 44642
log 5	= 1.60943 79124 34100 37460 07593	log 6	= 1.79175 94692 28055 00081 24773
log 7	= 1.94591 01490 55313 30510 54639	log 8	= 2.07944 15416 79835 92825 16964
log 9	= 2.19722 45773 36219 38279 04905	log 10	= 2.30258 50929 94045 68401 79914

All of these logarithms are computed from the above three series, with the exception of log 7, which can be found as follows. When in the last series given above, we let $x = \frac{1}{99}$, we obtain

$\log \frac{100}{98} = \log \frac{50}{49} = 0.0202027073175194484078230$. When this is subtracted from $\log 50 = 2\log 5 + \log 2 = 3.9120230054281460586187508$ we obtain $\log 49$. But $\log 7 = \frac{1}{2}\log 49$.

124. Let the natural logarithm of $1 + x$ be equal to y, then $y = \frac{x}{1} - \frac{x^2}{2} + \frac{x^3}{3} - \frac{x^4}{4} + \cdots$. Now let a be the base of a system of logarithms and let v be the logarithm of $1 + x$ in this system. Then as we have seen, $v = \frac{1}{k}\left(x - \frac{x^2}{2} + \frac{x^3}{3} - \frac{x^4}{4} + \cdots\right) = \frac{y}{k}$. It follows that $k = \frac{y}{v}$, and this is the most convenient method of calculating the value of k corresponding to the base a; it is given by the quotient of the natural logarithm of any number divided by the logarithms of that same number with the base a. Suppose the number is a, then $v = 1$ and k is equal to the natural logarithm of a. In the system of common logarithms, where the base is $a = 10$, then k is the natural logarithm of 10. It follows that $k = 2.3025850929940456840179914$, which is the value calculated not far above. If each natural logarithm is divided by this number k, or, which comes to the same thing, multiplied by the decimal fraction $0.4342944819032518276511289$, then the results are the common logarithms, with base $a = 10$.

125. Since $e^z = 1 + \frac{z}{1} + \frac{z^2}{1\cdot 2} + \frac{z^3}{1\cdot 2\cdot 3} + \cdots$, if we let $a^y = e^z$, then after taking natural logarithms, we have $y \log a = z$, since $\log e = 1$. We now substitute this value in the series to obtain

$$a^y = 1 + \frac{y\log a}{1} + \frac{y^2(\log a)^2}{1\cdot 2} + \frac{y^3(\log a)^3}{1\cdot 2\cdot 3} + \cdots .$$

In this way any exponential, with the aid of natural logarithms, can be expressed as an infinite series. Now let j be an infinitely large number, then both exponentials and logarithms can be expressed as powers. That is, $e^z = \left(1 + \frac{z}{j}\right)^j$ and so $a^y = \left(1 + \frac{y \log a}{j}\right)^j$. For natural logarithms, we have $\log(1 + x) = j((1 + x)^{\frac{1}{j}} - 1)$. Other uses of natural logarithms are discussed in integral calculus.

CHAPTER VIII

On Transcendental Quantities Which Arise from the Circle.

126. After having considered logarithms and exponentials, we must now turn to circular arcs with their sines and cosines. This is not only because these are further genera of transcendental quantities, but also since they arise from logarithms and exponentials when complex values are used. This will become clearer in the development to follow.

We let the radius, or total sine, of a circle be equal to 1, then it is clear enough that the circumference of the circle cannot be expressed exactly as a rational number. An approximation of half of the circumference of this circle is

3.14159265358979323846264338327950288419716939937510582097494459 2
3078164062862089986280348253421170679821480865132723066470938446+.

For the sake of brevity we will use the symbol π for this number. We say, then, that half of the circumference of a unit circle is π, or that the length of an arc of 180 degrees is π.

127. We always assume that the radius of the circle is 1 and let z be an arc of this circle. We are especially interested in the sine and cosine of this arc z. Henceforth we will signify the sine of the arc z by $\sin z$. Likewise, for the cosine of the arc z we will write $\cos z$. Since π is an arc of 180 degrees, $\sin 0\pi = 0$ and $\cos 0\pi = 1$. Also $\sin \frac{\pi}{2} = 1$, $\cos \frac{\pi}{2} = 0$, $\sin \pi = 0$, $\cos \pi = -1$,

$\sin \frac{3}{2}\pi = -1$, $\cos \frac{3}{2}\pi = 0$, $\sin 2\pi = 0$, and $\cos 2\pi = 1$. Every sine and cosine lies between +1 and -1. Further, we have $\cos z = \sin\left(\frac{\pi}{2} - z\right)$ and $\sin z = \cos\left(\frac{\pi}{2} - z\right)$. We also have $(\sin z)^2 + (\cos z)^2 = 1$. Besides these notations we mention also that $\tan z$ indicates the tangent of the arc z, $\cot z$ for the cotangent of the arc z. We agree that $\tan z = \frac{\sin z}{\cos z}$ and $\cot z = \frac{\cos z}{\sin z} = \frac{1}{\tan z}$, all of which is known from trigonometry.

128. We note further that if y and z are two arcs, then
$\sin(y + z) = \sin y \cos z + \cos y \sin z$ and
$\cos(y + z) = \cos y \cos z - \sin y \sin z$. Likewise
$\sin(y - z) = \sin y \cos z - \cos y \sin z$ and
$\cos(y - z) = \cos y \cos z + \sin y \sin z$. Now we substitute the arcs $\frac{\pi}{2}$, π, $\frac{3}{2}\pi$, etc. for y in the previous formulas:

$$\sin\left(\frac{\pi}{2} + z\right) = +\cos z \qquad \sin\left(\frac{\pi}{2} - z\right) = +\cos z$$

$$\cos\left(\frac{\pi}{2} + z\right) = -\sin z \qquad \cos\left(\frac{\pi}{2} - z\right) = +\sin z$$

$$\sin(\pi + z) = -\sin z \qquad \sin(\pi - z) = +\sin z$$

$$\cos(\pi + z) = -\cos z \qquad \cos(\pi - z) = -\cos z$$

$$\sin\left(\frac{3}{2}\pi + z\right) = -\cos z \qquad \sin\left(\frac{3}{2}\pi - z\right) = -\cos z$$

$$\cos\left(\frac{3}{2}\pi + z\right) = +\sin z \qquad \cos\left(\frac{3}{2}\pi - z\right) = -\sin z$$

$\sin(2\pi + z) = +\sin z \qquad \sin(2\pi - z) = -\sin z$

$\cos(2\pi + z) = +\cos z \qquad \cos(2\pi - z) = +\cos z.$

It follows that if n is any integer, then

$$\sin\left(\frac{4n+1}{2}\pi + z\right) = +\cos z \qquad \sin\left(\frac{4n+1}{2}\pi - z\right) = +\cos z$$

$$\cos\left(\frac{4n+1}{2}\pi + z\right) = -\sin z \qquad \cos\left(\frac{4n+1}{2}\pi - z\right) = -\sin z$$

$$\sin\left(\frac{4n+2}{2}\pi + z\right) = -\sin z \qquad \sin\left(\frac{4n+2}{2}\pi - z\right) = +\sin z$$

$$\cos\left(\frac{4n+2}{2}\pi + z\right) = -\cos z \qquad \cos\left(\frac{4n+2}{2}\pi - z\right) = -\cos z$$

$$\sin\left(\frac{4n+3}{2}\pi + z\right) = -\cos z \qquad \sin\left(\frac{4n+3}{2}\pi - z\right) = -\cos z$$

$$\cos\left(\frac{4n+3}{2}\pi + z\right) = +\sin z \qquad \cos\left(\frac{4n+3}{2}\pi - z\right) = -\sin z$$

$$\sin\left(\frac{4n+4}{2}\pi + z\right) = +\sin z \qquad \sin\left(\frac{4n+4}{2}\pi - z\right) = -\sin z$$

$$\cos\left(\frac{4n+4}{2}\pi + z\right) = +\cos z \qquad \cos\left(\frac{4n+4}{2}\pi - z\right) = +\cos z.$$

These formulas hold whether n is a positive or a negative integer.

129. Let $\sin z = p$ and $\cos z = q$, then $p^2 + q^2 = 1$; if also $\sin y = m$ and $\cos y = n$, then also $m^2 + n^2 = 1$. We have the following identities:

$\sin z = p \qquad \cos z = q$

$\sin(y + z) = mq + np \qquad \cos(y + z) = nq - mp$

$$\sin(2y + z) = 2mnq + (n^2 - m^2)p$$
$$\cos(2y + z) = (n^2 - m^2)q - 2mnp$$
$$\sin(3y + z) = (3mn^2 - m^3)q + (n^3 - 3m^2n)p$$
$$\cos(3y + z) = (n^3 - 3m^2n)q - (3mn^2 - m^3)p$$

etc. These arcs: $z, y + z, 2y + z, 3y + z \cdots$, form an arithmetic progression, however, both their sines and cosines form a recurrent progression which arises from the denominator $1 - 2nx + (m^2 + n^2)x^2$. This is seen from the following: $\sin(2y + z) = 2n \sin(y + z) - (m^2 + n^2)\sin z$, or

$\sin(2y + z) = 2 \cos y \sin(y + z) - \sin z$. In like manner

$\cos(2y + z) = 2 \cos y \cos(y + z) - \cos z$. Furthermore we have

$\sin(3y + z) = 2 \cos y \sin(2y + z) - \sin(y + z)$, and

$\cos(3y + z) = 2 \cos y \cos(2y + z) - \cos(y + z)$. Also

$\sin(4y + z) = 2 \cos y \sin(3y + z) - \sin(2y + z)$, and

$\cos(4y + z) = 2 \cos y \sin(3y + z) - \cos(2y + z)$, etc. The advantage of this law is that when the arcs form an arithmetic progression, then as many of the sines and cosines as may be desired can be expressed with little trouble.

130. Since $\sin(y + z) = \sin y \cos z + \cos y \sin z$, and

$\sin(y - z) = \sin y \cos z - \cos y \sin z$, when we add or subtract these expressions we obtain: $\sin y \cos z = \dfrac{\sin(y + z) + \sin(y - z)}{2}$, and

$\cos y \sin z = \dfrac{\sin(y + z) - \sin(y - z)}{2}$. Furthermore, since

$\cos(y + z) = \cos y \cos z - \sin y \sin z$, and

$\cos(y - z) = \cos y \cos z + \sin y \sin z$, by the same method we obtain:

$\cos y \cos z = \dfrac{\cos(y - z) + \cos(y + z)}{2}$, and

$\sin y \sin z = \frac{\cos(y - z) - \cos(y + z)}{2}$. Let $y = z = \frac{v}{2}$, then from these last formulas we obtain: $\left(\cos \frac{v}{2}\right)^2 = \frac{1 + \cos v}{2}$ so that $\cos \frac{v}{2} = \sqrt{(1 + \cos v)/2}$, and $\left(\sin \frac{v}{2}\right)^2 = \frac{1 - \cos v}{2}$ so that $\sin \frac{v}{2} = \sqrt{(1 - \cos v)/2}$. From this we see that if the cosine of an arc is given, then we can find the sine and cosine of the half arc.

131. Let the arcs $y + z = a$ and $y - z = b$, then $y = \frac{a + b}{2}$ and $z = \frac{a - b}{2}$. When we substitute these values in the formulas above we have the following equations, each of which is, as it were, a theorem:

$$\sin a + \sin b = 2 \sin \frac{a + b}{2} \cos \frac{a - b}{2}$$

$$\sin a - \sin b = 2 \cos \frac{a + b}{2} \sin \frac{a - b}{2}$$

$$\cos a + \cos b = 2 \cos \frac{a + b}{2} \cos \frac{a - b}{2}$$

$$\cos a - \cos b = 2 \sin \frac{a + b}{2} \sin \frac{a - b}{2}.$$

From these results we have, by division, the following theorems:

$$\frac{\sin a + \sin b}{\sin a - \sin b} = \tan \frac{a + b}{2} \cot \frac{a - b}{2} = \frac{\tan \frac{a + b}{2}}{\tan \frac{a - b}{2}},$$

$$\frac{\sin a + \sin b}{\cos a + \cos b} = \tan \frac{a + b}{2}, \qquad \frac{\sin a + \sin b}{\cos b - \cos a} = \cot \frac{a - b}{2},$$

$$\frac{\sin a - \sin b}{\cos a + \cos b} = \tan \frac{a - b}{2}, \qquad \frac{\sin a - \sin b}{\cos b - \cos a} = \cot \frac{a + b}{2},$$

$\frac{\cos a + \cos b}{\cos b - \cos a} = \cot \frac{a + b}{2} \cot \frac{a - b}{2}$. From these we deduce the following theorems:

$$\frac{\sin a + \sin b}{\cos a + \cos b} = \frac{\cos b - \cos a}{\sin a - \sin b},$$

$$\frac{\sin a + \sin b}{\sin a - \sin b} \cdot \frac{\cos a + \cos b}{\cos b - \cos a} = \left(\cot \frac{a - b}{2}\right)^2$$

$$\frac{\sin a + \sin b}{\sin a - \sin b} \cdot \frac{\cos b - \cos a}{\cos a + \cos b} = \left(\tan \frac{a + b}{2}\right)^2.$$

132. Since $(\sin z)^2 + (\cos z)^2 = 1$, we have the factors $(\cos z + i \sin z)(\cos z - i \sin z) = 1$. Although these factors are complex, still they are quite useful in combining and multiplying arcs. Consider the following product: $(\cos z + i \sin z)(\cos y + i \sin y)$, which results in $\cos y \cos z - \sin y \sin z + (\cos y \sin z + \sin y \cos z)i$. Since $\cos y \cos z - \sin y \sin z = \cos(y + z)$ and $\cos y \sin z + \sin y \cos z = \sin(y + z)$ we can express this product as $(\cos y + i \sin y)(\cos z + i \sin z) = \cos(y + z) + i \sin(y + z)$ and likewise

$$\begin{aligned}(\cos y - i \sin y)(\cos z - i \sin z) \\ = \cos(y + z) - i \sin(y + z)\end{aligned}$$

also

$$\begin{aligned}(\cos x \pm i \sin x)(\cos y \pm i \sin y)(\cos z \pm i \sin z) \\ = \cos(x + y + z) \pm i \sin(x + y + z).\end{aligned}$$

133. It now follows that $(\cos z \pm i \sin z)^2 = \cos 2z \pm i \sin 2z$ and $(\cos z \pm i \sin z)^3 = \cos 3z \pm i \sin 3z$. Generally we have $(\cos z \pm i \sin z)^n = \cos nz \pm i \sin nz$. It follows that

$$\cos nz = \frac{(\cos z + i \sin z)^n + (\cos z - i \sin z)^n}{2}$$ and

$$\sin nz = \frac{(\cos z + i \sin z)^n - (\cos z - i \sin z)^n}{2}.$$ Expanding the binomials we

obtain the following series:

$$\cos nz = (\cos z)^n - \frac{n(n-1)}{1\cdot 2}(\cos z)^{n-2}(\sin z)^2$$
$$+ \frac{n(n-1)(n-2)(n-3)}{1\cdot 2\cdot 3\cdot 4}(\cos z)^{n-4}(\sin z)^4$$
$$- \frac{n(n-1)(n-2)(n-3)(n-4)(n-5)}{1\cdot 2\cdot 3\cdot 4\cdot 5\cdot 6}(\cos z)^{n-6}(\sin z)^6$$
$$+ \cdots$$

and

$$\sin nz = \frac{n}{1}(\cos z)^{n-1}\sin z$$
$$- \frac{n(n-1)(n-2)}{1\cdot 2\cdot 3}(\cos z)^{n-3}(\sin z)^3$$
$$+ \frac{n(n-1)(n-2)(n-3)(n-4)}{1\cdot 2\cdot 3\cdot 4\cdot 5}(\cos z)^{n-5}(\sin z)^5$$
$$- \cdots .$$

134. Let the arc z be infinitely small, then $\sin z = z$ and $\cos z = 1$. If n is an infinitely large number, so that nz is a finite number, say $nz = v$, then, since $\sin z = z = \frac{v}{n}$, we have

$\cos v = 1 - \frac{v^2}{1\cdot 2} + \frac{v^4}{1\cdot 2\cdot 3\cdot 4} - \frac{v^6}{1\cdot 2\cdot 3\cdot 4\cdot 5\cdot 6} + \cdots$ and

$\sin v = v - \frac{v^3}{1\cdot 2\cdot 3} + \frac{v^5}{1\cdot 2\cdot 3\cdot 4\cdot 5} - \frac{v^7}{1\cdot 2\cdot 3\cdot 4\cdot 5\cdot 6\cdot 7} + \cdots$. It follows that if v is a given arc, by means of these series, the sine and cosine can be found. In order that the use of these formulas may become clearer, let us take v to be in the same ratio to the quarter circle, or 90 degrees, as m is to n. That is $v = \frac{m}{n}\frac{\pi}{2}$. Since the value of π is known, if we substitute this value we obtain

$$\sin \frac{m}{n}\frac{\pi}{2} = + \frac{m}{n}1.5707963267948966192313216916 \quad - \frac{m^3}{n^3}0.6459640975062462536557565636$$
$$+ \frac{m^5}{n^5}0.0796926262461670451205055488 \quad - \frac{m^7}{n^7}0.0046817541353186881006854632$$

$$+ \frac{m^9}{n^9} 0.000160441184787359821872660 5 \qquad - \frac{m^{11}}{n^{11}} 0.000003598843235212085340458 0$$

$$+ \frac{m^{13}}{n^{13}} 0.000000056921729219679268117 1 \qquad - \frac{m^{15}}{n^{15}} 0.000000000668803510981146722 4$$

$$+ \frac{m^{17}}{n^{17}} 0.000000000006066935731106195 0 \qquad - \frac{m^{19}}{n^{19}} 0.000000000000043770654673137 0$$

$$+ \frac{m^{21}}{n^{21}} 0.000000000000000257142289285 6 \qquad - \frac{m^{23}}{n^{23}} 0.000000000000000001253899540 3$$

$$+ \frac{m^{25}}{n^{25}} 0.000000000000000000005156455 0 \qquad - \frac{m^{27}}{n^{27}} 0.000000000000000000000018123 9$$

$$+ \frac{m^{29}}{n^{29}} 0.000000000000000000000000054 9$$

and

$$\cos \frac{m}{n} \frac{\pi}{2} = 1.000000000000000000000000000 0 \qquad - \frac{m^2}{n^2} 1.233700550136169827354311374 5$$

$$+ \frac{m^4}{n^4} 0.253669507901048013636563365 9 \qquad - \frac{m^6}{n^6} 0.020863480763352960873051636 4$$

$$+ \frac{m^8}{n^8} 0.000919260274839426580241715 8 \qquad - \frac{m^{10}}{n^{10}} 0.000025202042373060605481052 6$$

$$+ \frac{m^{12}}{n^{12}} 0.000000471087477881817150366 5 \qquad - \frac{m^{14}}{n^{14}} 0.000000006386603083791852240 8$$

$$+ \frac{m^{16}}{n^{16}} 0.000000000065659631149794723 0 \qquad - \frac{m^{18}}{n^{18}} 0.000000000000529440020073462 0$$

$$+ \frac{m^{20}}{n^{20}} 0.000000000000003437739179098 1 \qquad - \frac{m^{22}}{n^{22}} 0.000000000000000018359916521 2$$

$$+ \frac{m^{24}}{n^{24}} 0.000000000000000000082067532 7 \qquad - \frac{m^{26}}{n^{26}} 0.000000000000000000000311528 5$$

$$+ \frac{m^{28}}{n^{28}} 0.000000000000000000000001016 5 \qquad - \frac{m^{30}}{n^{30}} 0.000000000000000000000000002 6.$$

Since it is sufficient to know the sines and cosines of angles only to 45 degrees, the fraction $\frac{m}{n}$ will always be less than ½; because of the powers of the fraction

$\frac{m}{n}$, the series converge quickly. A few terms should be sufficient, especially if the number of decimal places is not so large.

135. Once sines and cosines have been computed, tangents and cotangents can be found in the ordinary way, however, since the multiplication and division of such gigantic numbers is so inconvenient, a different method of expressing these functions is desirable. Since $\tan v = \frac{\sin v}{\cos v}$

$$= \frac{v - \frac{v^3}{1\cdot 2\cdot 3} + \frac{v^5}{1\cdot 2\cdot 3\cdot 4\cdot 5} - \frac{v^7}{1\cdot 2\cdot 3\cdot 4\cdot 5\cdot 6\cdot 7} + \cdots}{1 - \frac{v^2}{1\cdot 2} + \frac{v^4}{1\cdot 2\cdot 3\cdot 4} - \frac{v^6}{1\cdot 2\cdot 3\cdot 4\cdot 5\cdot 6} + \cdots}$$

and

$$\cot v = \frac{\cos v}{\sin v}$$

$$= \frac{1 - \frac{v^2}{1\cdot 2} + \frac{v^4}{1\cdot 2\cdot 3\cdot 4} - \frac{v^6}{1\cdot 2\cdot 3\cdot 4\cdot 5\cdot 6} + \cdots}{v - \frac{v^3}{1\cdot 2\cdot 3} + \frac{v^5}{1\cdot 2\cdot 3\cdot 4\cdot 5} - \frac{v^7}{1\cdot 2\cdot 3\cdot 4\cdot 5\cdot 6\cdot 7} + \cdots}.$$

If the arc is $v = \frac{m}{n}\frac{\pi}{2}$, then as before

$$\tan v = \frac{2mn}{n^2 - m^2}0.6366197723675$$

$$+ \frac{m}{n}0.2975567820597 + \frac{m^3}{n^3}0.0186886502773$$

$$+ \frac{m^5}{n^5}0.0018424752034 + \frac{m^7}{n^7}0.0001975800714$$

$$+ \frac{m^9}{n^9}0.0000216977245 + \frac{m^{11}}{n^{11}}0.0000024011370$$

$$+ \frac{m^{13}}{n^{13}}0.0000002664132 + \frac{m^{15}}{n^{15}}0.0000000295864$$

$$+ \frac{m^{17}}{n^{17}}0.0000000032867 + \frac{m^{19}}{n^{19}}0.0000000003651$$

$$+\frac{m^{21}}{n^{21}}0.0000000000405+\frac{m^{23}}{n^{23}}0.0000000000045+\frac{m^{25}}{n^{25}}0.0000000000005$$

$$\cot v=\frac{n}{m}0.6366197723675$$

$$-\frac{4mn}{4n^2-m^2}0.3183098861837-\frac{m}{n}0.2052888894145$$

$$-\frac{m^3}{n^3}0.0065510747882-\frac{m^5}{n^5}0.0003450292554$$

$$-\frac{m^7}{n^7}0.0000202791060-\frac{m^9}{n^9}0.0000012366527$$

$$-\frac{m^{11}}{n^{11}}0.0000000764959-\frac{m^{13}}{n^{13}}0.0000000047597$$

$$-\frac{m^{15}}{n^{15}}0.0000000002969-\frac{m^{17}}{n^{17}}0.0000000000185-\frac{m^{19}}{n^{19}}0.0000000000011.$$

The basis for these series will be explained at length below in section 197.

136. From what we have seen previously, it is clear that when we know the sines and cosines of angles less than half a right angle, then we also have sines and cosines of greater angles. In fact, if we know the sines and cosines of angles less than only 30 degrees, then from these, by only addition and subtraction, we can find all sines and cosines of larger angles. Since $\sin\frac{\pi}{6}=\frac{1}{2}$, when we let $y=\frac{\pi}{6}$ in the formula from section 130, we have

$\cos z=\sin\left(\frac{\pi}{6}+z\right)+\sin\left(\frac{\pi}{6}-z\right)$ and

$\sin z=\cos\left(\frac{\pi}{6}-z\right)-\cos\left(\frac{\pi}{6}+z\right)$. It follows that from the sines and cosines of angles z and $\frac{\pi}{6}-z$ we obtain $\sin\left(\frac{\pi}{6}+z\right)=\cos z-\sin\left(\frac{\pi}{6}-z\right)$ and

$\cos\left(\frac{\pi}{6} + z\right) = \cos\left(\frac{\pi}{6} - z\right) - \sin z$. In this way we obtain sines and cosines of angles from 30 degrees to 60 degrees, and hence the sines and cosines are defined for all larger angles.

137. A similar strategy can be used to find tangents and cotangents. Since $\tan(a + b) = \dfrac{\tan a + \tan b}{1 - \tan a \tan b}$, we have $\tan 2a = \dfrac{2 \tan a}{1 - (\tan a)^2}$ and $\cot 2a = \dfrac{\cot a - \tan a}{2}$. It follows that from tangents and cotangents of arcs less than 30 degrees, we can find tangents and cotangents up to 60 degrees. Let $a = \frac{\pi}{6} - b$, then $2a = \frac{\pi}{3} - 2b$ and $\cot 2a = \tan\left(\frac{\pi}{6} + 2b\right)$. Then we have

$$\tan\left(\frac{\pi}{6} + 2b\right) = \frac{\cot\left(\frac{\pi}{6} - b\right) - \tan\left(\frac{\pi}{6} - b\right)}{2},$$

which gives tangents of arcs greater than 30 degrees. Secants and cosecants can be found from tangents by means of subtraction. Note that $\csc z = \cot \frac{z}{2} - \cot z$ and $\sec z = \cot\left(\frac{\pi}{4} - \frac{z}{2}\right) - \tan z$. From these remarks it should be very clear how tables of sines can be constructed.

138. Once again we use the formulas in section 133, where we let z be an infinitely small arc and let n be an infinitely large number j, so that jz has a finite value v. Now we have $nz = v$ and $z = \frac{v}{j}$, so that $\sin z = \frac{v}{j}$ and $\cos z = 1$. With these substitutions,

$$\cos v = \frac{\left(1 + \frac{iv}{j}\right)^j + \left(1 - \frac{iv}{j}\right)^j}{2} \quad \text{and} \quad \sin v = \frac{\left(1 + \frac{iv}{j}\right)^j - \left(1 - \frac{iv}{j}\right)^j}{2i}.$$

In the preceding chapter we saw that $(1 + z/j)^j = e^z$ where e is the base of the natural logarithms. When we let $z = iv$ and then $z = -iv$ we obtain $\cos v = \frac{e^{iv} + e^{-iv}}{2}$ and $\sin v = \frac{e^{iv} - e^{-iv}}{2i}$. From these equations we understand how complex exponentials can be expressed by real sines and cosines, since $e^{iv} = \cos v + i \sin v$ and $e^{-iv} = \cos v - i \sin v$.

139. Now let n be an infinitely small number in the formulas of section 130 or let $n = \frac{1}{j}$, where j is an infinitely large number. Then $\cos nz = \cos \frac{z}{j} = 1$ and $\sin nz = \sin \frac{z}{j} = \frac{z}{j}$, since the sine of a vanishing arc $\frac{z}{j}$ is equal to the arc itself, and the cosine of such an arc is equal to 1. With this hypothesis we have

$$1 = \frac{(\cos z + i \sin z)^{\frac{1}{j}} + (\cos z - i \sin z)^{\frac{1}{j}}}{2}$$ and

$$\frac{z}{j} = \frac{(\cos z + i \sin z)^{\frac{1}{j}} - (\cos z - i \sin z)^{\frac{1}{j}}}{2i}.$$ In section 125 we saw that $\log(1 + x) = j(1 + x)^{\frac{1}{j}} - j$ or, when we substitute y for $1 + x$, we have $y^{\frac{1}{j}} = 1 + \frac{1}{j}\log y$. Now we first substitute $\cos z + i \sin z$ for y, then substitute $\cos z - i \sin z$ for y to obtain

$$1 = \frac{1 + \frac{1}{j}\log(\cos z + i \sin z) + 1 + \frac{1}{j}\log(\cos z - i \sin z)}{2}.$$ Since the terms with logarithms vanish in this equation, nothing follows; however, from the other equation, for the sine, we obtain

$$\frac{z}{j} = \frac{\frac{1}{j}\log(\cos z + i \sin z) - \frac{1}{j}\log(\cos z - i \sin z)}{2i}.$$ From this we obtain

$z = \frac{1}{2i}\log\left(\frac{\cos z + i \sin z}{\cos z - i \sin z}\right)$, so that it becomes clear to what extent logarithms of complex numbers are related to circular arcs.

140. Since $\frac{\sin z}{\cos z} = \tan z$, the arc z can now be expressed through its tangent as follows: $z = \frac{1}{2i}\log\left(\frac{1 + i \tan z}{1 - i \tan z}\right)$. We have seen in section 123 that $\log\left(\frac{1 + x}{1 - x}\right) = \frac{2x}{1} + \frac{2x^3}{3} + \frac{2x^5}{5} + \frac{2x^7}{7} + \cdots$. When we substitute $i \tan z$ for x we obtain

$z = \frac{\tan z}{1} - \frac{(\tan z)^3}{3} + \frac{(\tan z)^5}{5} - \frac{(\tan z)^7}{7} + \cdots$. If we let $t = \tan z$ so that z is the arc whose tangent is t, which we will indicate by arctan t, then $z = \arctan t$. When we know the tangent of t, the corresponding arc z is given by $z = \frac{t}{1} - \frac{t^3}{3} + \frac{t^5}{5} - \frac{t^7}{7} + \frac{t^9}{9} - \cdots$. If the tangent of t is equal to the unit radius, then the arc z is equal to 45 degrees or $z = \frac{\pi}{4}$ and $\frac{\pi}{4} = 1 - \frac{1}{3} + \frac{1}{5} - \frac{1}{7} + \cdots$. This series, which was first discovered by Leibnitz, can be used to find the value of the circumference of the circle.

141. In order to see the ease with which the length of an arc can be found by means of this series, we should substitute a sufficiently small fraction for the tangent t. For example let us use this series to find the length of the arc z whose tangent is $\frac{1}{10}$. In this case the arc

$z = \frac{1}{10} - \frac{1}{3000} + \frac{1}{500000} - \cdots$, and the approximate value of this series is easily expressed by a decimal fraction. However, from such an arc, we cannot conclude anything about the whole circumference of the circle, since the ratio of the arc whose tangent is $\frac{1}{10}$ to the whole circumference is not given. For this reason, in order to find the circumference, we look for an arc such that not only is it some fractional part of the circumference, but also small and easily expressed. For this purpose it is customary to choose the arc to be 30 degrees, whose tangent is equal to $\frac{1}{\sqrt{3}}$, since smaller arcs have tangents which are extremely irrational. Wherefore, since an arc of 30 degrees has length $\frac{\pi}{6}$, we have $\frac{\pi}{6} = \frac{1}{\sqrt{3}} - \frac{1}{3 \cdot 3\sqrt{3}} + \frac{1}{5 \cdot 3^2\sqrt{3}} - \cdots$ and

$\pi = \frac{2\sqrt{3}}{1} - \frac{2\sqrt{3}}{3 \cdot 3} + \frac{2\sqrt{3}}{5 \cdot 3^2} - \frac{2\sqrt{3}}{7 \cdot 3^3} + \cdots$. By means of this series the value of π itself, which was previously exhibited, was determined with incredible labor.

142. The labor involved in this calculation is all the more since, first of all, each term is irrational, but also, since each succeeding term is only about one third of the preceding. In order to avoid these inconveniences, let us take the arc to be 45 degrees, that is of length $\frac{\pi}{4}$. Although this arc can be expressed by a series which hardly converges, $1 - \frac{1}{3} + \frac{1}{5} - \frac{1}{7} + \cdots$, still we keep this arc and express it by means of two arcs of lengths a and b so that $a + b = \frac{\pi}{4}$, that is 45 degrees. Since $\tan(a + b) = 1 = \frac{\tan a + \tan b}{1 - \tan a \tan b}$, we have $1 - \tan a \tan b = \tan a + \tan b$ and $\tan b = \frac{1 - \tan a}{1 + \tan a}$. If we let

$\tan a = \frac{1}{2}$, then $\tan b = \frac{1}{3}$ and both the arcs a and b can be expressed by rational series which converge much more rapidly than the series above. The sum of these two series gives the value of the arc $\frac{\pi}{4}$. It follows that

$$\pi = 4\left(\frac{1}{1\cdot 2} - \frac{1}{3\cdot 2^3} + \frac{1}{5\cdot 2^5} - \frac{1}{7\cdot 2^7} + \frac{1}{9\cdot 2^9} - \cdots\right)$$

$$+ 4\left(\frac{1}{1\cdot 3} - \frac{1}{3\cdot 3^3} + \frac{1}{5\cdot 3^5} - \frac{1}{7\cdot 3^7} + \frac{1}{9\cdot 3^9} - \cdots\right).$$ In this way we calculate the length of the semicircle, π, with much more ease than with the series mentioned before.

CHAPTER IX

On Trinomial Factors.

143. The means by which linear factors of any polynomial may be found, we have seen above, is through the solution of an equation. If the polynomial is $\alpha + \beta z + \gamma z^2 + \delta z^3 + \epsilon z^4 + \cdots$ and a linear factor is of the form $p - qz$, then it is clear that whenever $p - qz$ is a factor of the function $\alpha + \beta z + \gamma z^2 + \cdots$, and when we substitute $\frac{p}{q}$ for z, then the factor $p - qz$ becomes zero and the proposed function vanishes. It follows that $p - qz$ is a factor or divisor of the polynomial $\alpha + \beta z + \gamma z^2 + \delta z^3 + \epsilon z^4 + \cdots$ whenever $\alpha + \beta \frac{p}{q} + \gamma \frac{p^2}{q^2} + \delta \frac{p^3}{q^3} + \epsilon \frac{p^4}{q^4} + \cdots = 0$. Conversely, if all the roots $\frac{p}{q}$ of this equation have been extracted, they will give all of the linear factors of the proposed polynomial $\alpha + \beta z + \gamma z^2 + \delta z^3 + \cdots$, that is $p - qz$. It is clear now that the number of these linear factors is determined by the greatest power of z.

144. From time to time it happens that complex linear factors are found only with difficulty. It is for this reason that I present in this chapter a special method by which the complex linear factors can frequently be found. Since complex linear factors are so paired that the product of two of them is real. We will find those complex factors if we study the quadratic factors of the form

$p - qz + rz^2$ which are real, but whose linear factors are complex. If the function $\alpha + \beta z + \gamma z^2 + \delta z^3 + \cdots$ has only real quadratic factors in this form of a trinomial $p - qz + rz^2$, then all of the linear factors will be complex.

145. A trinomial $p - qz + rz^2$ has linear factors which are complex if $4pr > q^2$, that is if $\frac{q}{2\sqrt{pr}} < 1$. Since the sine and cosine of angles are less than 1, a trinomial $p - qz + rz^2$ has complex linear factors if $\frac{q}{2\sqrt{pr}}$ is equal to the sine or cosine of some angle. Now let $\frac{q}{2\sqrt{pr}} = \cos\phi$ or $q = 2\sqrt{pr}\cos\phi$, and the trinomial $p - qz + rz^2$ has complex linear factors. Lest some irrationality cause problems, we assume the trinomial has the form $p^2 - 2pqz\cos\phi + q^2z^2$, whose complex linear factors are $qz - p(\cos\phi + i\sin\phi)$ and $qz - p(\cos\phi - i\sin\phi)$. It is clear that if $\cos\phi = \pm 1$, then $\sin\phi = 0$ and both factors will be equal and real.

146. Given a polynomial $\alpha + \beta z + \gamma z^2 + \delta z^3 + \cdots$, the complex linear factors can be found if the values of p, q, and the arc ϕ are such that the trinomial $p^2 - 2pqz\cos\phi + q^2z^2$ is a factor of the function. In this case, the complex linear factors will be $qz - p(\cos\phi + i\sin\phi)$ and $qz - p(\cos\phi - i\sin\phi)$. For this reason, the given function vanishes if we substitute either $z = \frac{p}{q}(\cos\phi + i\sin\phi)$ or $z = \frac{p}{q}(\cos\phi - i\sin\phi)$. When each of these substitutions is made, we obtain two equations which can be solved for both the fraction $\frac{p}{q}$ and the arc ϕ.

147. It might seem at first that these substitutions for z would cause difficulties, but when we use some of the results treated in the preceding chapter, things go rather expeditiously. We have seen that $(\cos\phi \pm i\sin\phi)^n = \cos n\phi \pm i\sin n\phi$, so that the following formulas are used when substituting for the powers of z. In the first factor,

$$z = \frac{p}{q}(\cos\phi + i\sin\phi),\ z^2 = \frac{p^2}{q^2}(\cos 2\phi + i\sin 2\phi),$$

$$z^3 = \frac{p^3}{q^3}(\cos 3\phi + i\sin 3\phi),\ z^4 = \frac{p^4}{q^4}(\cos 4\phi + i\sin 4\phi), \text{ etc.}$$

In the second factor, $z = \frac{p}{q}(\cos\phi - i\sin\phi)$, $z^2 = \frac{p^2}{q^2}(\cos 2\phi - i\sin 2\phi)$,

$z^3 = \frac{p^3}{q^3}(\cos 3\phi - i\sin 3\phi)$, $z^4 = \frac{p^4}{q^4}(\cos 4\phi - i\sin 4\phi)$, etc. For the sake of brevity we let $\frac{p}{q} = r$ and then make the substitutions to obtain the two equations $0 = \alpha + \beta r\cos\phi + \gamma r^2\cos 2\phi + \delta r^3\cos 3\phi + \cdots$
$+ \beta ri\sin\phi + \gamma r^2 i\sin 2\phi + \delta r^3 i\sin 3\phi + \cdots$ and

$$0 = \alpha + \beta r\cos\phi + \gamma r^2\cos 2\phi + \delta r^3\cos 3\phi + \cdots$$
$$- \beta ri\sin\phi - \gamma r^2 i\sin 2\phi - \delta r^3 i\sin 3\phi - \cdots.$$

148. If these two equations are added and subtracted, and in the latter case also divided by $2i$ we obtain the two real equations $0 = \alpha + \beta r\cos\phi + \gamma r^2\cos 2\phi + \delta r^3\cos 3\phi + \cdots$ and $0 = \beta r\sin\phi + \gamma r^2\sin 2\phi + \delta r^3\sin 3\phi + \cdots$. In fact, given the polynomial $\alpha + \beta z + \gamma z^2 + \delta z^3 + \epsilon z^4 + \cdots$ we can immediately write down the two equations. In the first we put, for each power of z, $z^n = r^n\cos n\phi$ and in the second $z^n = r^n\sin n\phi$. Since $\sin 0\phi = 0$ and $\cos 0\phi = 1$, for z^0 in the first

equation we put 1 and in the second we put 0. If now we can find the two unknown quantities r and ϕ from the two equations, then, since $r = \frac{p}{q}$, we will have the trinomial factor, $p^2 - 2pqz \cos \phi + q^2z^2$, of the given function and so also the two complex linear factors.

149. If the first equation is multiplied by $\cos m\phi$ and the second by $\sin m\phi$, then by addition and subtraction the following equations result.

$0 = \alpha \cos m\phi + \beta r \cos(m - 1)\phi + \gamma r^2 \cos(m - 2)\phi$

$+ \delta r^3 \cos(m - 3)\phi + \cdots$ and

$0 = \alpha \cos m\phi + \beta r \cos(m + 1)\phi + \gamma r^2 \cos(m + 2)\phi$

$+ \delta r^3 \cos(m + 3)\phi + \cdots$. Any two equations of this kind determine the unknowns r and ϕ. Since frequently there are several different solutions, we obtain several different trinomial factors, indeed we obtain all such factors in this way.

150. In order that the use of these rules may become clearer, we will investigate trinomial factors of certain functions which occur rather frequently. Once we have these results, they will be ready at hand for future use. Let the first such function be $a^n + z^n$; we will determine the trinomial factors of the form $p^2 - 2pqz \cos \phi + q^2z^2$. When we let $r = \frac{p}{q}$ we have the following two equations: $0 = a^n + r^n \cos n\phi$ and $0 = r^n \sin n\phi$. The second of these equations gives $\sin n\phi = 0$, so that $n\phi = (2k + 1)\pi$, or $n\phi = 2k\pi$, where k is an integer. We will treat these two cases separately, since the cosines are different, being respectively $\cos(2k + 1)\pi = -1$ and $\cos 2k\pi = 1$. It should be clear that the choice will be $n\phi = (2k + 1)\pi$, since with $\cos n\phi = -1$, we have

$0 = a^n - r^n$. Since $r = a = \frac{p}{q}$, we have $p = a$, $q = 1$, and $\phi = \frac{(2k+1)\pi}{n}$.

It follows that a factor of $a^n + z^n$ will be $a^2 - 2az \cos\frac{(2k+1)}{n}\pi + z^2$. Since any integer can be substituted for k, several factors of this form will be produced, but not an infinite number. This is because when $2k + 1$ becomes larger than n the factors begin to reoccur. This is because $\cos(2\pi \pm \phi) = \cos\phi$, but this will become clearer from examples. If n is an odd number, when $2k + 1 = n$, then there is a quadratic factor $a^2 + 2az + z^2$. From this it does not follow that $(a + z)^2$ is a factor of $a^n + z^n$, since from section 148 we see that only one equation results. It is clear that only $a + z$ is the divisor of $a^n + z^n$. This rule applies whether $\cos\phi$ is equal to +1 or -1.

EXAMPLE

We will develop a few cases so that we can see more clearly what the factors are. In these cases we distinguish between the odd and even values of n. If $n = 1$ then the function is $a + z$ and the factor is $a + z$. If $n = 2$, then the function is $a^2 + z^2$ and the factor is $a^2 + z^2$. If $n = 3$ then the function is $a^3 + z^3$ and the factors are $a^2 - 2az \cos\frac{1}{3}\pi + z^2$ and $a + z$. If $n = 4$ then the function is $a^4 + z^4$ and the factors are $a^2 - 2az\cos\frac{1}{4}\pi + z^2$ and $a^2 - 2az\cos\frac{3}{4}\pi + z^2$. If $n = 5$ then the function is $a^5 + z^5$ and the factors are $a^2 - 2az\cos\frac{1}{5}\pi + z^2$, $a^2 - 2az\cos\frac{3}{5}\pi + z^2$, and $a + z$. If $n = 6$ then the function is $a^6 + z^6$ and the factors are $a^2 - 2az\cos\frac{1}{6}\pi + z^2$, $a^2 - 2az\cos\frac{3}{6}\pi + z^2$, and $a^2 - 2az\cos\frac{5}{6}\pi + z^2$. From these examples it is clear that all of the factors have been obtained when for $2k + 1$ all odd numbers less than n are substituted. In those cases when a perfect square is produced, only its square root is a factor.

151. If the given function is $a^n - z^n$, then a trinomial factor is $p^2 - 2pqz \cos \phi + q^2r^2$. If we let $r = \frac{p}{q}$, then $0 = a^n - r^n \cos n\phi$ and $0 = r^n \sin n\phi$. Once again $\sin n\phi = 0$, and $n\phi = (2k + 1)\pi$ or $n\phi = 2k\pi$. In this case, however, we make the second choice, so that $\cos n = 1$, with $0 = a^n - r^n$, and $r = \frac{p}{q} = a$. It follows that $p = a$, $q = 1$, and $\phi = \frac{2k\pi}{n}$, so that the trinomial factor will be $a^2 - 2az \cos \frac{2k\pi}{n} + z^2$. In this formula we let $2k$ be equal to all even integers no larger than n to obtain all factors. Concerning factors which are perfect squares, we follow the rule given above. First we let $k = 0$ to obtain $a^2 - 2az + z^2$ from which we take the square root, $a - z$. Likewise, if n is even and $2k = n$, then we obtain $a^2 + 2az + z^2$ and $a + z$ is a divisor of $a^n - z^n$.

EXAMPLE

As in the previous example we distinguish between the odd and even values of n. If $n = 1$, then the function is $a - z$ and the factor is $a - z$. If $n = 2$ then the function is $a^2 - z^2$ and the factors are $a - z$ and $a + z$. If $n = 3$, then the function is $a^3 - z^3$ and the factors are $a - z$ and $a^2 - 2az \cos \frac{2}{3}\pi + z^2$. If $n = 4$ then the function is $a^4 - z^4$ and the factors are $a - z$, $a^2 - 2az \cos \frac{2}{4}\pi + z^2$, and $a + z$. If $n = 5$ then the function is

$a^5 - z^5$ and the factors are $a - z$, $a^2 - 2az\cos\frac{2}{5}\pi + z^2$, and $a^2 - 2az\cos\frac{4}{5}\pi + z^2$. If $n = 6$, then the function is $a^6 - z^6$ and the factors are $a - z$, $a^2 - 2az\cos\frac{2}{6}\pi + z^2$, $a^2 - 2az\cos\frac{4}{6}\pi + z^2$, and $a + z$.

152. These examples confirm what had been stated earlier, namely, that every polynomial, can be expressed as the product of real linear factors and real quadratic factors. We have seen that functions with the form $a^n + z^n$ with any degree, can be expressed as a product of real quadratic factors and real linear factors. We progress to more complicated functions such as $\alpha + \beta z^n + \gamma z^{2n}$. If this function has two factors of the form $\eta + \theta z^n$, then the factorization is clear from what we have just considered. We will show how to resolve such a function $\alpha + \beta z^n + \gamma z^{2n}$ into real linear or real quadratic factor in the case where there are not two real factors of the form $\eta + \theta z^n$.

153. We consider this function $a^{2n} - 2a^n z^n \cos g + z^{2n}$ which cannot be expressed as the product of two real factors of the form $\eta + \theta z^n$. If we suppose one of the real quadratic factors to be $p^2 + 2pqz\cos\phi + q^2z^2$, when we let $r = \frac{p}{q}$ we obtain the following two equations:

$0 = a^{2n} - 2a^n r^n \cos g \cos n\phi + r^{2n}\cos 2n\phi$ and

$0 = -2a^n r^n \cos g \sin n\phi + r^{2n}\sin 2n\phi$. If instead of the first equation, we have from section 149, when $m = 2n$, $0 = a^{2n}\sin 2n\phi - 2a^n r^n \cos g \sin n\phi$. This equation with the second equation above give $r = a$. Then $\sin 2n\phi = 2\cos g \sin n\phi$. Since $\sin 2n\phi = 2\cos n\phi \sin n\phi$, it follows that $\cos n\phi = \cos g$. Since $\cos(2k\pi \pm g) = \cos g$, we have $n\phi = 2k\pi \pm g$ and

$\phi = \dfrac{2k\pi \pm g}{n}$. We now have the general quadratic factor of the proposed form $a^2 - 2az\cos\dfrac{2k\pi \pm g}{n} + z^2$, and all factors appear when we let $2k$ be all even integers no greater than n, as we shall see in the following.

EXAMPLE

We consider the cases in which n is 1, 2, 3, 4, etc. If the function is $a^2 - 2az\cos g + z^2$, then the factor is $a^2 - 2az\cos g + z^2$. If the function is $a^4 - 2az^2\cos g + z^4$, then the two factors are $a^2 - 2az\cos\dfrac{g}{2} + z^2$, and $a^2 - 2az\cos\dfrac{2\pi \pm g}{2} + z^2$, that is, $a^2 + 2az\cos\dfrac{g}{2} + z^2$. If the function is $a^6 - 2a^3z^3\cos g + z^6$, then the three factors are $a^3 - 2az\cos\dfrac{g}{3} + z^2$, $a^2 - 2az\cos\dfrac{2\pi - g}{3} + z^2$, and $a^2 - 2az\cos\dfrac{2\pi + g}{3} + z^2$. If the function is $a^8 - 2a^4z^4\cos g + z^8$, then the four factors are $a^2 - 2az\cos\dfrac{g}{4} + z^2$, $a^2 - 2az\cos\dfrac{2\pi - g}{4} + z^2$, $a^2 - 2az\cos\dfrac{2\pi + g}{4} + z^2$, and $a^2 - 2az\cos\dfrac{4\pi \pm g}{4} + z^2$, that is, $a^2 + 2az\cos\dfrac{g}{4} + z^2$. If the function is $a^{10} - 2a^5z^5\cos g + z^{10}$, then the five factors are $a^2 - 2az\cos\dfrac{g}{5} + z^2$, $a^2 - 2az\cos\dfrac{2\pi - g}{5} + z^2$, $a^2 - 2az\cos\dfrac{2\pi + g}{5} + z^2$, $a^2 - 2az\cos\dfrac{4\pi - g}{5} + z^2$, and $a^2 - 2az\cos\dfrac{4\pi + g}{5} + z^2$. Again it is confirmed in these examples that polynomials can be expressed as the product of real linear and real quadratic factors.

154. Now we can go further and consider a function of the form $\alpha + \beta z^n + \gamma z^{2n} + \delta z^{3n}$, which certainly has one factor of the form $\eta + \theta z^n$, and we have seen how to express this as a product of real linear and real quadratic factors. The other factor is of the form $\iota + \kappa z^n + \lambda z^{2n}$, which, according to the preceding section, can also be expressed as a product of real linear and real quadratic factors. Next we consider the function $\alpha + \beta z^n + \gamma z^{2n} + \delta z^{3n} + \epsilon z^{4n}$. This always has two real factors of the form $\eta + \theta z^n + \iota z^{2n}$ and these likewise can be expressed as products of real linear and real quadratic factors. Then we consider the function $\alpha + \beta z^n + \gamma z^{2n} + \delta z^{3n} + \epsilon z^{4n} + \zeta z^{5n}$, which always has one factor of the form $\eta + \theta z^n$, while the other factor is of the form just considered. It follows that this function can be expressed as a product of real linear and real quadratic factors. If there were any doubt that every polynomial can be expressed as a product of real linear and real quadratic factors, then that doubt by this time should be almost completely dissipated.

155. We can extend this factorization also to infinite series. For example, we have seen that $1 + \frac{x}{1} + \frac{x^2}{1 \cdot 2} + \frac{x^3}{1 \cdot 2 \cdot 3} + \frac{x^4}{1 \cdot 2 \cdot 3 \cdot 4} + \cdots = e^x$. We have also seen that $e^x = (1 + x/j)^j$, where j is an infinitely large number. It becomes clear now that the series $1 + \frac{x}{1} + \frac{x^2}{1 \cdot 2} + \frac{x^3}{1 \cdot 2 \cdot 3} + \cdots$ has an infinite number of linear factors, all of them equal, namely to $1 + \frac{x}{j}$. If we remove the first term from this series to obtain $\frac{x}{1} + \frac{x^2}{1 \cdot 2} + \frac{x^3}{1 \cdot 2 \cdot 3} + \cdots = e^x - 1 = (1 + x/j)^j - 1$. When we compare

this with the form in section 151, where we let $a = 1 + \frac{x}{j}$, $n = j$, and $z = 1$, each factor has the form $(1 + x/j)^2 - 2(1 + x/j)\cos\frac{2k\pi}{j} + 1$. When all even integers are substituted for $2k$ we obtain all of the factors. However, when $2k = 0$ we obtain the perfect square $\frac{x^2}{j^2}$ as a factor. For the reasons given before we take only the square root, $\frac{x}{j}$. It follows that x is a factor of the function $e^x - 1$, but that is already obvious. To find the other factors we have to note that the arc $\frac{2k}{j}\pi$ is infinitely small and according to section 134 we have $\cos\frac{2k}{j}\pi = 1 - 2\frac{k^2}{j^2}\pi^2$. The other terms in the series are neglected since j is infinitely large. It follows that each factor has the form $\frac{x^2}{j^2} + \frac{4k^2}{j^2}\pi^2 + \frac{4k^2}{j^3}\pi^2 x$ and $e^x - 1$ is divisible by $1 + \frac{x}{j} + \frac{x^2}{4k^2\pi^2}$. Therefore $e^x - 1 = x\left(1 + \frac{x}{1\cdot2} + \frac{x^2}{1\cdot2\cdot3} + \frac{x^3}{1\cdot2\cdot3\cdot4} + \cdots\right)$ and except for the factor x, it has the infinite product of factors

$$\left(1 + \frac{x}{j} + \frac{x^2}{4\pi^2}\right)\left(1 + \frac{x}{j} + \frac{x^2}{16\pi^2}\right)\left(1 + \frac{x}{j} + \frac{x^2}{36\pi^2}\right)$$

$$\left(1 + \frac{x}{j} + \frac{x^2}{64\pi^2}\right)\cdots.$$

156. Since all of these factors contain a term which is infinitely small $\frac{x}{j}$, which, since it is in each factor, and through the multiplication of all the factors which are $\frac{1}{2}j$ in number, there is produced a term $\frac{x}{2}$, so $\frac{x}{j}$ cannot be omitted. In order to avoid this inconvenience we consider the expression

$e^x - e^{-x} = (1 + x/j)^j - (1 - x/j)^j$

$= 2\left(\frac{x}{1} + \frac{x^3}{1\cdot 2\cdot 3} + \frac{x^5}{1\cdot 2\cdot 3\cdot 4\cdot 5} + \cdots\right)$, since

$e^{-x} = 1 - \frac{x}{1} + \frac{x^2}{1\cdot 2} - \frac{x^3}{1\cdot 2\cdot 3} + \cdots$. We compare this with the expression in section 151, with $n = j$, $a = 1 + \frac{x}{j}$, and $z = 1 - \frac{x}{j}$. It follows that the factor of this series will be

$a^2 - 2az\cos\frac{2k\pi}{n} + z^2 = 2 + \frac{2x^2}{j^2} - 2\left(1 - \frac{x^2}{j^2}\right)\cos\frac{2k}{j}\pi$

$= \frac{4x^2}{j^2} + \frac{4k^2}{j^2}\pi^2 - \frac{4k^2\pi^2x^2}{j^4}$, since $\cos\frac{2k}{j}\pi = 1 - \frac{2k^2\pi^2}{j^2}$. The function $e^x - e^{-x}$, therefore, is divisible by $1 + \frac{x^2}{k^2}\pi^2 - \frac{x^2}{j^2}$, however, we omit the term $\frac{x^2}{j^2}$, since even when multiplied by j, it remains infinitely small. Further, when $k = 0$, the factor will be x. For these reasons, the factors can be given in the order in which they are calculated:

$$\frac{e^x - e^{-x}}{2} = x\left(1 + \frac{x^2}{\pi^2}\right)\left(1 + \frac{x^2}{4\pi^2}\right)\left(1 + \frac{x^2}{9\pi^2}\right)\left(1 + \frac{x^2}{16\pi^2}\right)$$

$$\left(1 + \frac{x^2}{25\pi^2}\right)\cdots \qquad = x\left(1 + \frac{x^2}{1\cdot 2\cdot 3} + \frac{x^4}{1\cdot 2\cdot 3\cdot 4\cdot 5} + \frac{x^6}{1\cdot 2\cdot 3\cdot 4\cdot 5\cdot 6\cdot 7} + \cdots\right).$$

We have given each of the factors, multiplied by a constant of the same form so that when the factors are actually multiplied, the resulting first term will be x.

157. In the same way, $\frac{e^x + e^{-x}}{2} = 1 + \frac{x^2}{1\cdot 2} + \frac{x^4}{1\cdot 2\cdot 3\cdot 4} + \cdots$

$= \frac{(1 + x/j)^j + (1 - x/j)^j}{2}$. When this expression is compared to $a^n + z^n$, where we let $a = 1 + \frac{x}{j}$, $z = 1 - \frac{x}{j}$, and $n = j$, we obtain each factor as

$$a^2 - 2az \cos \frac{2ak + 1}{n} \pi + z^2$$

$$= 2 + \frac{2x^2}{j^2} - 2(1 - x^2/j^2) \cos \frac{2k + 1}{n} \pi.$$ Since

$\cos \frac{2k + 1}{j} \pi = 1 - \frac{(2k + 1)^2}{2j^2} \pi^2$, the factor takes the form

$\frac{4x^2}{j^2} + \frac{(2k + 1)^2}{j^2} \pi^2$, where we have omitted a term whose denominator is j^4.

Since each factor of $1 + \frac{x^2}{1 \cdot 2} + \frac{x^4}{1 \cdot 2 \cdot 3 \cdot 4} + \cdots$ should have the form $1 + \alpha x^2$,

we reduce the factor already found to the desired form when we divide by

$\frac{(2k + 1)^2}{j^2} \pi^2$. We then have the factors in the proper form $1 + \frac{4x^2}{(2k + 1)^2 \pi^2}$.

It follows from this that we can find the infinite product by substituting for $2k + 1$ successively all odd integers. Therefore we have

$$\frac{e^x + e^{-x}}{2} = 1 + \frac{x^2}{1 \cdot 2} + \frac{x^4}{1 \cdot 2 \cdot 3 \cdot 4} + \frac{x^6}{1 \cdot 2 \cdot 3 \cdot 4 \cdot 5 \cdot 6} + \cdots$$

$$= \left(1 + \frac{4x^2}{\pi^2} \right) \left(1 + \frac{4x^2}{9\pi^2} \right) \left(1 + \frac{4x^2}{25\pi^2} \right) \left(1 + \frac{4x^2}{49\pi^2} \right) \cdots .$$

158. If we let x be an imaginary number, then these exponential expressions can be represented by sines and cosines of a real arc. Let $x = zi$, then

$$\frac{e^{zi} - e^{-zi}}{2i} = \sin z \qquad = z - \frac{z^3}{1 \cdot 2 \cdot 3} + \frac{z^5}{1 \cdot 2 \cdot 3 \cdot 4 \cdot 5} - \frac{z^7}{1 \cdot 2 \cdot 3 \cdot 4 \cdot 5 \cdot 6 \cdot 7} + \cdots$$

which has an expression as an infinite product: $z(1 - z^2/\pi^2)(1 - z^2/4\pi^2)(1 - z^2/9\pi^2)(1 - z^2/16\pi^2)(1 - z^2/25\pi^2) \cdots$,

that is, we can write $\sin z = z(1 - z/\pi)(1 + z/\pi)(1 - z/2\pi)$ $(1 + z/2\pi)(1 - z/3\pi)(1 + z/3\pi) \cdots$. Whenever the arc z has a length such that any of the factors vanishes, that is when $z = 0, \pm \pi, \pm 2\pi$, etc. or generally when $z = \pm k\pi$, where k is any integer, then the sine of that arc must equal zero. But this is so obvious, that we might have found the factors from this fact. In like manner, since $\frac{e^{zi} + e^{-zi}}{2} = \cos z$ we also have $\cos z = (1 - 4z^2/\pi^2)(1 - 4z^2/9\pi^2)(1 - 4\pi^2/25\pi^2)(1 - 4z^2/49\pi^2) \cdots$, or when these factors are themselves factored, we obtain the expression $\cos z = (1 - 2z/\pi)(1 + 2z/\pi)(1 - 2z/3\pi)(1 + 2z/3\pi)$ $(1 - 2z/5\pi)(1 + 2z/5\pi) \cdots$. From this it again becomes obvious that when $z = \pm \frac{2k+1}{2}\pi$, then $\cos z = 0$, which is clear from the nature of the circle.

159. From section 152 we can also find the factors of the expression $e^x - 2\cos g + e^{-x} = 2\left(1 - \cos g + \frac{x^2}{1\cdot 2} + \frac{x^4}{1\cdot 2\cdot 3\cdot 4} + \cdots\right)$. This expression can also be written as $(1 + x/j)^j - 2\cos g + (1 - x/j)^j$, in which we let $2n = j$, $a = 1 + \frac{x}{j}$, and $z = 1 - \frac{x}{j}$. It follows that each of the factors has the form

$$a^2 - 2az\cos\frac{2k\pi \pm g}{n} + z^2 = 2 + \frac{2z^2}{j^2} - 2(1 - x^2/j^2)\cos\frac{2(2k\pi \pm g)}{j}.$$

Since $\cos\frac{2(2k\pi \pm g)}{j} = 1 - \frac{2(2k\pi \pm g)^2}{j^2}$, the factor has the form $\frac{4x^2}{j^2} + \frac{4(2k\pi \pm g)^2}{j^2}$ or the form $1 + \frac{x^2}{(2k\pi \pm g)^2}$. If the expression is divided by $2(1 - \cos g)$, so that in the resulting infinite series the constant term is 1, then we have the following infinite product: $\frac{e^x - 2\cos g + e^{-x}}{2(1 - \cos g)}$

$$= \left(1 + \frac{x^2}{g^2}\right)\left(1 + \frac{x^2}{(2\pi - g)^2}\right)\left(1 + \frac{x^2}{(2\pi + g)^2}\right)\left(1 + \frac{x^2}{(4\pi - g)^2}\right)$$

$$\left(1 + \frac{x^2}{(4\pi + g)^2}\right)\left(1 + \frac{x^2}{(6\pi - g)^2}\right)\left(1 + \frac{x^2}{(6\pi + g)^2}\right) \cdots .$$

Furthermore, if we substitute zi for x, then

$$\frac{\cos z - \cos g}{1 - \cos g} = \left(1 - \frac{z}{g}\right)\left(1 + \frac{z}{g}\right)\left(1 - \frac{z}{2\pi - g}\right)\left(1 + \frac{z}{2\pi - g}\right)$$

$$\left(1 - \frac{z}{2\pi + g}\right)\left(1 + \frac{z}{2\pi + g}\right)\left(1 - \frac{z}{4\pi - g}\right)\left(1 + \frac{z}{4\pi - g}\right) \cdots$$

$$= 1 - \frac{z^2}{1\cdot 2(1 - \cos g)} + \frac{z^4}{1\cdot 2\cdot 3\cdot 4(1 - \cos g)}$$

$$- \frac{z^6}{1\cdot 2\cdot 3\cdot 4\cdot 5\cdot 6(1 - \cos g)} + \cdots .$$

Now we have an infinite product expression for this infinite series.

160. It would be convenient to be able to find an infinite product expression for the function $e^{b+x} \pm e^{c-x}$. When we transform it into the form $\left(1 + \frac{b + x}{j}\right)^j \pm \left(1 + \frac{c - x}{j}\right)^j$ we can compare it with $a^j \pm z^j$, which has a factor $a^2 - 2az \cos \frac{m\pi}{j} + z^2$ where m is odd when the sign is positive and m is even when the sign is negative. Since j is infinitely large, $\cos \frac{m\pi}{j} = 1 - \frac{m^2\pi^2}{2j^2}$. Then the general factor has the form $(a - z)^2 + \frac{m^2\pi^2}{j^2} az$. In the present case we have $a = 1 + \frac{b + x}{j}$ and $z = 1 + \frac{c - x}{j}$, so that $(a - z)^2 = \frac{(b - c + 2x)^2}{j^2}$ and $az = 1 + \frac{b + c}{j} + \frac{bc + (c - b)x - x^2}{j^2}$. When these substitutions have been made and the result multiplied by j^2, we obtain

$(b - c)^2 + 4(b - c)x + 4x^2 + m^2\pi^2$, where we have neglected terms which have j or j^2 in the denominator, since they will vanish when compared with the remaining terms. We divide by the constant term which will make the constant term equal to 1 and obtain the factor $1 + \dfrac{4(b - c)x + 4x^2}{m^2\pi^2 + (b - c)^2}$.

161. Now, since each factor has 1 as its constant term, the function itself $e^{b+x} \pm e^{c-x}$ should be divided by that constant which will make the constant term equal 1. That is, the value should be 1 when $x = 0$. Such a divisor is $e^b \pm e^c$, so we have this expression: $\dfrac{e^{b+x} \pm e^{c-x}}{e^b \pm e^c}$ which can be written as an infinite product. If the sign is positive, so that m is odd, we have

$$\frac{e^{b+x} + e^{c-x}}{e^b + e^c} = \left(1 + \frac{4(b - c)x + 4x^2}{\pi^2 + (b - c)^2}\right)\left(1 + \frac{4(b - c)x + 4x^2}{9\pi^2 + (b - c)^2}\right)$$

$$\left(1 + \frac{4(b - c)x + 4x^2}{25\pi^2 + (b - c)^2}\right) \cdots .$$

If the sign is negative, then m is even, and in case $m = 0$, we take the square root of the factor. Hence we have

$$\frac{e^{b+x} - e^{c-x}}{e^b - e^c} = \left(1 + \frac{2x}{b - c}\right)\left(1 + \frac{4(b - c)x + 4x^2}{4\pi^2 + (b - c)^2}\right)$$

$$\left(1 + \frac{4(b - c)x + x^2}{16\pi^2 + (b - c)^2}\right)\left(1 + \frac{4(b - c)x + x^2}{36\pi^2 + (b - c)^2}\right) \cdots .$$

162. If we let $b = 0$, which we can do without loss of generality, then

$$\frac{e^x + e^c e^{-x}}{1 + e^c} = \left(1 - \frac{4cx + 4x^2}{\pi^2 + c^2}\right)\left(1 - \frac{4cx + 4x^2}{9\pi^2 + c^2}\right)$$

$$\left(1 - \frac{4cx + 4x^2}{25\pi^2 + c^2}\right) \cdots ,\text{ and}$$

$$\frac{e^x - e^c e^{-x}}{1 - e^c} = \left(1 - \frac{2x}{c}\right)\left(1 - \frac{4cx + 4x^2}{4\pi^2 + c^2}\right)\left(1 - \frac{4cx + 4x^2}{16\pi^2 + c^2}\right)$$

$$\left(1 - \frac{4cx + 4x^2}{36\pi^2 + c^2}\right) \cdots .$$

If now we let c be negative we obtain the two equations

$$\frac{e^x + e^{-c}e^{-x}}{1 + e^{-c}} = \left(1 + \frac{4cx + 4x^2}{\pi^2 + c^2}\right)\left(1 + \frac{4cx + 4x^2}{9\pi^2 + c^2}\right)$$

$$\left(1 + \frac{4cx + 4x^2}{25\pi^2 + c^2}\right) \cdots$$

and $\dfrac{e^x - e^{-c}e^{-x}}{1 - e^{-c}} = \left(1 + \dfrac{2x}{c}\right)\left(1 + \dfrac{4cx + 4x^2}{4\pi^2 + c^2}\right)\left(1 + \dfrac{4cx + 4x^2}{16\pi^2 + c^2}\right)$

$\left(1 + \dfrac{4cx + 4x^2}{36\pi^2 + c^2}\right) \cdots$. If the first of these forms is multiplied by the third form we obtain $\dfrac{e^{2x} + e^{-2x} + e^c + e^{-c}}{2 + e^c + e^{-c}}$. When we substitute y for $2x$ we obtain

$$\frac{e^y + e^{-y} + e^c + e^{-c}}{2 + e^c + e^{-c}} = \left(1 - \frac{2cy + y^2}{\pi^2 + c^2}\right)\left(1 + \frac{2cy + y^2}{\pi^2 + c^2}\right)$$

$$\left(1 - \frac{2cy + y^2}{9\pi^2 + c^2}\right)\left(1 + \frac{2cy + y^2}{9\pi^2 + c^2}\right)\left(1 - \frac{2cy + y^2}{25\pi^2 + c^2}\right)$$

$\left(1 + \dfrac{2cy + y^2}{25\pi^2 + c^2}\right) \cdots$. When the first form is multiplied by the fourth, we have $\dfrac{e^{2x} - e^{-2x} + e^c - e^{-c}}{e^c - e^{-c}}$. Again we substitute y for $2x$ to obtain

$$\frac{e^y - e^{-y} + e^c - e^{-c}}{e^c - e^{-c}} = \left(1 + \frac{y}{c}\right)\left(1 - \frac{2cy + y^2}{\pi^2 + c^2}\right)\left(1 + \frac{2cy + y^2}{4\pi^2 + c^2}\right)$$

$\left(1 - \dfrac{2cy + y^2}{9\pi^2 + c^2}\right)\left(1 + \dfrac{2cy + y^2}{16\pi^2 + c^2}\right)\left(1 - \dfrac{2cy + y^2}{25\pi^2 + c^2}\right) \cdots$. If the second form is multiplied by the third we obtain

$$\frac{e^c - e^{-c} - e^y + e^{-y}}{e^c - e^{-c}} = \left(1 - \frac{y}{c}\right)\left(1 + \frac{2cy + y^2}{\pi^2 + c^2}\right)\left(1 - \frac{2cy + y^2}{4\pi^2 + c^2}\right)$$

$$\left(1 + \frac{2cy + y^2}{9\pi^2 + c^2}\right)\left(1 - \frac{2cy + y^2}{16\pi^2 + c^2}\right)\left(1 + \frac{2cy + y^2}{25\pi^2 + c^2}\right)$$

$$\left(1 - \frac{2cy + y^2}{36\pi^2 + c^2}\right) \cdots .$$

Finally, if we multiply the second by the forth form, we have

$$\frac{e^y + e^{-y} - e^c - e^{-c}}{2 - e^c - e^{-c}} = \left(1 + \frac{y}{c}\right)\left(1 - \frac{2cy + y^2}{4\pi^2 + c^2}\right)\left(1 + \frac{2cy + y^2}{4\pi^2 + c^2}\right)$$

$$\left(1 - \frac{2cy + y^2}{16\pi^2 + c^2}\right)\left(1 + \frac{2cy + y^2}{16\pi^2 + c^2}\right)\left(1 - \frac{2cy + y^2}{36\pi^2 + c^2}\right)$$

$$\left(1 + \frac{2cy + y^2}{36\pi^2 + c^2}\right) \cdots .$$

163. These four formulas can now be conveniently written in terms of circular functions. We let $c = gi$, and $y = vi$ then $e^{vi} + e^{-vi} = 2 \cos v$, $e^{vi} - e^{-vi} = 2i \sin v$, $e^{gi} + e^{-gi} = 2 \cos g$, and $e^{gi} - e^{-gi} = 2i \sin g$. It follows that the first formula can be expressed as

$$\frac{\cos v + \cos g}{1 + \cos g} = 1 - \frac{v^2}{1 \cdot 2(1 + \cos g)} + \frac{v^4}{1 \cdot 2 \cdot 3 \cdot 4(1 + \cos g)}$$

$$- \frac{v^6}{1 \cdot 2 \cdot 3 \cdot 4 \cdot 5 \cdot 6(1 + \cos g)} + \cdots$$

$$= \left(1 + \frac{2gv - v^2}{\pi^2 - g^2}\right)\left(1 - \frac{2gv - v^2}{\pi^2 - g^2}\right)\left(1 + \frac{2gv - v^2}{9\pi^2 - g^2}\right)$$

$$\left(1 - \frac{2gv - v^2}{9\pi^2 - g^2}\right)\left(1 + \frac{2gv - v^2}{25\pi^2 - g^2}\right)\left(1 - \frac{2gv - v^2}{25\pi^2 - g^2}\right) \cdots$$

$$= \left(1 + \frac{v}{\pi - g}\right)\left(1 - \frac{v}{\pi + g}\right)\left(1 - \frac{v}{\pi - g}\right)\left(1 + \frac{v}{\pi + g}\right)$$

$$\left(1 + \frac{v}{3\pi - g}\right)\left(1 - \frac{v}{3\pi + g}\right)\left(1 - \frac{v}{3\pi - g}\right)\left(1 + \frac{v}{3\pi + g}\right) \cdots$$

$$= \left(1 - \frac{v^2}{(\pi - g)^2}\right)\left(1 - \frac{v^2}{(\pi + g)^2}\right)\left(1 - \frac{v^2}{(3\pi - g)^2}\right)$$

$$\left(1 - \frac{v^2}{(3\pi + g)^2}\right)\left(1 - \frac{v^2}{(5\pi - g)^2}\right) \cdots .$$

The fourth formula is expressed as

$$\frac{\cos v - \cos g}{1 - \cos g} = 1 - \frac{v^2}{1\cdot 2(1 - \cos g)} + \frac{v^4}{1\cdot 2\cdot 3\cdot 4(1 - \cos g)}$$

$$- \frac{v^6}{1\cdot 2\cdot 3\cdot 4\cdot 5\cdot 6(1 - \cos g)} + \cdots$$

$$= \left(1 - \frac{v^2}{g^2}\right)\left(1 + \frac{2gv - v^2}{4\pi^2 - g^2}\right)\left(1 - \frac{2gv - v^2}{4\pi^2 - g^2}\right)$$

$$\left(1 + \frac{2gv - v^2}{16\pi^2 - g^2}\right)\left(1 - \frac{2gv - v^2}{16\pi^2 - g^2}\right) \cdots$$

$$= \left(1 - \frac{v}{g}\right)\left(1 + \frac{v}{g}\right)\left(1 + \frac{v}{2\pi - g}\right)\left(1 - \frac{v}{2\pi + g}\right)$$

$$\left(1 - \frac{v}{2\pi - g}\right)\left(1 + \frac{v}{2\pi + g}\right)\left(1 + \frac{v}{4\pi - g}\right)\left(1 - \frac{v}{4\pi + g}\right) \cdots$$

$$= \left(1 - \frac{v^2}{g^2}\right)\left(1 - \frac{v^2}{(2\pi - g)^2}\right)\left(1 - \frac{v^2}{(2\pi + g)^2}\right)$$

$$\left(1 - \frac{v^2}{(4\pi - g)^2}\right)\left(1 - \frac{v^2}{(4\pi + g)^2}\right) \cdots .$$

The second formula is expressed as

$$\frac{\sin g + \sin v}{\sin g} = 1 + \frac{v}{\sin g} - \frac{v^3}{1\cdot 2\cdot 3 \sin g} + \frac{v^5}{1\cdot 2\cdot 3\cdot 4\cdot 5 \sin g} - \cdots$$

$$= \left(1 + \frac{v}{g}\right)\left(1 + \frac{2gv - v^2}{\pi^2 - g^2}\right)\left(1 - \frac{2gv - v^2}{4\pi^2 - g^2}\right)$$

$$\left(1 + \frac{2gv - v^2}{9\pi^2 - g^2}\right)\left(1 - \frac{2gv - v^2}{16\pi^2 - g^2}\right) \cdots$$

$$= \left(1 + \frac{v}{g}\right)\left(1 + \frac{v}{\pi - g}\right)\left(1 - \frac{v}{\pi + g}\right)\left(1 - \frac{v}{2\pi - g}\right)$$

$$\left(1 + \frac{v}{2\pi + g}\right)\left(1 + \frac{v}{3\pi - g}\right)\left(1 + \frac{v}{3\pi + g}\right)\left(1 - \frac{v}{5\pi - g}\right) \cdots .$$

When v is taken as negative, we obtain the third formula.

164. The same expressions we first saw in section 162 can be written in terms of circular arcs as follows. Since

$$\frac{e^x + e^c e^{-x}}{1 + e^c} = \frac{(1 + e^{-c})(e^x + e^c e^{-x})}{2 + e^c + e^{-c}} = \frac{e^x + e^{-x} + e^{c-x} + e^{-c+x}}{2 + e^c + e^{-c}},$$ if

we let $c = gi$ and $x = zi$, this expression becomes

$$\frac{\cos z + \cos(g - z)}{1 + \cos g} = \cos z + \frac{\sin g \sin z}{1 + \cos g}.$$

Since $\dfrac{\sin g}{1 + \cos g} = \tan \dfrac{g}{2}$, we have

$$\cos z + \tan \frac{g}{2} \sin z = 1 + \frac{z}{1} \tan \frac{g}{2} - \frac{z^2}{1\cdot 2} - \frac{z^3}{1\cdot 2\cdot 3} \tan \frac{g}{2}$$

$$+ \frac{z^4}{1\cdot 2\cdot 3\cdot 4} + \frac{z^5}{1\cdot 2\cdot 3\cdot 4\cdot 5} \tan \frac{g}{2} - \cdots$$

$$= \left(1 + \frac{4gz - 4z^2}{\pi^2 - g^2}\right)\left(1 + \frac{4gz - 4z^2}{9\pi^2 - g^2}\right)\left(1 + \frac{4gz - 4z^2}{25\pi^2 - g^2}\right) \cdots$$

$$= \left(1 + \frac{2z}{\pi - g}\right)\left(1 - \frac{2z}{\pi + g}\right)\left(1 + \frac{2z}{3\pi - g}\right)\left(1 - \frac{2z}{3\pi + g}\right)$$

$$\left(1 + \frac{2z}{5\pi - g}\right)\left(1 - \frac{2z}{5\pi + g}\right) \cdots .$$ Likewise the second expression, after multiplication of both numerator and denominator by $1 - e^{-c}$, becomes

$$\frac{e^x + e^{-x} - e^{c-x} - e^{x-c}}{2 - e^c - e^{-c}}.$$ When we let $c = gi$ and $x = zi$, we obtain

$$\frac{\cos z - \cos(g - z)}{1 - \cos g} = \cos z - \frac{\sin g \sin z}{1 - \cos g} = \cos z - \frac{\sin z}{\tan \frac{g}{2}}.$$ We have

$$\cos z - \cot \frac{g}{2} \sin z$$

$$= 1 - \frac{z}{1} \cot \frac{g}{2} - \frac{z^2}{1\cdot 2} + \frac{z^3}{1\cdot 2\cdot 3} \cot \frac{g}{2} + \frac{z^4}{1\cdot 2\cdot 3\cdot 4}$$

$$- \frac{z^5}{1\cdot 2\cdot 3\cdot 4\cdot 5} \cot \frac{g}{2} + \cdots$$

$$= \left(1 - \frac{2z}{g}\right)\left(1 + \frac{4gz - 4z^2}{4\pi^2 - g^2}\right)\left(1 + \frac{4gz - 4z^2}{16\pi^2 - g^2}\right)\left(1 + \frac{4gz - 4z^2}{36\pi^2 - g^2}\right)\cdots$$

$$= \left(1 - \frac{2z}{g}\right)\left(1 + \frac{2z}{2\pi - g}\right)\left(1 + \frac{2z}{2\pi + g}\right)\left(1 + \frac{2z}{4\pi - g}\right)$$

$$\left(1 - \frac{2z}{4\pi + g}\right)\cdots .$$

If we let $v = 2z$ or $z = \frac{v}{2}$, then we have $\dfrac{\cos\frac{g - v}{2}}{\cos\frac{g}{2}} = \cos\frac{v}{2} + \tan\frac{g}{2}\sin\frac{v}{2}$

$$= \left(1 + \frac{v}{\pi - g}\right)\left(1 - \frac{v}{\pi + g}\right)\left(1 + \frac{v}{3\pi - g}\right)\left(1 - \frac{v}{3\pi + g}\right)\cdots$$ and

$$\frac{\cos\frac{g + v}{2}}{\cos\frac{g}{2}} = \cos\frac{v}{2} - \tan\frac{g}{2}\sin\frac{v}{2}$$

$$= \left(1 - \frac{v}{\pi - g}\right)\left(1 + \frac{v}{\pi + g}\right)\left(1 - \frac{v}{3\pi - g}\right)\left(1 + \frac{v}{3\pi + g}\right)\cdots .$$

We also have $\dfrac{\sin\frac{g - v}{2}}{\sin\frac{g}{2}} = \cos\frac{v}{2} - \cot\frac{g}{2}\sin\frac{v}{2}$

$$= \left(1 - \frac{v}{g}\right)\left(1 + \frac{v}{2\pi - g}\right)\left(1 - \frac{v}{2\pi + g}\right)\left(1 + \frac{v}{4\pi - g}\right)\cdots$$ and

$$\frac{\sin\frac{g + v}{2}}{\sin\frac{g}{2}} = \cos\frac{v}{2} + \cot\frac{g}{2}\sin\frac{v}{2}$$

$$= \left(1 + \frac{v}{g}\right)\left(1 - \frac{v}{2\pi - g}\right)\left(1 + \frac{v}{2\pi + g}\right)\left(1 - \frac{v}{4\pi - g}\right)\cdots .$$ The law of

formation for these factors is sufficiently simple and uniform. Furthermore, from the multiplication of these expressions, there arise the expressions found in the previous section.

CHAPTER X

On the Use of the Discovered Factors to Sum Infinite Series.

165. If $1 + Az + Bz^2 + Cz^3 + Dz^4 + \cdots$ $= (1 + \alpha z)(1 + \beta z)(1 + \gamma z)(1 + \delta z) \cdots$, then these factors, whether they be finite or infinite in number, must produce the expression $1 + Az + Bz^2 + Cz^3 + Dz^4 + \cdots$, when they are actually multiplied. It follows then that the coefficient A is equal to the sum $\alpha + \beta + \gamma + \delta + \epsilon + \cdots$. The coefficient B is equal to the sum of the products taken two at a time. Hence $B = \alpha\beta + \alpha\gamma + \alpha\delta + \beta\gamma + \beta\delta + \gamma\delta + \cdots$. Also the coefficient C is equal to the sum of products taken three at a time, namely $C = \alpha\beta\gamma + \alpha\beta\delta + \beta\gamma\delta + \alpha\gamma\delta + \cdots$. We also have D as the sum of products taken four at a time, and E is the sum of products taken five at a time, etc. All of this is clear from ordinary algebra.

166. Since the sum $\alpha + \beta + \gamma + \delta + \cdots$ is given along with the sum of products taken two at a time, we can find the sum of the squares $\alpha^2 + \beta^2 + \gamma^2 + \delta^2 + \cdots$, since this is equal to the square of the sum diminished by two times the sum of the products taken two at a time. In a similar way the sums of the cubes, biquadratics, and higher powers can be found. If we let $P = \alpha + \beta + \gamma + \delta + \epsilon + \cdots$

$$Q = \alpha^2 + \beta^2 + \gamma^2 + \delta^2 + \epsilon^2 + \cdots$$
$$R = \alpha^3 + \beta^3 + \gamma^3 + \delta^3 + \epsilon^3 + \cdots$$
$$S = \alpha^4 + \beta^4 + \gamma^4 + \delta^4 + \epsilon^4 + \cdots$$
$$T = \alpha^5 + \beta^5 + \gamma^5 + \delta^5 + \epsilon^5 + \cdots$$
$$V = \alpha^6 + \beta^6 + \gamma^6 + \delta^6 + \epsilon^6 + \cdots .$$

Then P, Q, R, S, T, V, etc. can be found in the following way from A, B, C, D, etc. $P = A$, $Q = AP - 2B$, $R = AQ - BP + 3C$, $S = AR - BQ + CP - 4D$, $T = AS - BR + CQ - DP + 5E$, $V = AT - BS + CR - DQ + EP - 6F$, etc. The truth of these formulas is intuitively clear, but a rigorous proof will be given in the differential calculus.

167. Since we found above, in section 156, that

$$\frac{e^x - e^{-x}}{2} = x\left(1 + \frac{x^2}{1\cdot 2\cdot 3} + \frac{x^4}{1\cdot 2\cdot 3\cdot 4\cdot 5} + \frac{x^6}{1\cdot 2\cdots 7} + \cdots\right)$$
$$= x\left(1 + \frac{x^2}{\pi^2}\right)\left(1 + \frac{x^2}{4\pi^2}\right)\left(1 + \frac{x^2}{9\pi^2}\right)\left(1 + \frac{x^2}{16\pi^2}\right)\left(1 + \frac{x^2}{25\pi^2}\right)\cdots ,$$

it follows that

$$1 + \frac{x^2}{1\cdot 2\cdot 3} + \frac{x^4}{1\cdot 2\cdot 3\cdot 4\cdot 5} + \frac{x^6}{1\cdot 2\cdots 7} + \cdots$$
$$= \left(1 + \frac{x^2}{\pi^2}\right)\left(1 + \frac{x^2}{4\pi^2}\right)\left(1 + \frac{x^2}{9\pi^2}\right)\left(1 + \frac{x^2}{16\pi^2}\right)\cdots .$$

If we let $x^2 = \pi^2 z$,

$$1 + \frac{\pi^2}{1\cdot 2\cdot 3}z + \frac{\pi^4}{1\cdot 2\cdot 3\cdot 4\cdot 5}z^2 + \frac{\pi^6}{1\cdot 2\cdot 3\cdot 4\cdot 5\cdot 6\cdot 7}z^3 + \cdots$$
$$= (1 + z)(1 + z/4)(1 + z/9)(1 + z/16)(1 + z/25)\cdots .$$

We use the rules stated above where $A = \frac{\pi^2}{6}$, $B = \frac{\pi^4}{120}$, $C = \frac{\pi^6}{5040}$, $D = \frac{\pi^8}{362880}$, etc., and we also have

$$P = 1 + \frac{1}{4} + \frac{1}{9} + \frac{1}{16} + \frac{1}{25} + \frac{1}{36} + \cdots ,$$

$$Q = 1 + \frac{1}{4^2} + \frac{1}{9^2} + \frac{1}{16^2} + \frac{1}{25^2} + \frac{1}{36^2} + \cdots,$$

$$R = 1 + \frac{1}{4^3} + \frac{1}{9^3} + \frac{1}{16^3} + \frac{1}{25^3} + \frac{1}{36^3} + \cdots,$$

$$S = 1 + \frac{1}{4^4} + \frac{1}{9^4} + \frac{1}{16^4} + \frac{1}{25^4} + \frac{1}{36^4} + \cdots,$$

$$T = 1 + \frac{1}{4^5} + \frac{1}{9^5} + \frac{1}{16^5} + \frac{1}{25^5} + \frac{1}{36^5} + \cdots.$$

From the values of A, B, C, D, etc. we see that $P = \frac{\pi^2}{6}$, $Q = \frac{\pi^4}{90}$, $R = \frac{\pi^6}{945}$, $S = \frac{\pi^8}{9450}$, $T = \frac{\pi^{10}}{93555}$, etc.

168. It is clear that any infinite series of the form $1 + \frac{1}{2^n} + \frac{1}{3^n} + \frac{1}{4^n} + \cdots$, provided n is an even integer, can be expressed in terms of π, since it always has a sum equal to a fractional part of a power of π. In order that the values of these sums can be seen even more clearly, we set down in a convenient form some more sums of these series.

$$1 + \frac{1}{2^2} + \frac{1}{3^2} + \frac{1}{4^2} + \frac{1}{5^2} + \cdots = \frac{2^0}{1\cdot 2\cdot 3}\,\frac{1}{1}\pi^2$$

$$1 + \frac{1}{2^4} + \frac{1}{3^4} + \frac{1}{4^4} + \frac{1}{5^4} \cdots = \frac{2^2}{1\cdot 2\cdot 3\cdot 4\cdot 5}\,\frac{1}{3}\pi^4$$

$$1 + \frac{1}{2^6} + \frac{1}{3^6} + \frac{1}{4^6} + \frac{1}{5^6} + \cdots = \frac{2^4}{1\cdot 2\cdots 7}\,\frac{1}{3}\pi^6$$

$$1 + \frac{1}{2^8} + \frac{1}{3^8} + \frac{1}{4^8} + \frac{1}{5^8} + \cdots = \frac{2^6}{1\cdot 2\cdot 3\cdots 9}\,\frac{3}{5}\pi^8$$

$$1 + \frac{1}{2^{10}} + \frac{1}{3^{10}} + \frac{1}{4^{10}} + \frac{1}{5^{10}} + \cdots = \frac{2^8}{1\cdot 2\cdot 3\cdots 11}\,\frac{5}{3}\pi^{10}$$

$$1 + \frac{1}{2^{12}} + \frac{1}{3^{12}} + \frac{1}{4^{12}} + \frac{1}{5^{12}} + \cdots = \frac{2^{10}}{1\cdot 2\cdot 3\cdots 13}\,\frac{691}{105}\pi^{12}$$

$$1 + \frac{1}{2^{14}} + \frac{1}{3^{14}} + \frac{1}{4^{14}} + \frac{1}{5^{14}} + \cdots = \frac{2^{12}}{1 \cdot 2 \cdot 3 \cdots 15} \frac{35}{1} \pi^{14}$$

$$1 + \frac{1}{2^{16}} + \frac{1}{3^{16}} + \frac{1}{4^{16}} + \frac{1}{5^{16}} + \cdots = \frac{2^{14}}{1 \cdot 2 \cdot 3 \cdots 17} \frac{3617}{15} \pi^{16}$$

$$1 + \frac{1}{2^{18}} + \frac{1}{3^{18}} = \frac{1}{4^{18}} + \frac{1}{5^{18}} + \cdots = \frac{2^{16}}{1 \cdot 2 \cdot 3 \cdots 19} \frac{43867}{21} \pi^{18}$$

$$1 + \frac{1}{2^{20}} + \frac{1}{3^{20}} + \frac{1}{4^{20}} + \frac{1}{5^{20}} + \cdots = \frac{2^{18}}{1 \cdot 2 \cdot 3 \cdots 21} \frac{1222277}{55} \pi^{20}$$

$$1 + \frac{1}{2^{22}} + \frac{1}{3^{22}} + \frac{1}{4^{22}} + \frac{1}{5^{22}} + \cdots = \frac{2^{20}}{1 \cdot 2 \cdot 3 \cdots 23} \frac{854513}{3} \pi^{22}$$

$$1 + \frac{1}{2^{24}} + \frac{1}{3^{24}} + \frac{1}{4^{24}} + \frac{1}{5^{24}} + \cdots = \frac{2^{22}}{1 \cdot 2 \cdot 3 \cdots 25} \frac{1181820455}{273} \pi^{24}$$

$$1 + \frac{1}{2^{26}} + \frac{1}{3^{26}} + \frac{1}{4^{26}} + \frac{1}{5^{26}} + \cdots = \frac{2^{24}}{1 \cdot 2 \cdot 3 \cdots 27} \frac{76977927}{1} \pi^{26}.$$

We could continue with more of these, but we have gone far enough to see a sequence which at first seems quite irregular, $1, \frac{1}{3}, \frac{1}{3}, \frac{3}{5}, \frac{5}{3}, \frac{691}{105}, \frac{35}{1}, \ldots$, but it is of extraordinary usefulness in several places.

169. We now treat in the same manner the equation found in section 157. There we saw that

$$\frac{e^x + e^{-x}}{2} = 1 + \frac{x^2}{1 \cdot 2} + \frac{x^4}{1 \cdot 2 \cdot 3 \cdot 4} + \frac{x^6}{1 \cdot 2 \cdot 3 \cdot 4 \cdot 5 \cdot 6} + \cdots$$

$= \left(1 + \frac{4x^2}{\pi^2}\right)\left(1 + \frac{4x^2}{9\pi^2}\right)\left(1 + \frac{4x^2}{25\pi^2}\right)\left(1 + \frac{4x^2}{49\pi^2}\right) \cdots$. We let $x^2 = \frac{\pi^2 z}{4}$,

then $1 + \frac{\pi^2}{1 \cdot 2 \cdot 4} z + \frac{\pi^4}{1 \cdot 2 \cdot 3 \cdot 4 \cdot 4^2} z^2 + \frac{\pi^6}{1 \cdot 2 \cdots 6 \cdot 4^3} z^3 + \cdots$

$= (1 + z)(1 + z/9)(1 + z/25)(1 + z/49) \cdots$. We now use the formulas, where $A = \frac{\pi^2}{1 \cdot 2 \cdot 4}$, $B = \frac{\pi^4}{1 \cdot 2 \cdot 3 \cdot 4 \cdot 4^2}$, $C = \frac{\pi^6}{1 \cdot 2 \cdot 3 \cdots 6 \cdot 4^3}$, etc., and

$$P = 1 + \frac{1}{9} + \frac{1}{25} + \frac{1}{49} + \frac{1}{81} + \cdots$$

$$Q = 1 + \frac{1}{9^2} + \frac{1}{25^2} + \frac{1}{49^2} + \frac{1}{81^2} + \cdots$$

$$R = 1 + \frac{1}{9^3} + \frac{1}{25^3} + \frac{1}{49^3} + \frac{1}{81^3} + \cdots$$

$$S = 1 + \frac{1}{9^4} + \frac{1}{25^4} + \frac{1}{49^4} + \frac{1}{81^4} + \cdots .$$

It follows that $P = \frac{1}{1}\frac{\pi^2}{2^3}$, $Q = \frac{2}{1\cdot 2\cdot 3}\frac{\pi^4}{2^5}$, $R = \frac{16}{1\cdot 2\cdot 3\cdot 4\cdot 5}\frac{\pi^6}{2^7}$,

$S = \frac{272}{1\cdot 2\cdot 3\cdots 7}\frac{\pi^8}{2^9}$, $T = \frac{7936}{1\cdot 2\cdot 3\cdots 9}\frac{\pi^{10}}{2^{11}}$, $V = \frac{353792}{1\cdot 2\cdot 3\cdots 11}\frac{\pi^{12}}{2^{13}}$,

$W = \frac{22368256}{1\cdot 2\cdot 3\cdots 13}\frac{\pi^{14}}{2^{15}}$.

170. The same sums of powers of odd numbers can be found from the preceding sums in which all numbers occur. If we let $M = 1 + \frac{1}{2^n} + \frac{1}{3^n} + \frac{1}{4^n} + \frac{1}{5^n} + \cdots$ and multiply both sides by $\frac{1}{2^n}$, we obtain $\frac{M}{2^n} = \frac{1}{2^n} + \frac{1}{4^n} + \frac{1}{6^n} + \frac{1}{8^n} + \cdots$. This series contains only even numbers, which, when subtracted from the previous series, leaves the series with only odd numbers. Hence,

$M - \frac{M}{2^n} = \frac{2^n - 1}{2^n}M = 1 + \frac{1}{3^n} + \frac{1}{5^n} + \frac{1}{7^n} + \frac{1}{9^n} + \cdots$. If 2 times the series $\frac{M}{2^n}$ is subtracted from M an alternating series is produced:

$M - \frac{2M}{2^n} = \frac{2^{n-1} - 1}{2^{n-1}}M = 1 - \frac{1}{2^n} + \frac{1}{3^n} - \frac{1}{4^n} + \frac{1}{5^n} - \frac{1}{6^n} + \cdots$. In this way we can sum the series

$$1 \pm \frac{1}{2^n} + \frac{1}{3^n} \pm \frac{1}{4^n} + \frac{1}{5^n} \pm \frac{1}{6^n} + \frac{1}{7^n} \pm \cdots$$

$$1 + \frac{1}{3^n} + \frac{1}{5^n} + \frac{1}{7^n} + \frac{1}{9^n} + \frac{1}{11^n} + \cdots .$$

If n is an even number and the sum is $A\pi^n$, then A will be a rational number.

171. Furthermore, the expressions found in section 164 supply in the same way sums of series which are worthy of note. Since

$$\cos\frac{v}{2} + \tan\frac{g}{2}\sin\frac{v}{2} = \left(1 + \frac{v}{\pi - g}\right)\left(1 - \frac{v}{\pi + g}\right)\left(1 + \frac{v}{3\pi - g}\right)\cdots,$$

if we let $v = \frac{x}{n}\pi$ and $g = \frac{m}{n}\pi$, then

$$\left(1 + \frac{x}{n - m}\right)\left(1 - \frac{x}{n + m}\right)\left(1 + \frac{x}{3n - m}\right)\left(1 - \frac{x}{3n + m}\right)\left(1 + \frac{x}{5n - m}\right)$$
$$\left(1 - \frac{x}{5n + m}\right)\cdots = \cos\frac{x\pi}{2n} + \tan\frac{m\pi}{2n}\sin\frac{x\pi}{2n}$$
$$= 1 + \frac{\pi x}{2n}\tan\frac{m\pi}{2n} - \frac{\pi^2 x^2}{2\cdot 4n^2} - \frac{\pi^3 x^3}{2\cdot 4\cdot 6n^3}\tan\frac{m\pi}{2n}$$
$$+ \frac{\pi^4 x^4}{2\cdot 4\cdot 6\cdot 8n^4} + \cdots .$$

Using the expression in section 165, we have $A = \frac{\pi}{2n}\tan\frac{m\pi}{2n}$, $B = \frac{-\pi^2}{2\cdot 4n^2}$,

$C = \frac{-\pi^3}{2\cdot 4\cdot 6n^3}\tan\frac{m\pi}{2n}$, $D = \frac{\pi^4}{2\cdot 4\cdot 6\cdot 8n^4}$, $E = \frac{\pi^5}{2\cdot 4\cdot 6\cdot 8\cdot 10n^5}\tan\frac{m\pi}{2n}$, etc.

Further, since $\alpha = \frac{1}{n - m}$, $\beta = -\frac{1}{n + m}$, $\gamma = \frac{1}{3n - m}$,

$\delta = -\frac{1}{3n + m}$, $\epsilon = \frac{1}{5n - m}$, $\zeta = -\frac{1}{5n + m}$, etc.

172. When we follow the procedure given in section 166, we obtain the following.

$$P = \frac{1}{n - m} - \frac{1}{n + m} + \frac{1}{3n - m} - \frac{1}{3n + m} + \frac{1}{5n - m} - \frac{1}{5n + m} + \cdots$$

$$Q = \frac{1}{(n - m)^2} + \frac{1}{(n + m)^2} + \frac{1}{(3n - m)^2} + \frac{1}{(3n + m)^2} + \frac{1}{(5n - m)^2} + \cdots$$

$$R = \frac{1}{(n - m)^3} - \frac{1}{(n + m)^3} + \frac{1}{(3n - m)^3} - \frac{1}{(3n + m)^3} + \frac{1}{(5n - m)^3} - \cdots$$

$$S = \frac{1}{(n - m)^4} + \frac{1}{(n + m)^4} + \frac{1}{(3n - m)^4} + \frac{1}{(3n + m)^4} + \frac{1}{(5n - m)^4} + \cdots$$

$$T = \frac{1}{(n - m)^5} - \frac{1}{(n + m)^5} + \frac{1}{(3n - m)^5} - \frac{1}{(3n + m)^5} + \frac{1}{(5n - m)^5} - \cdots$$

$$V = \frac{1}{(n - m)^6} + \frac{1}{(n + m)^6} + \frac{1}{(3n - m)^6} + \frac{1}{(3n + m)^6} + \frac{1}{(5n - m)^6} + \cdots .$$

When we let $\tan \dfrac{m\pi}{2n} = k$, we obtain, as we have shown,

$$P = A = \frac{k\pi}{2n} = \frac{1}{2}\,\frac{k\pi}{n}$$

$$Q = \frac{(k^2 + 1)\pi^2}{4n^2} = \frac{(2k^2 + 2)\pi^2}{2\cdot 4n^2}$$

$$R = \frac{(k^3 + k)\pi^3}{8n^3} = \frac{(6k^3 + 6k)\pi^3}{2\cdot 4\cdot 6n^3}$$

$$S = \frac{(3k^4 + 4k^2 + 1)\pi^4}{48n^4} = \frac{(24k^4 + 32k^2 + 8)\pi^4}{2\cdot 4\cdot 6\cdot 8n^4}$$

$$T = \frac{(3k^5 + 5k^3 + 2k)\pi^5}{96n^5} = \frac{(120k^5 + 200k^3 + 80k)\pi^5}{2\cdot 4\cdot 6\cdot 8\cdot 10n^5}.$$

173. Likewise from the last form in section 164, we obtain

$$\cos\frac{v}{2}+\cot\frac{g}{2}\sin\frac{v}{2}=\left(1+\frac{v}{g}\right)\left(1-\frac{v}{2\pi-g}\right)\left(1+\frac{v}{2\pi+g}\right)$$
$$\left(1-\frac{v}{4\pi-g}\right)\left(1+\frac{v}{4\pi+g}\right)\cdots.$$

If we let $v=\frac{x}{n}\pi$, $g=\frac{m}{n}\pi$, and $\tan\frac{m\pi}{2n}=k$, so that $\cot\frac{g}{2}=\frac{1}{k}$ and

$$\cos\frac{\pi x}{2n}+\frac{1}{k}\sin\frac{\pi x}{2n}=1+\frac{\pi x}{2nk}-\frac{\pi^2x^2}{2\cdot4n^2}-\frac{\pi^3x^3}{2\cdot4\cdot6n^3k}+\frac{\pi^4x^4}{2\cdot4\cdot6\cdot8n^4}$$
$$+\frac{\pi^5x^5}{2\cdot4\cdot6\cdot8\cdot10n^5k}-\cdots=\left(1+\frac{x}{m}\right)\left(1-\frac{x}{2n-m}\right)\left(1+\frac{x}{2n+m}\right)$$
$$\left(1-\frac{x}{4n-m}\right)\left(1+\frac{x}{4n+m}\right)\cdots.$$

When we compare this with the general formula given in section 165, we find

$$A=\frac{\pi}{2nk},\quad B=\frac{-\pi^2}{2\cdot4n^2},\quad C=\frac{-\pi^3}{2\cdot4\cdot6n^3k},\quad D=\frac{\pi^4}{2\cdot4\cdot6\cdot8n^4},\quad E=\frac{\pi^5}{2\cdot4\cdot6\cdot8\cdot10n^5k},$$

etc. From the factors we obtain $\alpha=\frac{1}{m}$, $\beta=\frac{-1}{2n-m}$, $\gamma=\frac{1}{2n+m}$,

$\delta=\frac{-1}{4n-m}$, $\epsilon=\frac{1}{4n+m}$, etc.

174. Again we follow the procedure given in section 166 in order to obtain the sums of the following series.

$$P=\frac{1}{m}-\frac{1}{2n-m}+\frac{1}{2n+m}-\frac{1}{4n-m}+\frac{1}{4n+m}-\cdots$$

$$Q=\frac{1}{m^2}+\frac{1}{(2n-m)^2}+\frac{1}{(2n+m)^2}$$
$$+\frac{1}{(4n-m)^2}+\frac{1}{(4n+m)^2}+\cdots$$

$$R = \frac{1}{m^3} - \frac{1}{(2n-m)^3} + \frac{1}{(2n+m)^3} - \frac{1}{(4n-m)^3} + \frac{1}{(4n+m)^3} - \cdots$$

$$S = \frac{1}{m^4} + \frac{1}{(2n-m)^4} + \frac{1}{(2n+m)^4} + \frac{1}{(4n-m)^4} + \frac{1}{(4n+m)^4} + \cdots$$

$$T = \frac{1}{m^5} - \frac{1}{(2n-m)^5} + \frac{1}{(2n+m)^5} - \frac{1}{(4n-m)^5} + \frac{1}{(4n+m)^5} - \cdots .$$

We obtain the following sums:

$$P = A = \frac{\pi}{2nk} = \frac{1\pi}{2nk}$$

$$Q = \frac{(k^2+1)\pi^2}{4n^2k^2} = \frac{(2+2k^2)\pi^2}{2\cdot 4n^2k^2}$$

$$R = \frac{(k^2+1)\pi^3}{8n^3k^3} = \frac{(6+6k^2)\pi^3}{2\cdot 4\cdot 6n^3k^3}$$

$$S = \frac{(k^4+4k^2+3)\pi^4}{48n^4k^4} = \frac{(24+32k^2+3k^4)\pi^4}{2\cdot 4\cdot 6\cdot 8n^4k^4}$$

$$T = \frac{(2k^4+5k^2+3)\pi^5}{96n^5k^5} = \frac{(120+200k^2+80k^4)\pi^5}{2\cdot 4\cdot 6\cdot 8\cdot 10n^5k^5}$$

$$V = \frac{(2k^6+17k^4+30k^2+15)\pi^6}{960n^6k^6} = \frac{(720+1440k^2+816k^4+96k^6)\pi^6}{2\cdot 4\cdot 6\cdot 8\cdot 10\cdot 12n^6k^6}$$

175. These general series deserve to be particularized by giving special values to m and n. If $m = 1$ and $n = 2$, then $k = \tan \frac{\pi}{4} = 1$, and both of the series become the same:

$$\frac{\pi}{4} = 1 - \frac{1}{3} + \frac{1}{5} - \frac{1}{7} + \frac{1}{9} - \cdots$$

$$\frac{\pi^2}{8} = 1 + \frac{1}{3^2} + \frac{1}{5^2} + \frac{1}{7^2} + \frac{1}{9^2} + \cdots$$

$$\frac{\pi^3}{32} = 1 - \frac{1}{3^3} + \frac{1}{5^3} - \frac{1}{7^3} + \frac{1}{9^3} - \cdots$$

$$\frac{\pi^4}{96} = 1 + \frac{1}{3^4} + \frac{1}{5^4} + \frac{1}{7^4} + \frac{1}{9^4} + \cdots$$

$$\frac{\pi^6}{960} = 1 + \frac{1}{3^6} + \frac{1}{5^6} + \frac{1}{7^6} + \frac{1}{9^6} + \cdots .$$

The first of these series was seen before in section 140. The other series, which have equal exponents were discussed in section 169. The remaining series, in which the exponents are odd, we see here for the first time. It is clear that each of these series

$1 - \frac{1}{3^{2n+1}} + \frac{1}{5^{2n+1}} - \frac{1}{7^{2n+1}} + \frac{1}{9^{2n+1}} - \cdots$ has a sum which is some function of π.

176. Now we let $m = 1$, $n = 3$, then $k = \tan \frac{\pi}{6} = \frac{1}{\sqrt{3}}$ and the series in section 172 become

$$\frac{\pi}{6\sqrt{3}} = \frac{1}{2} - \frac{1}{4} + \frac{1}{8} - \frac{1}{10} + \frac{1}{14} - \frac{1}{16} + \cdots$$

$$\frac{\pi^2}{27} = \frac{1}{2^2} + \frac{1}{4^2} + \frac{1}{8^2} + \frac{1}{10^2} + \frac{1}{14^2} + \frac{1}{16^2} + \cdots$$

$$\frac{\pi^3}{162\sqrt{3}} = \frac{1}{2^3} - \frac{1}{4^3} + \frac{1}{8^3} - \frac{1}{10^3} + \frac{1}{14^3} - \frac{1}{16^3} + \cdots$$

etc., or

$$\frac{\pi}{3\sqrt{3}} = 1 - \frac{1}{2} + \frac{1}{4} - \frac{1}{5} + \frac{1}{7} - \frac{1}{8} + \cdots$$

$$\frac{4\pi^2}{27} = 1 + \frac{1}{2^2} + \frac{1}{4^2} + \frac{1}{5^2} + \frac{1}{7^2} + \frac{1}{8^2} + \cdots$$

$$\frac{4\pi^3}{81\sqrt{3}} = 1 - \frac{1}{2^3} + \frac{1}{4^3} - \frac{1}{5^3} + \frac{1}{7^3} - \frac{1}{8^3} + \cdots$$

In these series there is no term which is divisible by $\frac{1}{3}$. We can find the series which contain these terms, at least those series with even exponents, as follows. Since

$$\frac{\pi^2}{6} = 1 + \frac{1}{2^2} + \frac{1}{3^2} + \frac{1}{4^2} + \frac{1}{5^2} + \cdots,$$

it follows that

$$\frac{\pi^2}{6\cdot 9} = \frac{1}{3^2} + \frac{1}{6^2} + \frac{1}{9^2} + \frac{1}{12^2} + \cdots = \frac{\pi^2}{54}.$$

This last series contains only those terms which are divisible by $\frac{1}{3}$, and if it is subtracted from the previous series, there remains a series which contains all terms not divisible by $\frac{1}{3}$. Then

$$\frac{8\pi}{54} = \frac{4\pi}{27} = 1 + \frac{1}{2^2} + \frac{1}{4^2} + \frac{1}{5^2} + \frac{1}{7^2} + \cdots$$, as we have already seen.

177. With the same hypothesis, that is, $m = 1$, $n = 3$ and $k = \frac{1}{\sqrt{3}}$, from section 174 we obtain

$$\frac{\pi}{2\sqrt{3}} = 1 - \frac{1}{5} + \frac{1}{7} - \frac{1}{11} + \frac{1}{13} - \frac{1}{17} + \frac{1}{19} - \cdots$$

$$\frac{\pi^2}{9} = 1 + \frac{1}{5^2} + \frac{1}{7^2} + \frac{1}{11^2} + \frac{1}{13^2} + \frac{1}{17^2} + \frac{1}{19^2} + \cdots$$

$$\frac{\pi^3}{18\sqrt{3}} = 1 - \frac{1}{5^3} + \frac{1}{7^3} - \frac{1}{11^3} + \frac{1}{13^3} - \frac{1}{17^3} + \frac{1}{19^3} - \cdots$$

In these series, the denominators are all odd numbers, and the terms divisible by $\frac{1}{3}$ are missing. The sum of the even powers of these missing terms can be found from what we already know. Since

$$\frac{\pi^2}{8} = 1 + \frac{1}{3^2} + \frac{1}{5^2} + \frac{1}{7^2} + \frac{1}{9^2} + \cdots,$$ it follows that

$$\frac{\pi^2}{8\cdot 9} = \frac{1}{3^2} + \frac{1}{9^2} + \frac{1}{15^2} + \frac{1}{21^2} + \cdots = \frac{\pi^2}{72}.$$ If this series, which contains all the terms with odd denominators divisible by three, is subtracted from the series above it, there remains the series of squares of odd numbers not divisible by three, so that

$$\frac{\pi^2}{9} = 1 + \frac{1}{5^2} + \frac{1}{7^2} + \frac{1}{11^2} + \frac{1}{13^2} + \cdots.$$

178. If the series found in sections 172 and 174 are either added or subtracted, we obtain other series which are worthy of note. We have

$$\frac{k\pi}{2n} + \frac{\pi}{2nk} = \frac{1}{m} + \frac{1}{n-m} - \frac{1}{n+m} - \frac{1}{2n-m} + \frac{1}{2n+m} + \cdots$$
$$= \frac{(k^2+1)\pi}{2nk}.$$

If we let $k = \tan\frac{m\pi}{2n} = \frac{\sin\frac{m\pi}{2n}}{\cos\frac{m\pi}{2n}}$, then $1 + k^2 = \frac{1}{\left(\cos\frac{m\pi}{2n}\right)^2}$, so that

$\frac{2k}{1+k^2} = 2\sin\frac{m\pi}{2n}\cos\frac{m\pi}{2n} = \sin\frac{m\pi}{n}$. When we substitute these values, we obtain

$$\frac{\pi}{n\sin\frac{m\pi}{n}} = \frac{1}{m} + \frac{1}{n-m} - \frac{1}{n+m} - \frac{1}{2n-m} + \frac{1}{2n+m}$$
$$+ \frac{1}{3n-m} - \frac{1}{3n+m} - \cdots.$$ In a similar way, by subtraction, we obtain

$$\frac{\pi}{2nk} - \frac{k\pi}{2n} = \frac{(1-k^2)\pi}{2nk} = \frac{1}{m} - \frac{1}{n-m} + \frac{1}{n+m} - \frac{1}{2n-m}$$
$$+ \frac{1}{2n+m} - \frac{1}{3n-m} + \frac{1}{3n+m} - \cdots.$$ If we let

$$\frac{2k}{1-k^2} = \tan\frac{2m\pi}{2n} = \tan\frac{m\pi}{n} = \frac{\sin\frac{m\pi}{n}}{\cos\frac{m\pi}{n}}, \text{ then}$$

$$\frac{\pi\cos\frac{m\pi}{n}}{n\sin\frac{m\pi}{n}} = \frac{1}{m} - \frac{1}{n-m} + \frac{1}{n+m} - \frac{1}{2n-m}$$

$+ \frac{1}{2n+m} - \frac{1}{3n-m} + \cdots$. Series with squares and higher powers which arise in this way are more easily derived through differentiation, which we will do later.

179. Since we have already considered the results when $m = 1$ and $n = 2$, or 3, we now let $m = 1$ and $n = 4$. In this case $\sin\frac{m\pi}{n} = \sin\frac{\pi}{4} = \frac{1}{\sqrt{2}}$ and $\cos\frac{\pi}{4} = \frac{1}{\sqrt{2}}$. It follows that

$$\frac{\pi}{2\sqrt{2}} = 1 + \frac{1}{3} - \frac{1}{5} - \frac{1}{7} + \frac{1}{9} + \frac{1}{11} - \frac{1}{13} - \frac{1}{15} + \cdots \text{ and}$$

$\frac{\pi}{4} = 1 - \frac{1}{3} + \frac{1}{5} - \frac{1}{7} + \frac{1}{9} - \frac{1}{11} + \frac{1}{13} - \frac{1}{15} + \cdots$. If $m = 1$ and $n = 8$, then $\frac{m\pi}{n} = \frac{\pi}{8}$, $\sin\frac{\pi}{8} = \left(\frac{1}{2} - \frac{1}{2\sqrt{2}}\right)^{\frac{1}{2}}$, $\cos\frac{\pi}{8} = \left(\frac{1}{2} + \frac{1}{2\sqrt{2}}\right)^{\frac{1}{2}}$, and $\frac{\cos\frac{\pi}{8}}{\sin\frac{\pi}{8}} = 1 + \sqrt{2}$. From these we have

$$\frac{\pi}{4\left(2-\sqrt{2}\right)^{\frac{1}{2}}} = 1 + \frac{1}{7} - \frac{1}{9} - \frac{1}{15} + \frac{1}{17} + \frac{1}{23} - \cdots$$

$$\frac{\pi}{8\left(\sqrt{2}-1\right)} = 1 - \frac{1}{7} + \frac{1}{9} - \frac{1}{15} + \frac{1}{17} - \frac{1}{23} + \cdots.$$

Now we let $m = 3$ and $n = 8$, then $\frac{m\pi}{n} = \frac{3\pi}{8}$, $\sin\frac{3\pi}{8} = \left(\frac{1}{2} + \frac{1}{2\sqrt{2}}\right)^{\frac{1}{2}}$,

$\cos\frac{3\pi}{8} = \left(\frac{1}{2} - \frac{1}{2\sqrt{2}}\right)^{\frac{1}{2}}$, and $\dfrac{\cos\frac{3\pi}{8}}{\sin\frac{3\pi}{8}} = \frac{1}{\sqrt{2}+1}$. It follows that

$$\frac{\pi}{4\left(2+\sqrt{2}\right)^{\frac{1}{2}}} = \frac{1}{3} + \frac{1}{5} - \frac{1}{11} - \frac{1}{13} + \frac{1}{19} + \frac{1}{21} - \cdots$$

$$\frac{\pi}{8\left(\sqrt{2}+1\right)} = \frac{1}{3} - \frac{1}{5} + \frac{1}{11} - \frac{1}{13} + \frac{1}{19} - \frac{1}{21} + \cdots .$$

180. Through combinations of the above series we obtain

$$\frac{\pi\left(2+\sqrt{2}\right)^{\frac{1}{2}}}{4} = 1 + \frac{1}{3} + \frac{1}{5} + \frac{1}{7} - \frac{1}{9} - \frac{1}{11} - \frac{1}{13} - \frac{1}{15} + \frac{1}{17} + \frac{1}{19} + \cdots$$

$$\frac{\pi\left(2-\sqrt{2}\right)^{\frac{1}{2}}}{4} = 1 - \frac{1}{3} - \frac{1}{5} + \frac{1}{7} - \frac{1}{9} + \frac{1}{11} + \frac{1}{13} - \frac{1}{15} + \frac{1}{17} - \frac{1}{19} + \cdots$$

$$\frac{\pi\left(\left(4+2\sqrt{2}\right)^{\frac{1}{2}} + \sqrt{2} - 1\right)}{8} = 1 + \frac{1}{3} - \frac{1}{5} + \frac{1}{7} - \frac{1}{9} + \frac{1}{11} - \frac{1}{13} - \frac{1}{15} + \frac{1}{17} + \frac{1}{19} + \cdots$$

$$\frac{\pi\left(\left(4+2\sqrt{2}\right)^{\frac{1}{2}} - \sqrt{2} + 1\right)}{8} = 1 - \frac{1}{3} + \frac{1}{5} + \frac{1}{7} - \frac{1}{9} - \frac{1}{11} + \frac{1}{13} - \frac{1}{15} + \frac{1}{17} - \frac{1}{19} + \cdots$$

$$\frac{\pi\left(\sqrt{2}+1+\left(4-2\sqrt{2}\right)^{\frac{1}{2}}\right)}{8} = 1 + \frac{1}{3} + \frac{1}{5} - \frac{1}{7} + \frac{1}{9} - \frac{1}{11}$$

$$- \frac{1}{13} - \frac{1}{15} + \frac{1}{17} + \frac{1}{19} + \cdots$$

$$\frac{\pi\left(\sqrt{2}+1-\left(4-2\sqrt{2}\right)^{\frac{1}{2}}\right)}{8} = 1 - \frac{1}{3} - \frac{1}{5} - \frac{1}{7} + \frac{1}{9} + \frac{1}{11}$$

$$+ \frac{1}{13} - \frac{1}{15} + \frac{1}{17} - \frac{1}{19} - \cdots .$$

In the same way we could let $n = 16$ and $m = 1,3,5,$ or 7 which would show the sums of series in which the terms are $1, \frac{1}{3}, \frac{1}{5}, \frac{1}{7}, \frac{1}{9}, \cdots$ and in which the various changes of positive and negative signs are different from those already seen.

181. If in the series discussed in section 178, the terms are combined two by two, we obtain the following:

$$\frac{\pi}{n \sin \frac{m\pi}{n}} = \frac{1}{m} + \frac{2m}{n^2 - m^2} - \frac{2m}{4n^2 - m^2} + \frac{2m}{9n^2 - m^2}$$

$$- \frac{2m}{16n^2 - m^2} + \cdots .$$

From this it follows that

$$\frac{1}{n^2 - m^2} - \frac{1}{4n^2 - m^2} + \frac{1}{9n^2 - m^2} - \cdots = \frac{\pi}{2mn \sin \frac{m\pi}{n}} - \frac{1}{2m^2}.$$

The other series gives us

$$\frac{\pi}{n \tan \frac{m\pi}{n}} = \frac{1}{m} - \frac{2m}{n^2 - m^2} - \frac{2m}{4n^2 - m^2} - \frac{2m}{9n^2 - m^2} - \cdots .$$

From this we have

$$\frac{1}{n^2 - m^2} + \frac{1}{4n^2 - m^2} + \frac{1}{9n^2 - m^2} + \cdots = \frac{1}{2m^2} - \frac{\pi}{2mn \tan \frac{m\pi}{n}}.$$

When these two series are added, we obtain

$$\frac{1}{n^2 - m^2} + \frac{1}{9n^2 - m^2} + \frac{1}{25n^2 - m^2} + \cdots = \frac{\pi \tan \frac{m\pi}{2n}}{4mn}.$$

If we let $n = 1$ and let m be any even number $2k$ except zero, since $\tan k\pi = 0$, we always have

$\frac{1}{1 - 4k^2} + \frac{1}{9 - 4k^2} + \frac{1}{25 - 4k^2} + \frac{1}{49 - 4k^2} + \cdots = 0$. However, if in this series $n = 2$ and m is any odd number $2k + 1$,

since $\frac{1}{\tan \frac{m\pi}{n}} = 0$, we have

$$\frac{1}{4 - (2k + 1)^2} + \frac{1}{16 - (2k + 1)^2} + \frac{1}{36 - (2k + 1)^2} + \cdots = \frac{1}{2(2k + 1)^2}.$$

182. If we multiply the series by n^2 and let $\frac{m}{n} = p$,

then they take the form

$$\frac{1}{1 - p^2} - \frac{1}{4 - p^2} + \frac{1}{9 - p^2} - \frac{1}{16 - p^2} + \cdots = \frac{\pi}{2p \sin p\pi} - \frac{1}{2p^2},$$

$$\frac{1}{1 - p^2} + \frac{1}{4 - p^2} + \frac{1}{9 - p^2} + \frac{1}{16 - p^2} + \cdots = \frac{1}{p^2} - \frac{\pi}{2p \sin p\pi}.$$

If we let $p^2 = a$, then we obtain the series

$$\frac{1}{1 - a} - \frac{1}{4 - a} + \frac{1}{9 - a} - \frac{1}{16 - a} + \cdots = \frac{\pi\sqrt{a}}{2a \sin \pi\sqrt{a}} - \frac{1}{2a},$$

$$\frac{1}{1 - a} + \frac{1}{4 - a} + \frac{1}{9 - a} + \frac{1}{16 - a} + \cdots = \frac{1}{2a} - \frac{\pi\sqrt{a}}{2a \tan \pi\sqrt{a}}.$$

Provided a is not negative nor the square of an integer, then the sum of these series can be represented in terms of the circle.

183. By means of the reduction of complex exponentials to sines and cosines of circular arcs, which has been treated, we can assign negative values to a in the series just discussed. Since $e^{xi} = \cos x + i \sin x$ and $e^{-xi} = \cos x - i \sin x$, when we substitute yi for x, we obtain $\cos yi - 1 = \dfrac{e^{-y} + e^{y}}{2}$ and $\sin yi = \dfrac{e^{-y} - e^{y}}{2i}$. Now if $a = -b$ and $y = \pi\sqrt{b}$, then

$$\cos \pi\sqrt{-b} = \frac{e^{-\pi\sqrt{b}} + e^{\pi\sqrt{b}}}{2} \text{ and } \sin \pi\sqrt{-b} = \frac{e^{-\pi\sqrt{b}} - e^{\pi\sqrt{b}}}{2i}.$$

It follows that

$$\tan \pi\sqrt{-b} = \frac{e^{-\pi\sqrt{b}} - e^{\pi\sqrt{b}}}{\left(e^{-\pi\sqrt{b}} + e^{\pi\sqrt{b}}\right)i}.$$ Then we have

$$\frac{\pi\sqrt{-b}}{\sin \pi\sqrt{-b}} = \frac{-2\pi\sqrt{b}}{e^{-\pi\sqrt{b}} - e^{\pi\sqrt{b}}}$$ and

$$\frac{\pi\sqrt{-b}}{\tan \pi\sqrt{-b}} = \frac{\left(e^{-\pi\sqrt{b}} + e^{\pi\sqrt{b}}\right)\pi\sqrt{b}}{e^{-\pi\sqrt{b}} - e^{\pi\sqrt{b}}}.$$ From these remarks it follows that

$$\frac{1}{1+b} - \frac{1}{4+b} + \frac{1}{9+b} - \frac{1}{16+b} + \cdots$$

$$= \frac{1}{2b} - \frac{\pi\sqrt{b}}{\left(e^{\pi\sqrt{b}} - e^{-\pi\sqrt{b}}\right)b},$$

$$\frac{1}{1+b} + \frac{1}{4+b} + \frac{1}{9+b} + \frac{1}{16+b} + \cdots$$

$$= \frac{\left(e^{\pi\sqrt{b}} + e^{-\pi\sqrt{b}}\right)\pi\sqrt{b}}{2b\left(e^{\pi\sqrt{b}} - e^{-\pi\sqrt{b}}\right)} - \frac{1}{2b}.$$ These same series can be derived from section 162, using the same method which was used in this chapter. However, I have preferred to treat it in this way, since it is a nice illustration of the reduction of sines and cosines of complex arcs to real exponentials.

CHAPTER XI

On Other Infinite Expressions for Arcs and Sines.

184. We have already seen, in section 158, where we let z be any circular arc, that $\sin z = z\left(1 - \frac{z^2}{\pi^2}\right)\left(1 - \frac{z^2}{4\pi^2}\right)\left(1 - \frac{z^2}{9\pi^2}\right)\left(1 - \frac{z^2}{16\pi^2}\right) \cdots$ and $\cos z = \left(1 - \frac{4z^2}{\pi^2}\right)\left(1 - \frac{4z^2}{9\pi^2}\right)\left(1 - \frac{4z^2}{25\pi^2}\right)\left(1 - \frac{4z^2}{49\pi^2}\right) \cdots$. We let the arc z be equal to $\frac{m\pi}{n}$, then

$$\sin \frac{m\pi}{n} = \frac{m\pi}{n}\left(1 - \frac{m^2}{n^2}\right)\left(1 - \frac{m^2}{4n^2}\right)\left(1 - \frac{m^2}{9n^2}\right)\left(1 - \frac{m^2}{16n^2}\right) \cdots \text{ and}$$

$$\cos \frac{m\pi}{n} = \left(1 - \frac{4m^2}{n^2}\right)\left(1 - \frac{4m^2}{9n^2}\right)\left(1 - \frac{4m^2}{25n^2}\right)\left(1 - \frac{4m^2}{49n^2}\right) \cdots .$$

If we substitute $2n$ for n, then

$$\sin \frac{m\pi}{2n} = \frac{m\pi}{2n}\left(\frac{4n^2 - m^2}{4n^2}\right)\left(\frac{16n^2 - m^2}{16n^2}\right)\left(\frac{36n^2 - m^2}{36n^2}\right) \text{ and}$$

$$\cos \frac{m\pi}{2n} = \left(\frac{n^2 - m^2}{n^2}\right)\left(\frac{9n^2 - m^2}{9n^2}\right)\left(\frac{25n^2 - m^2}{25n^2}\right)\left(\frac{49n^2 - m^2}{49n^2}\right) \cdots .$$

These can be expressed by linear factors as follows

$$\sin \frac{m\pi}{2n} = \frac{m\pi}{2n}\left(\frac{2n - m}{2n}\right)\left(\frac{2n + m}{2n}\right)\left(\frac{4n - m}{4n}\right)\left(\frac{4n + m}{4n}\right)$$

$$\left(\frac{6n - m}{6n}\right) \cdots \text{ and}$$

$$\cos \frac{m\pi}{2n} = \left(\frac{n - m}{n}\right)\left(\frac{n + m}{n}\right)\left(\frac{3n - m}{3n}\right)\left(\frac{3n + m}{3n}\right)\left(\frac{5n - m}{5n}\right)$$

$\left(\frac{5n + m}{5n}\right) \cdots$. Since $\sin(n - m)\frac{\pi}{2n} = \cos \frac{m\pi}{2n}$ and

$\cos(n - m)\frac{\pi}{2n} = \sin \frac{m\pi}{2n}$, if we substitute $n - m$ for m, then we have

$$\cos \frac{m\pi}{2n} = \left((n - m)\frac{\pi}{2n}\right)\left(\frac{n + m}{2n}\right)\left(\frac{3n - m}{2n}\right)\left(\frac{3n + m}{4n}\right)\left(\frac{5n - m}{4n}\right)$$

$\left(\frac{5n + m}{6n}\right) \cdots$ and

$$\sin \frac{m\pi}{2n} = \frac{m}{n}\left(\frac{2n - m}{n}\right)\left(\frac{2n + m}{3n}\right)\left(\frac{4n - m}{3n}\right)\left(\frac{4n + m}{5n}\right)$$

$\left(\frac{6n - m}{5n}\right) \cdots$.

185. Since we have two expressions for each of the sine and cosine of $\frac{m\pi}{2n}$, when we divide one by the other we obtain

$1 = \frac{\pi}{2} \cdot \frac{1}{2} \cdot \frac{3}{2} \cdot \frac{3}{4} \cdot \frac{5}{4} \cdot \frac{5}{6} \cdot \frac{7}{6} \cdot \frac{7}{8} \cdot \frac{9}{8} \cdots$. It follows that $\frac{\pi}{2} = \frac{2\cdot2\cdot4\cdot4\cdot6\cdot6\cdot8\cdot8\cdot10\cdot10\cdot12\cdot12}{1\cdot3\cdot3\cdot5\cdot5\cdot7\cdot7\cdot9\cdot9\cdot11\cdot11\cdot13} \cdots$. And this is the exprssion for π which Wallis found in his *Arithmetic of the Infinite*. We could set down a great number of similar expressions derived from the first expression for the sine. For example, from it we deduce that

$$\frac{\pi}{2} = \frac{n}{m} \sin \frac{m\pi}{2n}\left(\frac{2n}{2n - m}\right)\left(\frac{2n}{2n + m}\right)\left(\frac{4n}{4n - m}\right)\left(\frac{4n}{4n + m}\right)$$

$\left(\frac{6n}{6n - m}\right) \cdots$ where, when we let $\frac{m}{n} = 1$, we obtain the Wallis product. If we let $\frac{m}{n} = \frac{1}{2}$, since $\sin \frac{1}{4}\pi = \frac{1}{\sqrt{2}}$, we have

$\frac{\pi}{2} = \frac{\sqrt{2}}{1} \cdot \frac{4}{3} \cdot \frac{4}{5} \cdot \frac{8}{7} \cdot \frac{8}{9} \cdot \frac{12}{11} \cdot \frac{12}{13} \cdot \frac{16}{15} \cdot \frac{16}{17} \cdots$. If we let $\frac{m}{n} = \frac{1}{3}$, since $\sin\frac{\pi}{6} = \frac{1}{2}$, we have

$\frac{\pi}{2} = \frac{3}{2} \cdot \frac{6}{5} \cdot \frac{6}{7} \cdot \frac{12}{11} \cdot \frac{12}{13} \cdot \frac{18}{17} \cdot \frac{18}{19} \cdot \frac{24}{23} \cdots$. We note that if the Wallis product is divided by the expression obtained when $\frac{m}{n} = \frac{1}{2}$, then we have

$$\sqrt{2} = \frac{2\cdot 2\cdot 6\cdot 6\cdot 10\cdot 10\cdot 14\cdot 14\cdot 18\cdot 18}{1\cdot 3\cdot 5\cdot 7\cdot 9\cdot 11\cdot 13\cdot 15\cdot 17\cdot 19} \cdots .$$

186. Since the tangent of any angle is equal to the quotient of the sine by the cosine of that angle, it follows that the tangent can also be expressed by infinite products. If the first expression for the sine is divided by the second for the cosine, we obtain the following expression.

$$\tan\frac{m\pi}{2n} = \frac{m}{n-m}\left(\frac{2n-m}{n+m}\right)\left(\frac{2n+m}{3n-m}\right)\left(\frac{4n-m}{3n+m}\right)\left(\frac{4n+m}{5n-m}\right)\cdots, \text{ and}$$

$$\cot\frac{m\pi}{2n} = \frac{n-m}{m}\left(\frac{n+m}{2n-m}\right)\left(\frac{3n-m}{2n+m}\right)\left(\frac{3n+m}{4n-m}\right)\left(\frac{5n-m}{4n+m}\right)\cdots .$$

In a similar way the secants and cosecants have the following expressions:

$$\sec\frac{m\pi}{2n} = \left(\frac{n}{n-m}\right)\left(\frac{n}{n+m}\right)\left(\frac{3n}{3n-m}\right)\left(\frac{3n}{3n+m}\right)\left(\frac{5n}{5n-m}\right)\left(\frac{5n}{5n+m}\right)\cdots \text{ and}$$

$$\csc\frac{m\pi}{2n} = \frac{n}{m}\left(\frac{n}{2n-m}\right)\left(\frac{3n}{2n+m}\right)\left(\frac{3n}{4n-m}\right)\left(\frac{5n}{4n+m}\right)\left(\frac{5n}{6n-m}\right)\cdots .$$

If the second expressions for the sines and cosines are used we obtain the following:

$$\tan\frac{m\pi}{2n} = \frac{\pi}{2} \cdot \frac{m}{n-m} \cdot \frac{1}{2}\frac{(2n-m)}{(n+m)} \cdot \frac{3}{2}\frac{(2n+m)}{(3n-m)} \cdot \frac{3}{4}\frac{(4n-m)}{(3n+m)} \cdots ,$$

$$\cot \frac{m\pi}{2n} = \frac{\pi}{2} \cdot \frac{n - m}{m} \cdot \frac{1}{2}\frac{(n + m)}{(2n - m)} \cdot \frac{3}{2}\frac{(3n - m)}{(2n + m)} \cdot \frac{3}{4}\frac{(3n + m)}{(4n - m)} \cdots ,$$

$$\sec \frac{m\pi}{2n} = \frac{2}{\pi} \cdot \frac{n}{n - m} \cdot \frac{2n}{n + m} \cdot \frac{2n}{3n - m} \cdot \frac{4n}{3n + m} \cdot \frac{4n}{5n - m} \cdots ,$$

$$\csc \frac{m\pi}{2n} = \frac{2}{\pi} \cdot \frac{n}{m} \cdot \frac{2n}{2n - m} \cdot \frac{2n}{2n + m} \cdot \frac{4n}{4n - m} \cdot \frac{4n}{4n + m} \cdots .$$

187. If we substitute k for m in a similar way the sine and cosine of the angle $\frac{k\pi}{2n}$ is defined. When the previous expressions are divided by these new expressions we obtain the following formulas.

$$\frac{\sin \frac{m\pi}{2n}}{\sin \frac{k\pi}{2n}} = \frac{m}{k}\left(\frac{2n - m}{2n - k}\right)\left(\frac{2n + m}{2n + k}\right)\left(\frac{4n - m}{4n - k}\right)\left(\frac{4n + m}{4n + k}\right) \cdots ,$$

$$\frac{\sin \frac{m\pi}{2n}}{\cos \frac{k\pi}{2n}} = \frac{m}{n - k}\left(\frac{2n - m}{n + k}\right)\left(\frac{2n + m}{3n - k}\right)\left(\frac{4n - m}{3n + k}\right)\left(\frac{4n + m}{5n - k}\right) \cdots ,$$

$$\frac{\cos \frac{m\pi}{2n}}{\cos \frac{k\pi}{2n}} = \left(\frac{n - m}{n - k}\right)\left(\frac{n + m}{n + k}\right)\left(\frac{3n - m}{3n - k}\right)\left(\frac{3n + m}{3n + k}\right)$$

$$\left(\frac{5n - m}{5n - k}\right) \cdots , \text{ and}$$

$$\frac{\cos \frac{m\pi}{2n}}{\cos \frac{k\pi}{2n}} = \left(\frac{n - m}{k}\right)\left(\frac{n + m}{2n - k}\right)\left(\frac{3n - m}{2n + k}\right)\left(\frac{3n + m}{4n - k}\right)$$

$\left(\frac{5n - m}{4n + k}\right) \cdots$. If we take $\frac{k\pi}{2n}$ as an angle whose sine and cosine are known, by means of the above formulas, we can find the sine and cosine of any other angle $\frac{m\pi}{2n}$.

188. Expressions of this type, which consist of infinite products, can be used to obtain accurate values of π or sines and cosines of given angles, and this is rather important, since even now we have no better methods for obtaining these values. There are other infinite products of little practical value which could be used to find approximate values for π or sines and cosines. Indeed, the following factors:

$\frac{\pi}{2} = 2\left(1 - \frac{1}{9}\right)\left(1 - \frac{1}{25}\right)\left(1 - \frac{1}{49}\right)\cdots$ can be expressed with little difficulty as decimal fractions, however, too many terms are required to obtain an accurate value of π even to only ten decimal places.

189. The principal use of these infinite products, however, is in the calculation of logarithms. Without these expressions the calculation of logarithms would be very difficult. In the first place, since

$\pi = 4\left(1 - \frac{1}{9}\right)\left(1 - \frac{1}{25}\right)\left(1 - \frac{1}{49}\right)\cdots$, when we take logarithms we find

$$\log \pi = \log 4 + \log\left(1 - \frac{1}{9}\right) + \log\left(1 - \frac{1}{25}\right) + \log\left(1 - \frac{1}{49}\right) + \cdots \qquad \text{or}$$

$$\log \pi = \log 2 - \log\left(1 - \frac{1}{4}\right) - \log\left(1 - \frac{1}{16}\right) - \log\left(1 - \frac{1}{36}\right) - \cdots ,$$

whether common logarithms or natural logarithms are used. Since it it easy to find the common logarithms from the natural logarithms, there is a significant advantage in using natural logarithms to find the logarithm of π.

190. Since, when we use natural logarithms,

$\log(1 - x) = -x - \frac{x^2}{2} - \frac{x^3}{3} - \frac{x^4}{4} - \cdots$, if each of the terms in the expression for $\log \pi$ are expressed in this form, we have

$$\log \pi = \log 4 + \left(-\frac{1}{9} - \frac{1}{2\cdot 9^2} - \frac{1}{3\cdot 9^3} - \frac{1}{4\cdot 9^4} - \cdots \right)$$

$$+ \left(-\frac{1}{25} - \frac{1}{2\cdot 25^2} - \frac{1}{3\cdot 25^3} - \frac{1}{4\cdot 25^4} - \cdots \right)$$

$$+ \left(-\frac{1}{49} - \frac{1}{2\cdot 49^2} - \frac{1}{3\cdot 49^3} - \frac{1}{4\cdot 49^4} - \cdots \right) + \cdots$$ In this infinite series

we notice that when we descend vertically we find infinite series whose sums we have already found. For the sake of brevity we use the following notation. Let

$$A = 1 + \frac{1}{3^2} + \frac{1}{5^2} + \frac{1}{7^2} + \frac{1}{9^2} + \cdots,$$

$$B = 1 + \frac{1}{3^4} + \frac{1}{5^4} + \frac{1}{7^4} + \frac{1}{9^4} + \cdots,$$

$$C = 1 + \frac{1}{3^6} + \frac{1}{5^6} + \frac{1}{7^6} + \frac{1}{9^6} + \cdots,$$

$$D = 1 + \frac{1}{3^8} + \frac{1}{5^8} + \frac{1}{7^8} + \frac{1}{9^8} + \cdots,$$ then

$$\log \pi = \log 4 - (A - 1) - \frac{1}{2}(B - 1) - \frac{1}{3}(C - 1) \quad - \frac{1}{4}(D - 1) - \cdots.$$

But we have, using the approximate results found above,

$A = 1.2337005501361698273543 1$ $\qquad B = 1.0146780316041920545462 5$

$C = 1.0014470766409421219064 7$ $\qquad D = 1.0001551790252961193029 8$

$E = 1.0000170413630448255081 6$ $\qquad F = 1.0000018858485831195759 0$

$G = 1.0000002092405192115001 0$ $\qquad H = 1.0000000232371573791567 0$

$I = 1.0000000025814375566597 7$ $\qquad K = 1.0000000002868076974555 8$

$L = 1.0000000000318667751404 4$ $\qquad M = 1.0000000000035407229439 2$

$N = 1.0000000000003934124669 1$ $\qquad O = 1.0000000000000437124485 9$

$P = 1.000000000000000485693682$ $\quad Q = 1.000000000000000053965957$

$R = 1.000000000000000005996217$ $\quad S = 1.000000000000000000666246$

$T = 1.000000000000000000074027$ $\quad V = 1.000000000000000000008225$

$W = 1.000000000000000000000913$ $\quad X = 1.000000000000000000000101.$

In this way with little tedium we find that the natural logarithm of π has the value 1.1447298858494001741434 2. If we multiply this value by $0.43429 \cdots$, we find the value of the common logarithms of π to be 0.49714987269413385435126.

191. Since we can also express both the sine and cosine of an angle equal to $\frac{m\pi}{2n}$ by means of infinite products, we can conveniently express the logarithms of both. From the formulas found earlier we have

$$\log \sin \frac{m\pi}{2n} = \log \pi + \log \frac{m}{2n} + \log\left(1 - \frac{m^2}{4n^2}\right) + \log\left(1 - \frac{m^2}{16n^2}\right)$$

$$+ \log\left(1 - \frac{m^3}{36n^2}\right) + \cdots \text{ and}$$

$$\log \cos \frac{m\pi}{2n} = \log\left(1 - \frac{m^2}{n^2}\right) + \log\left(1 - \frac{m^2}{9n^2}\right) + \log\left(1 - \frac{m^2}{25n^2}\right)$$

$$+ \log\left(1 - \frac{m^2}{49n^2}\right) + \cdots.$$ It follows, as before, that by using natural logarithms these can easily be expressed in rapidly converging series. Lest we have to multiply infinite series unnecessarily, we leave the first terms in the form of logarithms.

$$\log \sin \frac{m\pi}{2n} = \log \pi + \log m + \log(2n - m)$$

$$+ \log(2n + m) - \log 8 - 3\log n$$

$$- \frac{m^2}{16n^2} - \frac{m^4}{2 \cdot 16^2 n^4} - \frac{m^6}{3 \cdot 16^3 n^6} - \frac{m^8}{4 \cdot 16^4 n^8} - \cdots$$

$$- \frac{m^2}{36n^2} - \frac{m^4}{2 \cdot 36^2 n^4} - \frac{m^6}{3 \cdot 36^3 n^6} - \frac{m^8}{4 \cdot 36^4 n^8} - \cdots$$

$$- \frac{m^2}{64n^2} - \frac{m^4}{2 \cdot 64^2 n^4} - \frac{m^6}{3 \cdot 64^3 n^6} - \frac{m^8}{4 \cdot 64^4 n^8} - \cdots .$$

$$\log \cos \frac{m\pi}{2n} = \log(n - m) + \log(n + m) - 2 \log n$$

$$- \frac{m^2}{9n^2} - \frac{m^4}{2 \cdot 9^2 n^4} - \frac{m^6}{3 \cdot 9^3 n^6} - \frac{m^8}{4 \cdot 9^4 n^8} - \cdots$$

$$- \frac{m^2}{25n^2} - \frac{m^4}{2 \cdot 25^2 n^4} - \frac{m^6}{3 \cdot 25^3 n^6} - \frac{m^8}{4 \cdot 25^4 n^8} - \cdots$$

$$- \frac{m^2}{49n^2} - \frac{m^4}{2 \cdot 49^2 n^4} - \frac{m^6}{3 \cdot 49^3 n^6} - \frac{m^8}{4 \cdot 49^4 n^8} - \cdots$$

192. There occur in these series all even powers of $\frac{m}{n}$, which are multiplied by series whose sums we have already found. Therefore, we have

$$\log \sin \frac{m\pi}{2n} = \log m + \log(2n - m) + \log(2n + m) - 3 \log n$$

$$+ \log \pi - \log 8 - \frac{m^2}{n^2}\left(\frac{1}{4^2} + \frac{1}{6^2} + \frac{1}{8^2} + \frac{1}{10^2} + \frac{1}{12^2} + \cdots\right)$$

$$\frac{m^4}{2n^4}\left(\frac{1}{4^4} + \frac{1}{6^4} + \frac{1}{8^4} + \frac{1}{10^4} + \frac{1}{12^4} + \cdots\right)$$

$$- \frac{m^6}{3n^6}\left(\frac{1}{4^6} + \frac{1}{6^6} + \frac{1}{8^6} + \frac{1}{10^6} + \frac{1}{12^6} + \cdots\right)$$

$$- \frac{m^8}{4n^8}\left(\frac{1}{4^8} + \frac{1}{6^8} + \frac{1}{8^8} + \frac{1}{10^8} + \frac{1}{12^8} + \cdots\right) - \cdots$$

and

$$\log \cos \frac{m\pi}{2n} = \log(n - m) + \log(n + m) - 2 \log n$$

$$- \frac{m^2}{n^2}\left(\frac{1}{3^2} + \frac{1}{5^2} + \frac{1}{7^2} + \frac{1}{9^2} + \cdots\right)$$

$$-\frac{m^4}{2n^4}\left(\frac{1}{3^4}+\frac{1}{5^4}+\frac{1}{7^4}+\frac{1}{9^4}+\cdots\right)$$

$$-\frac{m^6}{3n^6}\left(\frac{1}{3^6}+\frac{1}{5^6}+\frac{1}{7^6}+\frac{1}{9^6}+\cdots\right)$$

$$-\frac{m^8}{4n^8}\left(\frac{1}{3^8}+\frac{1}{5^8}+\frac{1}{7^8}+\frac{1}{9^8}+\cdots\right)-\cdots$$

The sums of the latter series were given in section 190, while the sums of the former can be derived from these. In order that they may be more easily used, however, we will append at least partial results in what follows.

193. For the sake of brevity we let $\alpha = \frac{1}{2^2}+\frac{1}{4^2}+\frac{1}{6^2}+\frac{1}{8^2}+\cdots$

$\beta = \frac{1}{2^4}+\frac{1}{4^4}+\frac{1}{6^4}+\frac{1}{8^4}+\cdots$ $\gamma = \frac{1}{2^6}+\frac{1}{4^6}+\frac{1}{6^6}+\frac{1}{8^6}+\cdots$

$\delta = \frac{1}{2^8}+\frac{1}{4^8}+\frac{1}{6^8}+\frac{1}{8^8}+\cdots$. The sums are approximated as follows:

$\alpha = 0.411233516712056609118 10$ $\beta = 0.067645202106946136969 75$

$\gamma = 0.015895985343507017808 04$ $\delta = 0.003922177172648220075 70$

$\epsilon = 0.000977533764773259848 98$ $\zeta = 0.000244200704724928722 74$

$\eta = 0.000061038894539493329 15$ $\theta = 0.000015259022251272699 77$

$\iota = 0.000003814711827443180 08$ $\kappa = 0.000000953675226175340 53$

$\lambda = 0.000000238418635952591 54$ $\mu = 0.000000059604648328315 55$

$\nu = 0.000000014901161415898 13$ $\xi = 0.000000003725290312339 86$

$o = 0.000000000931322575482 84$ $\pi = 0.000000000232830643708 07$

$ro = 0.000000000058207660916 85$ $\sigma = 0.000000000014551915228 58$

$\tau = 0.000000000003637978807 10$ $\upsilon = 0.000000000000909494701 77$

$\phi = 0.000000000000227373675 44$ $\chi = 0.000000000000056843418 86$

$\psi = 0.000000000000014210854 71$ $\omega = 0.000000000000003552713 67.$

The remaining sums decrease by about one fourth per sum.

194. When we use these results we have

$$\log \sin \frac{m\pi}{2n} = \log n + \log(2n - m) + \log(2n + m) - 3 \log n + \log \pi$$

$$- \log 8 - \frac{m^2}{n^2}\left(\alpha - \frac{1}{2^2}\right) - \frac{m^4}{2n^4}\left(\beta - \frac{1}{2^4}\right) - \frac{m^6}{3n^6}\left(\gamma - \frac{1}{2^6}\right) - \cdots$$ and

$$\log \cos \frac{m\pi}{2n} = \log(n - m) + \log(n + m) - 2 \log n$$

$$- \frac{m^2}{n^2}(A - 1) - \frac{m^4}{2n^4}(B - 1) - \frac{m^6}{3n^6}(C - 1) - \cdots.$$ Since we know $\log \pi$ and $\log 8$, the natural logarithm of the sine of the angle $\frac{m}{n}.\frac{\pi}{2}$ is equal to

$$\log m + \log(2n - m) + \log(2n + m) - 3 \log n$$

$$- 0.93471165583043575410 \qquad - \frac{m^2}{n^2}0.16123351671205660911$$

$$- \frac{m^4}{n^4}0.00257260105347306848 \qquad - \frac{m^6}{n^6}0.00009032844783567260$$

$$- \frac{m^8}{n^8}0.00000398179316205501 \qquad - \frac{m^{10}}{n^{10}}0.00000019425295465196$$

$$- \frac{m^{12}}{n^{12}}0.00000001001328748812 \qquad - \frac{m^{14}}{n^{14}}0.00000000053404135618$$

$$- \frac{m^{16}}{n^{16}}0.00000000002914859658 \qquad - \frac{m^{18}}{n^{18}}0.00000000000161797979$$

$$- \frac{m^{20}}{n^{20}}0.00000000000009097690 \qquad - \frac{m^{22}}{n^{22}}0.00000000000000516827$$

$$- \frac{m^{24}}{n^{24}}0.00000000000000029607 \qquad - \frac{m^{26}}{n^{26}}0.00000000000000001708$$

$$- \frac{m^{28}}{n^{28}}0.00000000000000000099 \qquad - \frac{m^{30}}{n^{30}}0.00000000000000000005.$$

The natural logarithms of the cosine of the angle $\frac{m}{n}.\frac{\pi}{2}$ is equal to

$$\log(n-m)+\log(n+m)-2\log n \quad -\frac{m^2}{n^2}0.23370055013616982735$$

$$-\frac{m^4}{n^4}0.00733901580209602727 \quad -\frac{m^6}{n^6}0.00048235888031404063$$

$$-\frac{m^8}{n^8}0.00003879475632402982 \quad -\frac{m^{10}}{n^{10}}0.00000340827260896510$$

$$-\frac{m^{12}}{n^{12}}0.00000031430809718659 \quad -\frac{m^{14}}{n^{14}}0.00000002989150274450$$

$$-\frac{m^{16}}{n^{16}}0.00000000290464467239 \quad -\frac{m^{18}}{n^{18}}0.00000000028682639518$$

$$-\frac{m^{20}}{n^{20}}0.00000000002868076974 \quad -\frac{m^{22}}{n^{22}}0.00000000000289697956$$

$$-\frac{m^{24}}{n^{24}}0.00000000000029506024 \quad -\frac{m^{26}}{n^{26}}0.00000000000003026249$$

$$-\frac{m^{28}}{n^{28}}0.00000000000000312232 \quad -\frac{m^{30}}{n^{30}}0.00000000000000032379$$

$$-\frac{m^{32}}{n^{32}}0.00000000000000003373 \quad -\frac{m^{34}}{n^{34}}0.00000000000000000352$$

$$-\frac{m^{36}}{n^{36}}0.00000000000000000037 \quad -\frac{m^{38}}{n^{38}}0.00000000000000000004.$$

195. If these natural logarithms of sines and cosines are multiplied by 0.4342944819 etc., the common logarithms of these functions are produced. Since it is customary in tables of common logarithms of sines and cosines to add 10, this is done in what follows, after this multiplication. The tabular logarithm of the sine of the angle $\frac{m}{n}.\frac{\pi}{2}$ is equal to

$$\log m+\log(2n-m)+\log(2n+m)-3\log n \quad +9.594059885702190$$

$$-\frac{m^2}{n^2}0.070022826605901 \quad -\frac{m^4}{n^4}0.001117266441661 \quad -\frac{m^6}{n^6}0.000039229146453$$

$$-\frac{m^8}{n^8}0.000001729270798 \quad -\frac{m^{10}}{n^{10}}0.000000084362986$$

$$- \frac{m^{12}}{n^{12}} 0.000000004348715 \quad - \frac{m^{14}}{n^{14}} 0.000000000231931$$

$$- \frac{m^{16}}{n^{16}} 0.000000000012659 \quad - \frac{m^{18}}{n^{18}} 0.000000000000702$$

$$- \frac{m^{20}}{n^{20}} 0.000000000000039.$$

The tabular logarithm of the cosine of the angle $\frac{m}{n} \cdot \frac{\pi}{2}$ is equal to

$$\log(n - m) + \log(n + m) - 2 \log n + 10.000000000000000$$

$$- \frac{m^2}{n^2} 0.101494859341892 - \frac{m^4}{n^4} 0.003187294065451 - \frac{m^6}{n^6} 0.000209485800017$$

$$- \frac{m^8}{n^8} 0.000016848348597 \quad - \frac{m^{10}}{n^{10}} 0.000001480193986$$

$$- \frac{m^{12}}{n^{12}} 0.000000136502272 \quad - \frac{m^{14}}{n^{14}} 0.000000012981715$$

$$- \frac{m^{16}}{n^{16}} 0.000000001261471 \quad - \frac{m^{18}}{n^{18}} 0.000000000124567$$

$$- \frac{m^{20}}{n^{20}} 0.000000000012456 \quad - \frac{m^{22}}{n^{22}} 0.000000000001258$$

$$- \frac{m^{24}}{n^{24}} 0.000000000000128 - \frac{m^{26}}{n^{26}} 0.000000000000013$$

196. With these formulas we can find both the natural and common logarithms of the sine and cosine of any angle, even without knowing the sines and cosines. Furthermore, from the logarithms of the sine and cosine we can find the logarithms of the tangent, cotangent, secant and cosecant by simple subtraction. For this reason there is no need for special formulas for the other functions. We also note that in the formulas, when we find the logarithms of $m, n, n - m, n + m$, etc., they must be natural or common logarithms depending on which formula we use. Furthermore, the ratio $\frac{m}{n}$ indicates that

part of the right angle, which represents the angle. Therefore, we recall that the sine of an angle larger than a right angle is equal to the cosine of an angle less than a right angle and vice versa. The fraction $\frac{m}{n}$ need never be greater then $\frac{1}{2}$, and for this reason the terms converge much more quickly.

197. Before we leave this topic, we show a method for finding the tangents and secants of any angle, which is better than the methods we have seen so far in this chapter. Although the tangent and secant are determined by the sine and cosine, nevertheless this is accomplished through division, and when there are many digits, this is extremely tedious. Furthermore, in section 136 we gave formulas for the tangents and cotangents without the rationale, which we have saved for this chapter.

We recall from section 181 an expression for the tangent of the angle $\frac{m}{2n}\pi$. Since $\frac{1}{n^2 - m^2} + \frac{1}{9n^2 - m^2} + \frac{1}{25n^2 - m^2} + \cdots = \frac{\pi}{4mn} \tan \frac{m}{2n}\pi$, we have $\tan \frac{m}{2n}\pi = \frac{4mn}{\pi}\left(\frac{1}{n^2 - m^2} + \frac{1}{9n^2 - m^2} + \frac{1}{25n^2 - m^2} + \ldots\right)$ and also $\frac{1}{n^2 - m^2} + \frac{1}{4n^2 - m^2} + \frac{1}{9n^2 - m^2} + \cdots = \frac{1}{2m^2} - \frac{\pi}{2mn} \cot \frac{m}{n}\pi$. If we substitute $2n$ for n, then

$$\cot \frac{m}{2n}\pi = \frac{2n}{m\pi} - \frac{4mn}{\pi}\left(\frac{1}{4n^2 - m^2} + \frac{1}{16n^2 - m^2} + \frac{1}{36n^2 - m^2} + \cdots\right).$$

Each of the fractions, except for the first, can be expressed as infinite series which can easily be summed. Hence,

$$\tan \frac{m}{2n}\pi = \frac{mn}{n^2 - m^2}\cdot\frac{4}{\pi} + \frac{4}{\pi}\left(\frac{m}{3^2 n} + \frac{m^3}{3^4 n^3} + \frac{m^5}{3^6 n^5} + \cdots\right)$$

$$+ \frac{4}{\pi}\left(\frac{m}{5^2 n} + \frac{m^3}{5^4 n^3} + \frac{m^5}{5^6 n^5} + \cdots\right)$$

$$+ \frac{4}{\pi}\left(\frac{m}{7^2 n} + \frac{m^3}{7^4 n^3} + \frac{m^5}{7^6 n^5} + \cdots\right) + \cdots,$$

and

$$\cot\frac{m}{2n}\pi = \frac{n}{m}\cdot\frac{2}{\pi} - \frac{mn}{4n^2 - m^2}\cdot\frac{4}{\pi} - \frac{4}{\pi}\left(\frac{m}{4^2 n} + \frac{m^3}{4^4 n^3} + \frac{m^5}{4^6 n^5} + \cdots\right)$$

$$- \frac{4}{\pi}\left(\frac{m}{6^2 n} + \frac{m^3}{6^4 n^3} + \frac{m^5}{6^6 n^5} + \cdots\right)$$

$$- \frac{4}{\pi}\left(\frac{m}{8^2 n} + \frac{m^3}{8^4 n^3} + \frac{m^5}{8^6 n^5} + \cdots\right) + \cdots.$$

198. Since we know the value of π, we compute the value of $\frac{1}{\pi}$ to be 0.318309886183790671537767926745028724. We have already found the sums of the series, which we designated by A, B, C, D, etc. and $\alpha, \beta, \gamma, \delta$, etc. With this notation we have

$$\tan\frac{m}{2n}\pi = \frac{mn}{n^2 - m^2}\cdot\frac{4}{\pi} + \frac{m}{n}\cdot\frac{4}{\pi}(A - 1) + \frac{m^3}{n^3}\cdot\frac{4}{\pi}(B - 1)$$

$$+ \frac{m^5}{n^5}\cdot\frac{4}{\pi}(C - 1) + \frac{m^7}{n^7}\cdot\frac{4}{\pi}(D - 1) + \cdots,$$

and

$$\cot\frac{m}{2n}\pi = \frac{n}{m}\cdot\frac{2}{\pi} - \frac{4mn}{4n^2 - m^2}\cdot\frac{1}{\pi} - \frac{m}{n}\cdot\frac{4}{\pi}\left(\alpha - \frac{1}{2^2}\right)$$

$$- \frac{m^3}{n^3}\cdot\frac{4}{\pi}\left(\beta - \frac{1}{2^4}\right) - \frac{m^5}{n^5}\cdot\frac{4}{\pi}\left(\gamma - \frac{1}{2^6}\right) - \cdots$$

From these formulas are derived the expressions given above in section 135 for the tangent and cotangent. In section 137 we have shown that from the tangent and cotangent, we find the secant and cosecant by simple addition and subtraction. With the use of these rules the entire table of sines, cosines, tangents, and secants, as well as their logarithms can much more easily be found, than was the case in previous times.

CHAPTER XII

On the Development of Real Rational Functions

199. It was in the second chapter that we gave a method for expressing any rational function by means of just as many partial fractions as the denominator had linear factors, indeed those linear factors became the denominators of the partial fractions. From this it is clear that if the linear factors are complex, then the partial fractions to which they give rise will also be complex. In this case it will be of little use to express a real rational function in terms of complex fractions. Since we have seen than every polynomial, which is what the denominator of a rational function is, can always be expressed by real quadratic factors, no matter how many complex linear factors it may contain. In this way we can express any real rational function in real partial fractions by allowing the denominators to be quadratic.

200. We let the given rational function be $\frac{M}{N}$, and for each of the real linear factors of the denominator N we find partial fractions by the method already given. Instead of the complex linear factors we use the factor of N with the form $p^2 - 2pqz \cos \phi + q^2z^2$. Since in this process it will be helpful to look at the numerator and denominator of the function at this stage of development, we suppose the form to be

$\frac{A + Bz + Cz^2 + Dz^3 + Ez^4 + \cdots}{(p^2 - 2pqz \cos \phi + q^2z^2)(\alpha + \beta z + \gamma z^2 + \delta z^3 + \cdots)}$. The partial fraction whose denominator is the factor $p^2 - 2pqz \cos \phi + q^2z^2$ has the form $\frac{P + Qz}{p^2 - 2pqz \cos \phi + q^2z^2}$. Since the denominator is a second degree polynomial, the numerator must be a first degree polynomial and not of higher degree, since otherwise the fraction would contain a polynomial, which should have been removed.

201. For the sake of brevity, we let, M be the numerator $A + Bz + Cz^2 + \cdots$, and we let Z be the other factor of the denominator, $\alpha + \beta z + \gamma z^2 + \cdots$. We let the rational function which comes from the factor Z be $\frac{Y}{Z}$, where $Y = \frac{M - PZ - QZz}{p^2 - 2pqz \cos \phi + q^2z^2}$. Since Y should be a polynomial, it is necessary that $M - PZ - QZz$ be divisible by $p^2 - 2pqz \cos \phi + q^2z^2$. It follows that $M - PZ - QZz$ vanishes if we let $p^2 - 2pqz \cos \phi + q^2z^2 = 0$, that is when we let $z = \frac{p}{q}(\cos \phi + i \sin \phi)$ or when $z = \frac{p}{q}(\cos \phi - i \sin \phi)$. We let $\frac{p}{q} = f$, then $z^n = f^n(\cos n\phi \pm i \sin n\phi)$. When we make the two substitutions for z we obtain two equations from which we can find the two unknown quantities P and Q.

202. After we make the two substitutions in the equation $M = PZ + QZz$ we obtain the following two equations:

$$A + Bf \cos \phi + Cf^2 \cos 2\phi + Df^3 \cos 3\phi + \cdots$$
$$\pm (Bf \sin \phi + Cf^2 \sin 2\phi + Df^3 \sin 3\phi + \cdots)i$$
$$= P(\alpha + \beta f \cos \phi + \gamma f^2 \cos 2\phi + \delta f^3 \cos 3\phi + \cdots)$$
$$\pm P(\beta f \sin \phi + \gamma f^2 \sin 2\phi + \delta f^3 \sin 3\phi + \cdots)i$$

$$+ Q(\alpha f \cos\phi + \beta f^2\cos 2\phi + \gamma f^3\cos 3\phi + \cdots)$$

$$\pm Q(\alpha f \sin\phi + \beta f^2\sin 2\phi + \gamma f^3\sin 3\phi + \cdots)i.$$

For the sake of brevity in calculation we let

$$A + Bf\cos\phi + Cf^2\cos 2\phi + Df^3\cos 3\phi + \cdots = R$$

$$Bf\sin\phi + Cf^2\sin 2\phi + Df^3\sin 3\phi + \cdots = r$$

$$\alpha + \beta f\cos\phi + \gamma f^2\cos 2\phi + \delta f^3\cos 3\phi + \cdots = S$$

$$\beta f\sin\phi + \gamma f^2\sin 2\phi + \delta f^3\sin 3\phi + \cdots = s$$

$$\alpha f\cos\phi + \beta f^2\cos 2\phi + \gamma f^3\cos 3\phi + \cdots = T$$

$$\alpha f\sin\phi + \beta f^2\sin 2\phi + \gamma f^3\sin 3\phi + \cdots = t.$$

With these substitutions, we obtain the equations

$$R \pm ri = PS \pm Psi + QT \pm Qti.$$

203. Since we have the plus and minus signs, we obtain the equations $R = PS + QT$, $r = Ps + Qt$ and from these we find $P = \frac{Rt - rT}{St - sT}$ and $Q = \frac{Rs - rS}{sT - St}$. It follows that the rational function $\frac{M}{(p^2 - 2pqz\cos\phi + q^2z^2)Z}$ has the partial fraction $\frac{P + Qz}{p^2 - 2pqz\cos\phi + q^2z^2}$ which we calculate by the following rules. Let $f = \frac{p}{q}$ and express each of the terms as follows: when $z^n = f^n\cos n\phi$, then $M = R$, when $z^n = f^n\sin n\phi$, then $M = r$, when $z^n = f^n\cos n\phi$, then $Z = S$, when $z^n = f^n\sin n\phi$, then $Z = s$, when $z^n = f^n\cos n\phi$, then $zZ = T$, when $z^n = f^n\sin n\phi$, then $zZ = t$. When R, r, S, s, T, and t have been found, then $P = \frac{Rt - rT}{St - sT}$ and $Q = \frac{rS - Rs}{St - sT}$.

EXAMPLE I

Let the given rational function be $\dfrac{z^2}{(1 - z + z^2)(1 + z^4)}$. We have to find $\dfrac{P + Qz}{1 - z + z^2}$ which arises from the factor $1 - z + z^2$ in the denominator. In the first place we recall the general form $p^2 - 2pqz \cos \phi + q^2z^2$, which, when compared with the given factor, give $p = 1$, $q = 1$, and $\cos \phi = \dfrac{1}{2}$. From this it follows that $\phi = \dfrac{\pi}{3}$. Since $M = z^2$, $Z = 1 + z^4$, and $f = 1$, we have $R = \cos \dfrac{2}{3}\pi = -\dfrac{1}{2}$, $r = \dfrac{\sqrt{3}}{2}$, $S = 1 + \cos \dfrac{4}{3}\pi = \dfrac{1}{2}$, $s = -\dfrac{\sqrt{3}}{2}$, $T = \cos \dfrac{\pi}{3} + \cos \dfrac{5}{3}\pi = 1$, $t = 0$. From these we find that $P = -1$ and $Q = 0$, so that the partial fraction is $\dfrac{-1}{1 - z + z^2}$ and its complement is $\dfrac{1 + z + z^2}{1 + z^4}$. The real factors of the denominator $1 + z^4$ are $1 + \sqrt{2}z + z^2$ and $1 - \sqrt{2}z + z^2$, so that this rational function can also be expressed in partial fractions, where $\phi = \dfrac{\pi}{4}$ and in one case $f = -1$ and in the other $f = +1$.

EXAMPLE II

Let the given rational function be $\dfrac{1 + z + z^2}{(1 + \sqrt{2}z + z^2)(1 - \sqrt{2}z + z^2)}$, so that $M = 1 + z + z^2$. For the first factor we have $f = -1$, $\phi = \dfrac{\pi}{4}$, and $Z = 1 - z\sqrt{2} + z^2$ so that $R = 1 - \cos \dfrac{\pi}{4} + \cos \dfrac{2\pi}{4} = \dfrac{\sqrt{2} - 1}{\sqrt{2}}$, $r = -\sin \dfrac{\pi}{4} + \sin \dfrac{2\pi}{4} = \dfrac{\sqrt{2} - 1}{\sqrt{2}}$, $S = 1 + \sqrt{2} \cos \dfrac{\pi}{4} + \cos \dfrac{2\pi}{4} = 2$,

$s = \sqrt{2}\sin\frac{\pi}{4} + \sin\frac{2\pi}{4} = 2$, $T = -\cos\frac{\pi}{4} - \sqrt{2}\cos\frac{2\pi}{4} - \cos\frac{3\pi}{4} = 0$,

$t = -\sin\frac{\pi}{4} - \sqrt{2}\sin\frac{2\pi}{4} - \sin\frac{3\pi}{4} = -2\sqrt{2}$. From these we have

$St - sT = -4\sqrt{2}$, $P = \frac{\sqrt{2}-1}{2\sqrt{2}}$, and $Q = 0$. It follows that the partial

fraction which arises from the factor $1 + z\sqrt{2} + z^2$ is $\frac{(\sqrt{2}-1)/2\sqrt{2}}{1+\sqrt{2}z+z^2}$. In a

similar way we find the other partial fraction to be $\frac{(\sqrt{2}+1)/2\sqrt{2}}{1-\sqrt{2}z+z^2}$. Now we see

that the original function $\frac{z^2}{(1-z+z^2)(1+z^4)}$ is expressed as

$$\frac{-1}{1-z+z^2} + \frac{(\sqrt{2}-1)/2\sqrt{2}}{1+\sqrt{2}z+z^2} + \frac{(\sqrt{2}+1)/2\sqrt{2}}{1-\sqrt{2}z+z^2}.$$

EXAMPLE III

Let the given function be $\frac{1+2z+z^2}{\left(-\frac{8}{5}z+z^2\right)\left(1+2z+3z^2\right)}$. From the factor in the denominator $1 - \frac{8}{5}z + z^2$ there will arise a partial fraction

$\frac{P+Qz}{1-\frac{8}{5}z+z^2}$. Here $r = 1$, $s = 1$, and $\cos\phi = \frac{4}{5}$, so that

$M = 1 + 2z + z^2$, $Z = 1 + 2z + 3z^2$. Since this angle is not a fractional part of a right angle, we have to investigate the sine and cosine of its multiples. Since

$\cos\phi = \frac{4}{5}$, we have $\sin\phi = \frac{3}{5}$. Further, $\cos 2\phi = \frac{7}{25}$, $\sin 2\phi = \frac{24}{25}$,

$\cos 3\phi = \frac{-44}{125}$, $\sin 3\phi = \frac{117}{125}$. From these we have

$R = 1 + 2.\frac{4}{5} + \frac{7}{25} = \frac{72}{25}$, $\qquad r = 2.\frac{3}{5} + \frac{24}{25} = \frac{54}{25}$,

$$S = 1 + \frac{2.4}{5} + \frac{3.7}{25} = \frac{86}{25}, \qquad s = 2.\frac{3}{5} + 3.\frac{24}{25} = \frac{102}{25},$$

$$T = \frac{4}{5} + 2.\frac{7}{25} - 3.\frac{44}{125} = \frac{38}{125}, \qquad t = \frac{3}{5} + \frac{2.24}{25} + \frac{3.117}{125} = \frac{666}{125}.$$

It follows that $St - sT = \dfrac{53400}{25 \cdot 125} = \dfrac{2136}{125}$, $P = \dfrac{1836}{2136} = \dfrac{153}{178}$, $Q = -\dfrac{540}{2136} = -\dfrac{45}{178}$, and so the partial fraction arising from $1 - \frac{8}{5}z + z^2$ is $\dfrac{9(17 - 5z)/178}{1 - \frac{8}{5}z + z^2}$. We seek in like manner the partial fraction corresponding to the other factor. Here $r = 1$, $s = -\sqrt{3}$ and $\cos\phi = \dfrac{1}{\sqrt{3}}$, so that $f = -\dfrac{1}{\sqrt{3}}$, $M = 1 + 2z + z^2$ and $Z = 1 - \frac{8}{5}z + z^2$. Since $\cos\phi = \dfrac{1}{\sqrt{3}}$, we have $\sin\phi = \dfrac{\sqrt{2}}{\sqrt{3}}$, $\cos 2\phi = -\dfrac{1}{3}$, $\sin 2\phi = \dfrac{2\sqrt{2}}{3}$, $\cos 3\phi = -\dfrac{5}{3\sqrt{3}}$, $\sin 3\phi = \dfrac{\sqrt{3}}{3\sqrt{3}}$. From this we have

$$R = 1 - \frac{2}{\sqrt{3}}.\frac{1}{\sqrt{3}} + \frac{1}{3}.\left(-\frac{1}{3}\right) = \frac{2}{9}$$

$$r = -\frac{2}{\sqrt{3}}.\frac{\sqrt{2}}{\sqrt{3}} + \frac{1}{3}.\frac{2\sqrt{2}}{3} = -\frac{4\sqrt{2}}{9}$$

$$S = 1 + \frac{8}{5\sqrt{3}}.\frac{1}{\sqrt{3}} + \frac{1}{3}.\left(-\frac{1}{3}\right) = \frac{64}{45}$$

$$s = \frac{8}{5\sqrt{3}}.\frac{\sqrt{2}}{\sqrt{3}} + \frac{1}{3}.\frac{2\sqrt{2}}{3} = \frac{34\sqrt{2}}{45}$$

$$T = -\frac{1}{\sqrt{3}}.\frac{1}{\sqrt{3}} - \frac{8}{5 \cdot 3}.\left(-\frac{1}{3}\right) - \frac{1}{3\sqrt{3}}.\frac{(-5)}{3\sqrt{3}} = \frac{4}{135}$$

$$t = -\frac{1}{\sqrt{3}}.\frac{\sqrt{2}}{\sqrt{3}} - \frac{8}{5 \cdot 3}.\frac{2\sqrt{2}}{3} - \frac{1}{3\sqrt{3}}.\frac{\sqrt{2}}{3\sqrt{3}} = \frac{-98\sqrt{2}}{135}.$$ It then follows

that $St - sT = \frac{-712\sqrt{2}}{675}$, $P = \frac{100}{712} = \frac{25}{178}$, $Q = \frac{540}{712} = \frac{135}{178}$. The given rational function $\frac{1 + 2z + z^2}{\left(1 - \frac{8}{5}z + z^2\right)\left(1 + 2z + 3z^2\right)}$ is expressed by the partial fractions $\frac{9(17 - 5z)/178}{1 - \frac{8}{5}z + z^2} + \frac{5(5 + 27z)/178}{1 + 2z + 3z^2}$.

204. If we know the values of S and s, we can calculate the values of T and t. Since $S = \alpha + \beta f \cos \phi + \gamma f^2 \cos 2\phi + \delta f^3 \cos 3\phi + \cdots$ $s = \beta f \sin \phi + \gamma f^2 \sin 2\phi + \delta f^3 \sin 3\phi + \cdots$ we have

$S \cos \phi - s \sin \phi = \alpha \cos \phi + \beta f \cos 2\phi + \gamma f^2 \cos 3\phi + \cdots$ so that $T = f(S \cos \phi - s \sin \phi)$.

Likewise $S \sin \phi + s \cos \phi = \alpha \sin \phi + \beta f \sin 2\phi + \gamma f^2 \sin 3\phi + \cdots$ so that $t = f(S \sin \phi + s \cos \phi)$. Furthermore $St - sT = (S^2 + s^2)f \sin \phi$ and $Rt - rT = (RS - rs)f \sin \phi + (Rs - rS)f \cos \phi$. It follows that $P = \frac{RS + rs}{S^2 + s^2} + \frac{Rs - rS}{S^2 + s^2} \cdot \frac{\cos \phi}{\sin \phi}$ and $Q = -\frac{Rs + rS}{(S^2 + s^2)f \sin \phi}$. Hence from a factor $p^2 - 2pqz \cos \phi + q^2z^2$ in the denominator there arises a partial fraction with the form $\frac{(RS + rs)f \sin \phi + (Rs - rS)(f \cos \phi - z)}{(p^2 - 2pqz \cos \phi + q^2z^2)(S^2 + s^2)f \sin \phi}$ or since $f = \frac{p}{q}$, we have the form

$$\frac{(RS + rs)p \sin \phi + (Rs - rS)(p \cos \phi - qz)}{(p^2 - 2pqz \cos \phi + q^2z^2)(S^2 + s^2)p \sin \phi}.$$

205. This partial fraction has arisen from the rational function $\frac{M}{(p^2 - 2pqz \cos \phi + q^2z^2)Z}$ with factor $p^2 - 2pqz \cos \phi + q^2z^2$ in the denomi-

nator. The values of R, r, S, and s can be determined in the following way from M and Z. We substitute $\frac{p^n}{q^n}\cos n\phi$ for z^n in M to obtain R and in Z to obtain S. We substitute $\frac{p^n}{q^n}\sin n\phi$ for z^n in M to obtain r and in Z to obtain s. It is to be noted that the functions M and Z should be written as polynomials before the substitutions, so that they have the form

$M = A + Bz + Cz^2 + Dz^3 + Ez^4 + \cdots$ and

$Z = \alpha + \beta z + \gamma z^2 + \delta z^3 + \epsilon z^4 + \cdots$ so that

$$R = A + B\frac{p}{q}\cos\phi + C\frac{p^2}{q^2}\cos 2\phi + D\frac{p^3}{q^3}\cos 3\phi + \cdots$$

$$r = B\frac{p}{q}\sin\phi + C\frac{p^2}{q^2}\sin 2\phi + D\frac{p^3}{q^3}\sin 3\phi + \cdots$$

$$S = \alpha + \beta\frac{p}{q}\cos\phi + \gamma\frac{p^2}{q^2}\cos 2\phi + \delta\frac{p^3}{q^3}\cos 3\phi + \cdots$$

$$s = \beta\frac{p}{q}\sin\phi + \gamma\frac{p^2}{q^2}\sin 2\phi + \delta\frac{p^3}{q^3}\sin 3\phi + \cdots .$$

206. From what we have just done, it should be clear that this procedure will not succeed if the function Z still contains the same factor $p^2 - 2pqz\cos\phi + q^2z^2$. In this case after the substitution $z^n = f^n(\cos n\phi \pm i\sin\phi)$ is made in the equation $M = PZ + QZz$, the quantity Z vanishes and we cannot solve for P and Q. For this reason, if in the denominator of the rational function $\frac{M}{N}$ there is the factor $(p^2 - 2pqz\cos\phi + q^2z^2)^2$ or some higher power, we need a special form for the partial fractions. Let $N = (p^2 - 2pqz\cos\phi + q^2z^2)^2 Z$, then from the factor $(p^2 - 2pqz\cos\phi + q^2z^2)^2$ there arise these two partial fractions,

$\frac{P + Qz}{(p^2 - 2pqz \cos \phi + q^2z^2)^2} + \frac{K + Lz}{p^2 - 2pqz \cos \phi + q^2z^2}$, where P, Q, K, and L are to be determined.

207. With the above hypothesis, the expression $\frac{M - (P + Qz)Z - (K + Lz)Z(p^2 - 2pqz \cos \phi + q^2z^2)}{(p^2 - 2pqz \cos \phi + q^2z^2)^2}$ is a polynomial, so that the numerator is divisible by the denominator. In the first place, the expression $M - PZ - QzZ$ is divisible by $p^2 - 2pqz \cos \phi + q^2z^2$. Since this was true in the previous case, the same method will determine P and Q. Therefore we let $z^n = \frac{p^n}{q^n} \cos n\phi$ in M to obtain R and in Z to obtain N. We let $z^n = \frac{p^n}{q^n} \sin n\phi$ in M to obtain r and in Z to obtain n. Having calculated these values, we now have, by the rule derived above

$$P = \frac{RN + rn}{N^2 + n^2} + \frac{Rn - rN}{N^2 + n^2} \frac{\cos \phi}{\sin \phi} \text{ and } Q = \frac{-Rn + rN}{N^2 + n^2} \frac{q}{p \sin \phi}.$$

208. When we have found P and Q, then $\frac{M - (P + Qz)Z}{p^2 - 2pqz \cos \phi + q^2z^2}$ is a polynomial W, and $W - KZ - LzZ$ is divisible by $p^2 - 2pqz \cos \phi + q^2z^2$. Since this case is now like the first case, we substitute $\frac{p^n}{q^n} \cos n\phi$ for z^n in W to obtain T and $\frac{p^n}{q^n} \sin n\phi$ for z^n in W to obtain t. Then

$$K = \frac{TN + tn}{N^2 + n^2} + \frac{Tn - tN}{N^2 + n^2} \frac{\cos \phi}{\sin \phi} \text{ and } L = \frac{-Tn + tN}{N^2 + n^2} \frac{q}{p \sin \phi}.$$

209. Now we show how to proceed in the general case, when the rational function $\frac{M}{N}$ has a factor $(p^2 - 2pqz \cos \phi + q^2z^2)^k$ in the denominator. Let

$N = (p^2 - 2pqz \cos \phi + q^2 z^2)^k Z$, so that $\dfrac{M}{(p^2 - 2pqz \cos \phi + q^2 z^2)^k Z}$ is to be expressed in partial fractions. The factor $(p^2 - 2pqz \cos \phi + q^2 z^2)^k$ gives rise to the following partial fractions :

$$\frac{U + uz}{(p^2 - 2pqz \cos \phi + q^2 z^2)^k} + \frac{V + vz}{(p^2 - 2pqz \cos \phi + q^2 z^2)^{k-1}}$$

$$+ \frac{W + wz}{(p^2 - 2pqz \cos \phi + q^2 z^2)^{k-2}} + \frac{X + xz}{(p^2 - 2pqz \cos \phi + q^2 z^2)^{k-3}} + \cdots .$$

Now let $z^n = \dfrac{p^n}{q^n} \cos n\phi$ in M to obtain Y and in Z to obtain N. We let $z^n = \dfrac{p^n}{q^n} \sin n\phi$ in M to obtain Y and in Z to obtain n . Then

$U = \dfrac{YN + yn}{N^2 + n^2} + \dfrac{Yn - yN}{N^2 + n^2}\dfrac{\cos \phi}{\sin \phi}$ and $u = -\dfrac{Yn + yN}{N^2 + n^2}\dfrac{q}{p \sin \phi}$. Then we let $\dfrac{M - (U + uz)Z}{p^2 - 2pqz \cos \phi + q^2 z^2}$ be the polynomial F. When we let $z^n = \dfrac{p^n}{q^2} \cos n\phi$ in F we obtain P and when we let $z^n = \dfrac{p^n}{q^n} \sin n\phi$ in F we obtain p, $V = \dfrac{PN + pn}{N^2 + n^2} + \dfrac{Pn - pN}{N^2 + n^2}\dfrac{\cos \phi}{\sin \phi}$, and $v = -\dfrac{Pn + pN}{N^2 + n^2}\dfrac{q}{p \sin \phi}$. Next we let $\dfrac{F - (V + vz)Z}{p^2 - 2pqz \cos \phi + q^2 z^2}$ be the polynomial G. When we let $z^n = \dfrac{p^n}{q^n} \cos n\phi$ in G we obtain S and when we let $z^n = \dfrac{p^n}{q^n} \sin n\phi$ in G we obtain s. Then $W = \dfrac{SN + sn}{N^2 + n^2} + \dfrac{Sn - sN}{N^2 + n^2}\dfrac{\cos \phi}{\sin \phi}$, and $w = -\dfrac{Sn + sN}{N^2 + n^2}\dfrac{q}{p \sin \phi}$. Now we let $\dfrac{S - (W + wz)Z}{p^2 - 2pqz \cos \phi + q^2 z^2}$ be the polynomial H. We let $z^n = \dfrac{p^n}{q^n} \cos n\phi$ in H to obtain R and we let $z^n = \dfrac{p^n}{q^n} \sin n\phi$ in H to obtain r. Then $X = \dfrac{RN + rn}{N^2 + n^2} + \dfrac{Rn - rN}{N^2 + n^2}\dfrac{\cos \phi}{\sin \phi}$

and $x = -\frac{Rn + rN}{N^2 + n^2}\frac{q}{p \sin \phi}$. In this way we proceed until the numerator of the last partial fraction, whose denominator is $p^2 - 2pqz \cos \phi + q^2z^2$, shall have been determined.

EXAMPLE

We consider the rational function $\frac{z - z^3}{(1 + z^2)^4(1 + z^4)}$. From the factor $(1 + z^2)^4$ in the denominator there arise the partial fractions $\frac{U + uz}{(1 + z^2)^4} + \frac{V + vz}{(1 + z^2)^3} + \frac{W + wz}{(1 + z^2)^2} + \frac{X + xz}{1 + z^2}$. When we compare this factor with the general form we find that $p = 1$, $q = 1$, $\cos \phi = 0$, so that $\phi = \frac{1}{2}\pi$. Furthermore $M = z - z^3$ and $Z = 1 + z^4$. It follows that $Y = 0$, $y = 2$, $N = 2$, $n = 0$, and $\sin \phi = 1$. Then $U = -\frac{4}{4} \cdot 0 = 0$, $U = 1$ and $U + uz = Z$. Now the polynomial $F = \frac{z - z^3 - z - z^5}{1 + z^2} = -z^3$ so that $P = 0$, $p = 1$, $V = 0$, and $v = \frac{1}{2}$. Hence $V + vz = \frac{1}{2}z$ and the polynomial

$$G = \frac{-z^3 - \frac{1}{2}z - \frac{1}{2}z^5}{1 + z^2} = -\frac{1}{2}z - \frac{1}{2}z^3.$$

From this it follows that $S = 0$ and $s = 0$, so that $W = 0$, $w = 0$, and the polynomial

$$H = \frac{-\frac{1}{2}z - \frac{1}{2}z^3}{1 + z^2} = -\frac{1}{2}z.$$

Then $R = 0$, $r = -\frac{1}{2}$, $X = 0$, $x = -\frac{1}{4}$. From these calculations we have the required partial fractions $\frac{z}{(1 + z^2)^4} + \frac{z}{2(1 + z^2)^3} - \frac{z}{4(1 + z^2)}$. The numerator of the remaining fraction is $I = \frac{H - (X + xz)Z}{1 + z^2} = -\frac{1}{4}z + \frac{1}{4}z^3$, so that the remaining fraction is

$$\frac{-z+z^3}{4(1+z^4)}.$$

210. This method gives not only the partial fractions but also the complementary fraction, which when added to the partial fractions gives the original rational function. For example, if all of the partial fractions for $\dfrac{M}{(p^2-2pqz\cos\phi+q^2z^2)^k Z}$ arising from the factor $(p^2-2pqz\cos\phi+q^2z^2)^k$ have been found, we have used the sequence of polynomials F, G, H, I, and K if the sequence has to be continued further. It will be the last of these polynomials which will be the numerator of the complimentary fraction, which has Z for a denominator. For example, if $k=1$, then $\dfrac{F}{Z}$ is the complementary fraction; if $k=2$, it will be $\dfrac{G}{Z}$; if $K=3$, it will be $\dfrac{H}{Z}$; and so forth. Once this complementary fraction has been found, this fraction with denominator Z may itself be expressed in partial fractions.

CHAPTER XIII

On Recurrent Series.

211. We here refer to a type of series, called by DeMoivre *recurrent,* which arise from rational functions by actual division. Above we have shown that series of this type have the property that any term is determined by a certain number of preceding terms, according to some fixed law. This law depends on the denominator of the rational function. Now we have a method for expressing a rational function by simpler partial fractions, so that the corresponding recurrent series can be expressed in terms of simpler recurrent series. In this chapter we propose to express any recurrent series in terms of simpler recurrent series.

212. Let $\dfrac{a + bz + cz^2 + dz^3 + \cdots}{1 - \alpha z - \beta z^2 - \gamma z^3 - \delta z^4 - \cdots}$ be a proper rational function which by division is expressed by the recurrent series $A + Bz + Cz^2 + Dz^3 + Ez^4 + Fz^5 + \cdots$. The partial fractions, which we have already learned how to find, give partial series, whose property is easily perceived: all of the partial series taken together must give the original recurrent series.

213. Let the respective recurrent series from the partial fractions be

$a + bz + cz^2 + dz^3 + ez^4 + \cdots$

$a' + b'z + c'z^2 + d'z^3 + e'z^4 + \cdots$

$a'' + b''z + c''z^2 + d''z^3 + e''z^4 + \cdots$

$a''' + b'''z + c'''z^3 + d'''z^3 + e'''z^4 + \cdots$

Since these series taken together must be equal to

$A + Bz + Cz^2 + Dz^3 + Ez^4 + \cdots$, it follows that

$A = a + a' + a'' + a''' + \cdots, \quad B = b + b' + b'' + b''' + \cdots,$

$C = c + c' + c'' + c''' + \cdots, \quad D = d + d' + d'' + d''' + \cdots.$

If we can find the coefficients of z^n in each of the series arising from the partial fractions then their sum will be the coefficient of z^n in the series $A + Bz + Cz^2 + Dz^3 + \cdots$.

214. A doubt could arise as to whether the two series really are equal, that is, whether the coefficients of the same powers of z are equal. We can easily remove this doubt if we recall that equality must hold for every value of z. If we let $z = 0$, then it is clear that $A = a + a' + a'' + a''' + \cdots$. When we have seen this equality, we subtract these terms and divide by z. Then the constant terms will be B and $b + b' + b'' + b''' + \cdots$. From this it follows that $B = b + b' + b'' + b''' + \cdots$, $C = c + c' + c'' + c''' + \cdots$, $D = d + d' + d'' + d''' + \cdots$, etc.

215. Now we examine the series which arise from the partial fractions into which any rational function can be resolved. The first such fraction is $\frac{A}{1 - pz}$ which clearly gives rise to the series $A + Apz + Ap^2z^2 + Ap^3z^3 + \cdots$, whose general term is Ap^nz^n. This expression indeed is customarily called the *general term*, since from it, when n is given successive integral values, all of the terms of the series are produced. Then from $\frac{A}{(1 - pz)^2}$ arises the series

$A + 2Apz + 3Ap^2z^2 + 4Ap^3z^3 + \cdots$, whose general term is $(n+1)Ap^nz^n$.

Also from $\frac{A}{(1-pz)^3}$ we have the series

$A + 3Apz + 6Ap^2z^2 + 10Ap^3z^3 + \cdots$, whose general term is $\frac{(n+1)}{1}\frac{(n+2)}{2}Ap^nz^n$. In general the fraction $\frac{A}{(1-pz)^k}$ gives rise to the series $A + kApz + \frac{k}{1}\frac{(k+1)}{2}Ap^2z^2 + \frac{k}{1}\frac{(k+1)}{2}\frac{(k+2)}{3}Ap^3z^3 + \cdots$, whose general term is $\frac{(n+1)(n+2)(n+3)\cdots(n+k-1)}{1\cdot 2\cdot 3\cdots(k-1)}Ap^nz^n$. From the progression in the series itself we gather that this general term should be equal to $\frac{k(k+1)(k+2)\cdots(k+n-1)}{1\cdot 2\cdot 3\cdots n}Ap^nz^n$. This expression is equal to the other, which is clear from the cross multiplication, which gives the identical equation

$$1\cdot 2\cdot 3\cdots n(n+1)\cdots(n+k-1) = 1\cdot 2\cdot 3\cdots(k-1)k\cdots(k+n-1).$$

216. It follows that as often as we have, in the expression of a rational function, the partial fractions $\frac{A}{(1-pz)^k}$, then we can find the coefficient of each of the powers of z, since it will be the sum of the corresponding general terms of the partial fractions.

EXAMPLE I

Find the general terms of the recurrent series which arises from the rational function $\frac{1-z}{1-z-2z^2}$.

The series which arises from this function is $1 + 0z + 2z^2 + 2z^3 + 6z^4 + 10z^5 + 22z^6 + 42z^7 + 86z^8 + \cdots$. In order to find the coefficient of the general term, we express $\dfrac{1 - z}{1 - z - 2z^2}$ in the form

$\dfrac{\frac{2}{3}}{1 + z} + \dfrac{\frac{1}{3}}{1 - 2z}$. From this we obtain the desired general term

$\left(\frac{2}{3}(-1)^n + \frac{1}{3}2^n\right)z^n = \dfrac{2^n \pm 2}{3}z^n$ where the positive sign is used when n is

even, and the negative sign when n is odd.

EXAMPLE II

Find the general term of the recurrent series which arises from

$$\frac{1 - z}{1 - 5z + 6z^2} = 1 + 4z + 14z^2 + 46z^3 + 146z^4 + 454z^5 + \cdots.$$

Since the denominator is equal to $(1 - 2z)(1 - 3z)$, the function is

expressed as $\dfrac{-1}{1 - 2z} + \dfrac{2}{1 - 3z}$. From this we obtain the general term

$2 \cdot 3^n z^n - 2^n z^n = (2 \cdot 3^n - 2^n)z^n$.

EXAMPLE III

Find the general term of the series

$$1 + 3z + 4z^2 + 7z^3 + 11z^4 + 18z^5 + 29z^6 + 47z^7 + \cdots$$

which arises from $\dfrac{1 + 2z}{1 - z - z^2}$.

Since the factors of the denominator are $1 - \left(\dfrac{1 + \sqrt{5}}{2}\right)z$ and

$1 - \left(\dfrac{1 - \sqrt{5}}{2}\right)z$, the expression in partial fractions is

$\dfrac{\frac{\sqrt{5} + 1}{2}}{1 - \left(\frac{1 + \sqrt{5}}{2}\right)z} + \dfrac{\frac{1 - \sqrt{5}}{2}}{1 - \left(\frac{1 - \sqrt{5}}{2}\right)z}$. From this we obtain the general term

$$\left(\frac{1+\sqrt{5}}{2}\right)^{n+1} z^n + \left(\frac{1-\sqrt{5}}{2}\right)^{n+1} z^n.$$

EXAMPLE IV

Find the general term of the series $z + (\alpha a + b)z + (\alpha^2 a + \alpha b + \beta a)z^2 + (\alpha^3 a + \alpha^2 b + 2\alpha\beta a + \beta b)z^3 + \cdots$ *which arises from the function* $\dfrac{a + bz}{1 - \alpha z - \beta z^2}$.

For the partial fraction expression we have $\dfrac{\dfrac{\left(a\left(\alpha + \sqrt{\alpha^2 + 4\beta}\right) + 2b\right)}{2\sqrt{\alpha^2 + 4\beta}}}{1 - \left(\dfrac{\alpha + \sqrt{\alpha^2 + 4\beta}}{2}\right)z}$

$+ \dfrac{\dfrac{\left(a\left(\sqrt{\alpha^2 + 4\beta} - \alpha\right) - 2b\right)}{2\sqrt{\alpha^2 + 4\beta}}}{1 - \left(\dfrac{\alpha - \sqrt{\alpha^2 + 4\beta}}{2}\right)z}$. Hence the general term is

$\dfrac{a\left(\sqrt{\alpha^2 + 4\beta} + \alpha\right) + 2b}{2\sqrt{\alpha^2 + 4\beta}}\left(\dfrac{\alpha + \sqrt{\alpha^2 + 4\beta}}{2}\right)^n z^n$

$+ \dfrac{a\left(\sqrt{\alpha^2 + 4\beta} - \alpha\right) - 2b}{2\sqrt{\alpha^2 + 4\beta}}\left(\dfrac{\alpha - \sqrt{\alpha^2 + 4\beta}}{2}\right)^n z^n$. From this result it becomes reasonably easy to express the general term of any recurrent series in which each term is determined by the two preceding terms.

EXAMPLE V

To find the general term of the series

$1 + z + 2z^2 + 2z^3 + 3z^4 + 3z^5 + 4z^6 + 4z^7 + \cdots$

which arises from the function

$$\frac{1}{1 - z - z^2 + z^3} = \frac{1}{(1 - z)^2(1 + z)}.$$

Although the law of the progression seems at first quite obvious and in need of no explanation, nevertheless, the partial fraction expression is

$$\frac{\frac{1}{2}}{(1 - z)^2} + \frac{\frac{1}{4}}{1 - z} + \frac{\frac{1}{4}}{1 + z}.$$

From this we obtain the general term

$\frac{1}{2}(n + 1)z^n + \frac{1}{4}z^n + \frac{1}{4}(- 1)^n z^n = \frac{2n + 3 \pm 1}{4}z^n$, where the positive sign is used when n is even, and the negative when n is odd.

217. After having established this result, we note that the general term for any recurrent series can be found, since every rational function can be expressed as partial fractions with denominators which are powers of linear factors. However, if we wish to avoid complex expressions, we will frequently encounter partial fraction of the form

$\frac{A + Bpz}{1 - 2pz \cos \phi + p^2z^2}, \frac{A + Bpz}{(1 - 2pz \cos \phi + p^2z^2)^2}, \ldots,$

$\frac{A + Bpz}{(1 - 2pz \cos \phi + p^2z^2)^k}$, so we have to investigate the series which arise from such fractions. First we note that since

$\cos n \phi = 2 \cos \phi \cos(n - 1)\phi - \cos(n - 2)\phi$, the series corresponding to

$\frac{A}{1 - 2pz \cos \phi + p^2z^2}$ is $A + 2pz \cos \phi + 2Ap^2z^2\cos 2\phi + Ap^2z^2$ $+ 2Ap^3z^3\cos 3\phi + 2Ap^4z^4\cos 4\phi + 2Ap^3z^3\cos \phi + 2Ap^4z^4\cos 2\phi$ $+ Ap^4z^4 + \cdots$. The general term for this series is far from obvious.

218. In order that we may arrive at our goal, we consider the following two

series:

$$Ppz\sin\phi + Pp^2z^2\sin 2\phi + Pp^3z^3\sin 3\phi + Pp^4z^4\sin 4\phi + \cdots$$

$$Q + Qpz\cos\phi + Qp^2z^2\cos 2\phi + Qp^3z^3\cos 3\phi + Qp^4z^4\cos 4\phi + \cdots .$$

These two series arise from the partial fraction whose denominator is $1 - 2pz\cos\phi + p^2z^2$. The first arises from the fraction $\dfrac{Ppz\sin\phi}{1 - 2pz\cos\phi + p^2z^2}$, while the second from $\dfrac{Q - Qpz\cos\phi}{1 - 2pz\cos\phi + p^2z^2}$. When these two fractions are added, the sum is $\dfrac{Q + Ppz\sin\phi - Qpz\cos\phi}{1 - 2pz\cos\phi + p^2z^2}$ and the series arising from the sum has a general term equal to $(P\sin n\phi + Q\cos n\phi)p^nz^n$. If the given partial fraction is $\dfrac{A + Bpz}{1 - 2pz\cos\phi + p^2z^2}$, then $Q = A$, and $P = A\cot\phi + B\csc\phi$. It follows that the general term of the series arising from the given partial fraction is equal to

$$\frac{A\cos\phi\sin n\phi + B\sin n\phi + A\sin\phi\cos n\phi}{\sin\phi}p^nz^n$$
$$= \frac{A\sin(n+1)\phi + B\sin n\phi}{\sin\phi}p^nz^n.$$

219. In order to find the general term when the denominator of the partial fraction is a power, like $(1 - 2pz\cos\phi + p^2z^2)^k$, there is an advantage in expressing the fraction as the sum of two complex fractions $\dfrac{a}{(1 - (\cos\phi + i\sin\phi)pz)^k} + \dfrac{b}{(1 - (\cos\phi - i\sin\phi)pz)^k}$. If we take the sum of the general terms of the series arising from each of the fractions, we obtain

$$\frac{(n+1)}{1}\frac{(n+2)}{2}\frac{(n+3)}{3}\cdots\frac{(n+k-1)}{(k-1)}(\cos n\phi + i\sin n\phi)ap^nz^n$$
$$+ \frac{(n+1)}{1}\frac{(n+2)}{2}\frac{(n+3)}{3}\cdots\frac{(n+k-1)}{(k-1)}(\cos n\phi - i\sin\phi)bp^nz^n.$$ Let

$a + b = f$ and $a - b = \frac{g}{i}$, so that $a = \frac{fi - 1 + g}{2i}$ and $b = \frac{fi - g}{2i}$, then the expression for the general term of the series becomes

$$\frac{(n+1)}{1}\frac{(n+2)}{2}\frac{(n+3)}{3}\cdots\frac{(n+k-1)}{(k-1)}(f \cos n\phi + g \sin n\phi)p^n z^n$$ and

the fractions from which it arose become

$$\frac{\frac{1}{2}f + \frac{1}{2i}g}{(1 - (\cos\phi + i\sin\phi)pz)^k} + \frac{\frac{1}{2}f - \frac{1}{2i}g}{(1 - (\cos\phi - i\sin\phi)pz)^k}.$$ This sum can be expressed as a single fraction whose numerator is

$$f - kfpz\cos\phi + \frac{k}{1}\frac{(k-1)}{2}fp^2z^2\cos 2\phi$$
$$- \frac{k}{1}\frac{(k-1)}{2}\frac{(k-2)}{3}fp^3z^3\cos 3\phi + \cdots$$
$$+ kgpz\sin\phi - \frac{k}{1}\frac{(k-1)}{2}gp^2z^2\sin 2\phi$$
$$+ \frac{k}{1}\frac{(k-1)}{2}\frac{(k-2)}{3}gp^3z^3\sin 3\phi - \cdots$$

and whose denominator is $(1 - 2pz\cos\phi + p^2z^2)^k$

220. If $k = 2$, then the series arising from the fraction $\frac{f - 2pz(f\cos\phi - g\sin\phi) + p^2z^2(f\cos 2\phi - g\sin 2\phi)}{(1 - 2pz\cos\phi + p^2z^2)^2}$ has a general term equal to $(n+1)(f\cos n\phi + g\sin n\phi)p^n z^n$. The series which arises from $\frac{a}{1 - 2pz\cos\phi + p^2z^2}$ or $\frac{a - 2apz\cos\phi + ap^2z^2}{(1 - 2pz\cos\phi + p^2z^2)^2}$ has a general term equal to $\frac{a\sin(n+1)\phi}{\sin\phi}p^n z^n$. When these two fractions are added and we let

$a + f = A$, $\quad 2a\cos\phi + 2f\cos\phi - 2g\sin\phi = -B$, and

$a + f\cos 2\phi - g\sin 2\phi = 0$, then $\quad g = \frac{B + 2A\cos\phi}{2\sin\phi}$,

$a = \frac{A + B\cos\phi}{1 - \cos 2\phi} = \frac{A + B\cos\phi}{2(\sin\phi)^2}$, $\quad f = -\frac{A\cos 2\phi - B\cos\phi}{2(\sin\phi)^2}$, and

$g = \dfrac{B \sin\phi + A(\sin\phi)^2}{2(\sin\phi)^2}$. It follows that the general term of the series which arises from $\dfrac{A + Bpz}{(1 - 2pz\cos\phi + p^2z^2)^2}$ is equal to

$$\frac{A + B\cos\phi}{2(\sin\phi)^3}\sin(n+1)\phi p^n z^n$$

$$+ \frac{(n+1)(B\sin\phi\sin n\phi + A\sin 2\phi\sin n\phi)}{2(\sin\phi)^2}p^n z^n$$

$$+ \frac{(n+1)(-B\cos\phi\cos n\phi - A\cos 2\phi\cos n\phi)}{2(\sin\phi)^2}p^n z^n$$

$$= \frac{-(n+1)(A\cos(n+2)\phi + B\cos(n+1)\phi)p^n z^n}{2(\sin\phi)^2}$$

$$+ \frac{(A + B\cos\phi)\sin(n+1)\phi p^n z^n}{2(\sin\phi)^3}$$

$$= \frac{\left(\frac{1}{2}(n+3)\sin(n+1)\phi - \frac{1}{2}(n+1)\sin(n+3)\phi\right)}{2(\sin\phi)^3}Ap^n z^n$$

$$+ \frac{\left(\frac{1}{2}(n+2)\sin n\phi - \frac{1}{2}n\sin(n+2)\phi\right)}{2(\sin\phi)^3}Bp^n z^n.$$ Hence the general term for the series arising from $\dfrac{A + Bpz}{(1 - 2pz\cos\phi + p^2z^2)^2}$ is

$$\frac{(n+3)\sin(n+1)\phi - (n+1)\sin(n+3)\phi}{4(\sin\phi)^3}Ap^n z^n$$

$$+ \frac{(n+2)\sin n\phi - n\sin(n+2)\phi}{4(\sin\phi)^3}Bp^n z^n.$$

221. Let $k = 3$, then from the fraction

$$\frac{f - 3pz(f\cos\phi - g\sin\phi) + 3p^2z^2(f\cos 2\phi - g\sin 2\phi)}{(1 - 2pz\cos\phi + p^2z^2)^3}$$

$$- \frac{p^3z^3(f\cos 3\phi - g\sin 3\phi)}{(1 - 2pz\cos\phi + p^2z^2)^3},$$ the series which arises has a general term of the

form $\frac{(n+1)}{1}\frac{(n+2)}{2}(f\cos n\phi + g\sin n\phi)p^n z^n$. Further, the series arising from the fraction

$$\frac{a + bpz}{(1 - 2pz\cos\phi + p^2z^2)^2} = \frac{a - 2apz\cos\phi + ap^2z^2 + bpz - 2bp^2z^2\cos\phi + bp^3z^3}{(1 - 2pz\cos\phi + p^2z^2)^3}$$

has a general term of the form

$$\frac{(n+3)\sin(n+1)\phi - (n+1)\sin(n+3)\phi ap^n z^n}{4(\sin\phi)^3}$$

$$+ \frac{(n+2)\sin n\phi - n\sin(n+2)\phi bp^n z^n}{4(\sin\phi)^3}.$$ When these two fractions are added and the numerator is set equal to A, then $a + f = A$,

$3f\cos\phi - 3g\sin\phi + 2a\cos\phi - b = 0,$

$3f\cos 2\phi - 3g\sin 2\phi + a - 2b\cos\phi = 0$, and

$b = f\cos 3\phi - g\sin 3\phi.$

It follows that

$$a = \frac{f\cos 3\phi - g\sin 3\phi - 3f\cos\phi + 3g\sin\phi}{2\cos\phi}$$

$$= 2g(\sin\phi)^2\tan\phi - f - 2f(\sin\phi)^2.$$

We also have

$$\frac{f}{g} = \frac{\sin 5\phi - 2\sin 3\phi + \sin\phi}{\cos 5\phi - 2\cos 3\phi + \cos\phi}$$ and

$a + f = A = 2g(\sin\phi)^2\tan\phi - 2f(\sin\phi)^2$. It follows that

$\frac{A}{2(\sin\phi)^2} = \frac{g\sin\phi - f\cos\phi}{\cos\phi}$. From this, finally, we have

$$f = \frac{A(\sin\phi - 2\sin 3\phi + \sin 5\phi)}{16(\sin\phi)^5} \text{ and } g = \frac{A(\cos\phi - 2\cos 3\phi + \cos 5\phi)}{16(\sin\phi)^5}.$$

Since $16(\sin\phi)^5 = \sin 5\phi - 5\sin 3\phi + 10\sin\phi$, we have

$a = \dfrac{A(9 \sin \phi - 3 \sin 3\phi)}{16(\sin \phi)^5}$ and $b = \dfrac{A(-\sin 2\phi + \sin 2\phi)}{16(\sin \phi)^5} = 0.$ Since $3 \sin \phi - \sin 3\phi = 4(\sin \phi)^3$, we have $a = \dfrac{3A}{4(\sin \phi)^2}$. From this we have the general term equal to

$$\frac{(n+1)}{1}\frac{(n+2)}{2}p^n z^n A \frac{\sin(n+1)\phi - 2\sin(n+3)\phi + \sin(n+5)\phi}{16(\sin\phi)^5}$$

$$+ 3Ap^n z^n \frac{(n+3)\sin(n+1)\phi - (n+1)\sin(n+3)\phi}{16(\sin\phi)^5}$$

$$= \frac{Ap^n z^n}{16(\sin\phi)^5}\left(\frac{(n+4)}{1}\frac{(n+5)}{2}\sin(n+1)\phi\right)$$

$$- \frac{2Ap^n z^n}{16(\sin\phi)^5}\left(\frac{(n+1)}{1}\frac{(n+5)}{2}\sin(n+3)\phi\right)$$

$$+ \frac{Ap^n z^n}{16(\sin\phi)^5}\left(\frac{(n+1)}{1}\frac{(n+2)}{2}\sin(n+5)\phi\right)$$

222. The series which arises from $\dfrac{A + Bpz}{(1 - 2pz\cos\phi + p^2z^2)^3}$ has the general term

$$\frac{Ap^n z^n}{16(\sin\phi)^5}\left(\frac{(n+5)}{1}\frac{(n+4)}{2}\sin(n+1)\phi\right)$$

$$- \frac{Ap^n z^n}{16(\sin\phi)^5}\left(2\frac{(n+1)}{1}\frac{(n+5)}{2}\sin(n+3)\phi\right)$$

$$+ \frac{Ap^n z^n}{16(\sin\phi)^5}\left(\frac{(n+1)}{1}\frac{(n+2)}{2}\sin(n+5)\phi\right)$$

$$+ \frac{Bp^n z^n}{16(\sin\phi)^5}\left(\frac{(n+4)}{1}\frac{(n+3)}{2}\sin n\phi\right)$$

$$- \frac{Bp^n z^n}{16(\sin\phi)^5}\left(2\frac{n}{1}\frac{(n+4)}{2}\sin(n+2)\phi\right)$$

$$+ \frac{Bp^n z^n}{16(\sin\phi)^5}\left(\frac{n}{1}\frac{(n+1)}{2}\sin(n+4)\phi\right).$$ When we take the next step, the

general term of the series which arises from $\frac{A + Bpz}{(1 - 2pz\cos\phi + p^2z^2)^4}$ is equal to

$$\frac{Ap^n z^n}{64(\sin\phi)^7}\left(\frac{(n+7)}{1}\frac{(n+6)}{2}\frac{(n+5)}{3}\sin(n+1)\phi\right)$$

$$-\frac{Ap^n z^n}{64(\sin\phi)^7}\left(3\frac{(n+1)}{1}\frac{(n+7)}{2}\frac{(n+6)}{3}\sin(n+3)\phi\right)$$

$$+\frac{Ap^n z^n}{64(\sin\phi)^7}\left(3\frac{(n+1)}{1}\frac{(n+2)}{2}\frac{(n+7)}{3}\sin(n+5)\phi\right)$$

$$-\frac{Ap^n z^n}{64(\sin\phi)^7}\left(\frac{(n+1)}{1}\frac{(n+2)}{2}\frac{(n+3)}{3}\sin(n+7)\phi\right)$$

$$+\frac{Bp^n z^n}{64(\sin\phi)^7}\left(\frac{(n+6)}{1}\frac{(n+5)}{2}\frac{(n+4)}{3}\sin n\phi\right)$$

$$-\frac{Bp^n z^n}{64(\sin\phi)^7}\left(3\frac{n}{1}\frac{(n+6)}{2}\frac{(n+5)}{3}\sin(n+2)\phi\right)$$

$$+\frac{Bp^n z^n}{64(\sin\phi)^7}\left(3\frac{n}{1}\frac{(n+1)}{2}\frac{(n+6)}{3}\sin(n+4)\phi\right)$$

$$-\frac{Bp^n z^n}{64(\sin\phi)^7}\left(\frac{n}{1}\frac{(n+1)}{2}\frac{(n+2)}{3}\sin(n+6)\phi\right).$$

From these expressions it can easily be understood how to form the general term corresponding to higher powers of the denominator. It is important to note the following identities when discussing these general terms.

$$\sin\phi = \sin\phi$$
$$4(\sin\phi)^3 = 3\sin\phi - \sin 3\phi$$
$$16(\sin\phi)^5 = 10\sin\phi - 5\sin 3\phi + \sin 5\phi$$
$$64(\sin\phi)^7 = 35\sin\phi - 21\sin 3\phi + 7\sin 5\phi - \sin 7\phi$$
$$256(\sin\phi)^9 = 126\sin\phi - 84\sin 3\phi + 36\sin 5\phi - 9\sin 7\phi + \sin 9\phi$$

etc.

223. Now that we are able to find the general term for the series arising from any real partial fraction, we also are able to find the general term of the series of any real rational function, by summing the individual general terms. In order that this become perfectly clear, we add the following examples.

EXAMPLE I

From the fraction

$\frac{1}{(1 - z)(1 - z^2)(1 - z^3)} = \frac{1}{1 - z - z^2 + z^4 + z^5 - z^6}$, there arises the series $1 + z + 2z^2 + 3z^3 + 4z^4 + 5z^5 + 7z^6 + 8z^7 + 10z^8 + 12z^9 + \cdots$. We would like to find the general term of this series. The given function is expressed in terms of the factors of the denominator as $\frac{1}{(1 - z)^3(1 + z)(1 + z + z^2)}$. The expression in terms of partial fractions is

$$\frac{1}{6(1 - z)^3} + \frac{1}{4(1 - z)^2} + \frac{17}{72(1 - z)} + \frac{1}{8(1 + z)} + \frac{2 + z}{9(1 + z + z^2)}.$$

The general term corresponding to the first fraction is

$\frac{(n + 1)}{1}\frac{(n + 2)}{2}\frac{1}{6}z^n = \frac{n^2 + 3n + 2}{12}z^n$. The second fraction, $\frac{1}{4(1 - z)^2}$, gives rise to $\frac{(n + 1)}{4}z^n$; the third, $\frac{17}{72(1 - z)}$, gives $\frac{17}{72}z^n$; the fourth, $\frac{1}{8(1 + z)}$, gives $\frac{1}{8}(-1)^n z^n$; the fifth, $\frac{(2 + z)}{9(1 + z + z^2)}$, when compared to the form $\frac{A + Bz}{1 - 2pz \cos\phi + p^2z^2}$ gives $p = -1$, $\phi = \frac{\pi}{3}$, $A = \frac{2}{9}$, $B = -\frac{1}{9}$.

From section 218 we see that the general term has the form

$$\frac{2\sin(n+1)\phi - \sin n\phi}{9\sin\phi}(-1)^n z^n$$

$$= \frac{4\sin(n+1)\phi - 2\sin n\phi}{9\sqrt{3}}(-1)^n z^n$$

$$= \frac{4\sin(n+1)\frac{\pi}{3} - 2\sin n\frac{\pi}{3}}{9\sqrt{3}}(-1)^n z^n.$$

When we take the sum of all these general terms we have the general term of the series which we are seeking. This general term is

$\left(\frac{n^2}{12} + \frac{n}{2} + \frac{47}{72}\right)z^n \pm \frac{1}{8}z^n \pm \frac{4\sin(n+1)\frac{\pi}{3} - 2\sin n\frac{\pi}{3}}{9\sqrt{3}}z^n$, where the positive sign is used when n is even, and the negative when n is odd. We further note that when n is of the form $3m$, then

$\frac{4\sin\frac{1}{3}(n+1)\pi - 2\sin\frac{1}{3}n\pi}{9\sqrt{3}} = \pm\frac{2}{9}$; when $n = 3m + 1$, then the expression is equal to the negative of $\pm\frac{1}{9}$; when $n = 3m + 2$, then the expression is equal to the negative of $\pm\frac{1}{9}$, depending on whether n is even or odd. From these results we can explain the nature of the series as follows.

If	then the general term is
$n = 6m + 0$	$\left(\frac{n^2}{12} + \frac{n}{2} + 1\right)z^n$
$n = 6m + 1$	$\left(\frac{n^2}{12} + \frac{n}{2} + \frac{5}{12}\right)z^n$
$n = 6m + 2$	$\left(\frac{n^2}{12} + \frac{n}{2} + \frac{2}{3}\right)z^n$
$n = 6m + 3$	$\left(\frac{n^2}{12} + \frac{n}{2} + \frac{3}{4}\right)z^n$

$$n = 6m + 4 \qquad \left(\frac{n^2}{12} + \frac{n}{2} + \frac{2}{3}\right) z^n$$

$$n = 6m + 5 \qquad \left(\frac{n^2}{12} + \frac{n}{2} + \frac{5}{12}\right) z^n.$$

For example, if $n = 50$, then n has the form $6m + 2$, so that the term of the series is equal to $234z^{50}$.

EXAMPLE II

From the rational function $\dfrac{1 + z + z^2}{1 - z - z^4 + z^5}$ there arises the recurrent series

$1 + 2z + 3z^2 + 3z^3 + 4z^4 + 5z^5 + 6z^6 + 6z^7 + 7z^8 + \cdots$, and we would like to find the general term. The given function can be expressed as $\dfrac{1 + z + z^2}{(1 - z)^2(1 + z)(1 + z^2)}$, and this gives rise to the partial fraction expression $\dfrac{3}{4(1 - z)^2} + \dfrac{3}{8(1 - z)} + \dfrac{1}{8(1 + z)} - \dfrac{1 + z}{4(1 + z^2)}$. The first partial fraction $\dfrac{3}{4(1 - z)^2}$ gives the general term $\dfrac{3}{4}(n + 1)z^n$; the second fraction, $\dfrac{3}{8(1 - z)}$, gives $\dfrac{3}{8}z^n$; the third gives $\dfrac{1}{8}(-1)^n z^n$; the fourth, $-\dfrac{1 + z}{4(1 + z^2)}$, when compared to the form $\dfrac{A + Bz}{1 - 2pz \cos \phi + p^2z^2}$ gives

$p = 1$, $\cos \phi = 0$, $\phi = \dfrac{1}{2}\pi$, $A = -\dfrac{1}{4}$, $B = \dfrac{1}{4}$, so that the general term is $\left(-\dfrac{1}{4}\sin \dfrac{1}{2}(n + 1)\pi + \dfrac{1}{4}\sin \dfrac{1}{2}n\pi\right) z^n$. When all of these terms are collected to form the desired general term, we obtain

$\left(\dfrac{3}{4}n + \dfrac{9}{8}\right) z^n \pm \dfrac{1}{8}z^n - \dfrac{1}{4}\left(\sin \dfrac{1}{2}(n + 1)\pi - \sin \dfrac{1}{2}n\pi\right) z^n$. Hence,

if	then the general term is
$n = 4m + 0$	$\left(\frac{3}{4}n + 1\right)z^n$
$n = 4m + 1$	$\left(\frac{3}{4}n + \frac{5}{4}\right)z^n$
$n = 4m + 2$	$\left(\frac{3}{4}n + \frac{3}{2}\right)z^n$
$n = 4m + 3$	$\left(\frac{3}{4}n + \frac{3}{4}\right)z^n.$

Thus, if $n = 50$, then $n = 4m + 2$, so that the term is $39z^{50}$.

224. Suppose a recurrent series is given. Since the rational function from which the series arose is easily recognized, we can find the general term of the series from our previous discussion. From the law of the recurrent series, that is, the law by which each term is defined by its predecessors, we immediately know the denominator of the rational function. The factors of this denominator give the form of the general term, since the coefficients are determined by the numerator alone. Let
$A + Bz + Cz^2 + Dz^3 + Ez^4 + Fz^5 + \cdots$ be a recurrent series. Suppose the law of progression, by which each term is determined by a certain number of its predecessors, gives the denominator $1 - \alpha z - \beta z^2 - \gamma z^3$. Then $D = \alpha C + \beta B + \gamma A$, $E = \alpha D + \beta C + \gamma B$, $F = \alpha E + \beta D + \gamma C$, etc. These multipliers, α, β, and γ were called by De Moivre the *scale of the relation*. The law of progression is contained in the scale of the relation, and the scale of the relation immediately gives us the denominator of the rational function from which the series arose.

225. In order to find the general term, that is, the coefficient of any power z^n, we find the linear factors or quadratic factors, if we wish to avoid complex factors, of the denominator $1 - \alpha z - \beta z^2 - \gamma z^3$. If the factors are all distinct and real, namely, $(1 - pz)(1 - qz)(1 - rz)$, and the function generating the series is expressed as $\frac{A}{1 - pz} + \frac{B}{1 - qz} + \frac{C}{1 - rz}$, then the general term of the series will be $(Ap^n + Bq^n + Cr^n)z^n$. If two of the factors are equal, namely, $q = p$, then the general term will be $((An + B)p^n + Cr^n)z^n$. If $r = q = p$, then the general term will be $(An^2 + Bn + C)p^n z^n$. If the denominator $1 - \alpha z - \beta z^2 - \gamma z^3$ has a quadratic factor, for instance, $(1 - pz)(1 - 2qz \cos \phi + q^2 z^2)$, then the general term will be $\left(Ap^n + \frac{B \sin(n + 1)\phi + C \sin n\phi}{\sin \phi} q^n\right) z^n$. When we let n be equal to three successive integers, 0, 1, and 2, then we obtain A, Bz, Cz^2 and in this we find the values of A, B, and C.

226. If the scale of the relation has only two members, that is, each term is determined by two of its predecessors, so that $C = \alpha B - \beta A$, $D = \alpha C - \beta B$, $E = \alpha D - \beta C$, etc. It is clear that the series is recurrent with the form $A + Bz + Cz^2 + Dz^3 + Ez^4 + \cdots + Pz^n + Qz^{n+1} + \cdots$ which arises from a rational function with denominator $1 - \alpha z + \beta z^2$. If the factors of the denominator are $(1 - pz)(1 - qz)$, then $p + q = \alpha$ and $pq = \beta$. The general term of the series is $(Up^n + Vq^n)z^n$. We let $n = 0$, so that $A = U + V$. When $n = 1$, $B = Up + Vq$, and $Aq - B = U(p - q)$, so that $U = \frac{Aq - B}{q - p}$ and $V = \frac{Ap - B}{p - q}$. When we have found the values of U and

V, then $P = Up^n + Vq^n$ and $Q = Up^{n+1} + Vq^{n+1}$. Furthermore, we have

$$UV = \frac{B^2 - \alpha AB + \beta A^2}{4\beta - \alpha^2}.$$

227. It follows that we can find a method for forming any term of the series from the preceding term alone, although according to the law of progression two are required. Since $P = Up^n + Vq^n$ and $Q = Upp^n + Vqq^n$, we have $Pq - Q = U(q - p)p^n$ and $Pp - Q = V(p - q)q^n$. When these two expressions are multiplied we have $P^2pq - (p + q)PQ + Q^2 = - UV(p - q)^2p^nq^n$. Recall that $p + q = \alpha$, $pq = \beta$, $(p - q)^2 = (p + q)^2 - 4pq = \alpha^2 - 4\beta$, and $p^nq^n = \beta^n$. After we make these substitutions, we obtain $\beta P^2 - \alpha PQ + Q^2 = (\beta A^2 - \alpha AB + B^2)\beta^n$. It follows that

$\dfrac{q^2 - \alpha PQ + \beta P^2}{B^2 - \alpha AB + \beta A^2} = \beta^n$. This is the principal property of the recurrent series, that each term be determined by the two preceding terms. Note that if the term P is known, then

$Q = \dfrac{1}{2}\alpha P + \sqrt{((1/4)\alpha^2 - \beta)P^2 + (B^2 - \alpha AB + \beta A^2)\beta^n}$. This expression appears to contain an irrationality but it is always rational, since no irrational terms occur in the series.

228. Given any two successive terms, Pz^n and Qz^{n+1}, without much work we can find some remote term Xz^{2n}. Let $X = fP^2 + gPQ - hUV$, since $P = Up^n + Vq^n$, $Q = Upp^n + vqq^n$, and $X = Up^{2n} + Vq^n$. It follows then that $fP^2 = fu^2p^{2n} + fV^2q^{2n} + 2fUV\beta^n$,

$gPQ = gu^2pp^{2n} + gV^2qq^{2n} + gUV\alpha\beta^n$, $\quad - hUV = - hUV$, and so

$X = Up^{2n} + Vq^{2n}$. Hence $f + gp = \dfrac{1}{U}$, $f + gq = \dfrac{1}{V}$, and $h = 2f + g\alpha$.

It follows that $g = \frac{V - U}{UV(p - q)}$ and $f = \frac{Up - Vq}{UV(p - q)}$. Since $V - U = \frac{\alpha A - 2B}{p - q}$ and $Up - Vq = \frac{\alpha B - 2A\beta}{p - q}$, we have $f = \frac{\alpha B - 2A\beta}{UV(\alpha^2 - 4\beta)}$ and $g = \frac{\alpha A - 2B}{UV(\alpha^2 - 4\beta)}$ or $f = \frac{2A\beta - \alpha B}{B^2 - \alpha AB + \beta A^2}$ and $g = \frac{2B - \alpha A}{B^2 - \alpha AB + \beta A^2}$. Hence $h = \frac{(4\beta - \alpha^2)A}{B^2 - \alpha AB + \beta A^2}$. Finally we have

$$X = \frac{(2A\beta - \alpha B)P^2 + (2B - \alpha A)PQ - A\beta^n}{B^2 - \alpha AB + \beta A^2}.$$

In a similar way we also have

$$X = \frac{(\alpha\beta A - (\alpha^2 - 2\beta)B)p^2 + (2B - \alpha A)Q^2}{\alpha(B^2 - \alpha AB + \beta A^2)} - \frac{2B\beta^n}{\alpha}.$$

With these two expression for X we eliminate the β^n term to obtain

$$X = \frac{(\beta A - \alpha B)P^2 + 2BPQ - AQ^2}{B^2 - \alpha AB + \beta A^2}.$$

229. In a similar way, if we let the terms have the form $A + Bz + Cz^2 + \cdots + Pz^n + Qz^{n+1} + Rz^{n+2} + \cdots + Xz^{2n} + Yz^{2n+1} + Zz^{2n+2} \cdots$, Then

$$Z = \frac{(\beta A - \alpha B)Q^2 + 2BQR - AR^2}{B^2 - \alpha AB + \beta A^2},$$ and, since $R = \alpha Q - \beta P$,

$$Z = \frac{-\beta^2 AP^2 + 2\beta(A - B)PQ + (\alpha B - (\alpha^2 - \beta)A)Q^2}{B^2 - \alpha AB + \beta A^2}.$$

Since $Z = \alpha Y - \beta X$, it follows that $Y = \frac{Z + \beta X}{\alpha}$, so that $Y = \frac{-\beta BP^2 + 2\beta APQ + \alpha(B - \alpha A)Q^2}{B^2 - \alpha AB + \beta A^2}$. Furthermore, from X and Y, we can define the coefficients of z^{4n} and z^{4n+1} and also those of z^{8n}, z^{8n+1}, and so forth.

EXAMPLE

Let the recurrent series

$1 + 3z + 4z^2 + 7z^3 + 11z^4 + 18z^5 + \cdots + Pz^n + Qz^{n+1} + \cdots$ be such that the coefficient of each term is the sum of the coefficients of the two preceding terms, then the denominator of the rational function giving rise to this function is $1 - z - z^2$. Hence $\alpha = 1$, $\beta = -1$, $A = 1$, and $B = 3$, so that $B^2 - \alpha AB + \beta A^2 = 5$ and

$Q = \dfrac{P + \sqrt{5p^2 + 20(-1)^n}}{2} = \dfrac{P + \sqrt{5p^2 \pm 20}}{2}$ where the sign is positive when n is even and the sign is negative when n is odd. Thus, when $n = 4$, since $P = 11$, we have $Q = \dfrac{11 + \sqrt{5 \cdot 121 + 20}}{2} = \dfrac{11 + 25}{2} = 18$. Furthermore, if X is the coefficient of z^{2n}, then $X = \dfrac{-4P^2 + 6PQ - Q^2}{5}$. It follows that the coefficient of z^8 is equal to $\dfrac{-4 \cdot 121 + 6 \cdot 198 - 324}{5} = 76$. Since $Q = \dfrac{P + \sqrt{5p^2 \pm 20}}{2}$, we have $Q^2 = \dfrac{3P^2 \pm 10 + P\sqrt{5p^2 \pm 20}}{2}$ and $X = \dfrac{-P^2 \pm 2 + P\sqrt{5p^2 \pm 20}}{2}$. We have seen that from any term of this series Pz^n, we can obtain $\dfrac{P + \sqrt{5P^2 \pm 20}}{2} z^{n+1}$ and $\dfrac{-P^2 \pm 2 + P\sqrt{5p^2 \pm 20}}{2} z^{2n}$.

230. Likewise, for a recurrent series in which each term is determined by the three preceding terms, any term can actually be determined by only two preceding terms. Let such a series be

$A + Bz + Cz^2 + Dz^3 + \cdots + Pz^n + Qz^{n+1} + Rz^{n+2} + \cdots$ in which the scale of relation is α, $-\beta$, γ, that is, it arises from a rational function with denominator $1 - \alpha z + \beta z^2 - \gamma z^3$. When the coefficients P, Q, and R are expressed in terms of the factors of this denominator, say $(1 - pz)(1 - qz)(1 - rz)$, then $P = Up^n + Vq^n + Wr^n$, $Q = Upp^n + Vqq^n + Wrr^n$, and $R = Up^2p^n + Vq^2q^n + Wr^2r^n$. Since $p + q + r = \alpha$, $pq + pr + qr = \beta$ and $pqr = \gamma$, we have the following equation :

$$r^3 + (-2\alpha Q + \beta P)R^2 + ((\alpha^2 + \beta)Q^2 - (\alpha\beta + 3\gamma)PQ + \alpha\gamma P^2)R$$
$$+ (-(\alpha\beta - \gamma)Q^3 + (\alpha\gamma + \beta^2)PQ^2 - 2\beta\gamma P^2 Q + \gamma^2 P^3)$$
$$= \gamma^n(W^3 + (-2\alpha V + \beta U)W^2 + ((\alpha^3 + \beta)V^2 - (\alpha\beta + 3\gamma)UV + \alpha\gamma U^2)W$$
$$+ (-(\alpha\beta - \gamma)V^3 + (\alpha\gamma + \beta^2)UV^2 - 2\beta\gamma U^2 V + \gamma^2 U^3)).$$

It follows that the value of R can be found as a solution of a cubic equation in the two preceding terms P and Q.

231. Now that we have discussed the general term of a recurrent series, it remains to investigate the sum of such a series. The first thing to note is that the sum is equal to the rational function from which it arose. The denominator of that function is clear from the law of progression, so we have only to define the numerator. Let such a series be

$A + Bz + Cz^2 + Dz^3 + Ez^4 + Fz^5 + Gz^6 + \cdots$, with a law of progression which yields $1 - \alpha z + \beta z^2 - \gamma z^3 + \delta z^4$ as a denominator. We take the sum of the infinite series to be equal to $\dfrac{a + bz + cz^2 + dz^3}{1 - \alpha z + \beta z^2 - \gamma z^3 + \delta z^4}$. Since the proposed series is supposed to arise from this function, we see that

$$a = A$$
$$b = B - \alpha A$$
$$c = C - \alpha B + \beta A$$
$$d = D - \alpha C + \beta B - \gamma A.$$

It follows that the desired sum is equal to

$$\frac{A + (B - \alpha A)z + (C - \alpha B + \beta A)z^2 + (D - \alpha C + \beta B - \gamma A)z^3}{1 - \alpha z + \beta z^2 - \gamma z^3 + \delta z^4}.$$

232. With this result it is not difficult to understand how to find the sum of the series only up to a certain term. That is, we would like to find the sum up to only the term Pz^n. We let $s = A + Bz + Cz^3 + Dz^3 + Ez^4 + \cdots Pz^n$. Since we know the sum of the infinite series, we seek the sum of the infinite number of terms following Pz^n, and let $t = Qz^{n+1} + Rz^{n+2} + Sz^{n+3} + Tz^{n+4} + \cdots$. When this series is divided by z^{n+1}, we have a recurrent series like the original series, so that

$$t = \frac{Qz^{n+1} + (R - \alpha Q)z^{n+2} + (S - \alpha R + \beta Q)z^{n+3}}{1 - \alpha z + \beta z^2 - \gamma z^3 + \delta z^4}$$

$$+ \frac{(T - \alpha S + \beta R - \gamma Q)z^{n+4}}{1 - \alpha z + \beta z^2 - \gamma z^3 + \delta z^4}.$$ It follows that the desired sum S is equal to

$$\frac{A + (B - \alpha A)z + (C - \alpha B + \beta A)z^2 + (D - \alpha C + \beta B - \delta A)z^3}{1 - \alpha z + \beta z^2 - \gamma z^3 + \delta z^4}$$

$$- \frac{(R - \alpha Q)z^{n+2} - (S - \alpha R + \beta Q)z^{n+3} - (T - \alpha S + \beta R - \gamma Q)z^{n+4}}{1 - \alpha z + \beta z^2 - \gamma z^3 + \delta z^4}.$$

233. If the scale of the relation has two members, α, and $-\beta$, then the series $A + Bz + Cz^2 + Dz^3 + \cdots + Pz^n$, which arises from the fraction $\frac{A + (B - \alpha A)z}{1 - \alpha z + \beta z^2}$ has a sum equal to

$$\frac{A + (B - \alpha A)z - Qz^{n+1} - (R - \alpha Q)z^{n+2}}{1 - \alpha z + \beta z^2}.$$ Since from the nature of the

series, we have $R = \alpha Q - \beta P$, we can write the sum in the form

$$\frac{A + (B - \alpha A)z - Qz^{n+1} + \beta Pz^{n+2}}{1 - \alpha z + \beta z^2}.$$

EXAMPLE

Let the given series be $1 + 3z + 4z^2 + 7z^3 + \cdots + Pz^n$, where $\alpha = 1, \beta = -1, A = 1, B = 3$. Then the sum is $\dfrac{1 + 2z - Qz^{n+1} - Pz^{n+2}}{1 - z - z^2}$.

If we let $z = 1$, then the sum of the series is equal to

$1 + 3 + 4 + 7 + 11 + \cdots + P = P + Q - 3$. Since

$Q = \dfrac{P + \sqrt{5p^2 \pm 20}}{2}$, the sum of the series

$1 + 3 + 4 + 7 + 11 + \cdots + P$ is equal to $\dfrac{3P - 6 + \sqrt{5p^2 \pm 20}}{2}$. It follows that the sum can be expressed in terms of the last term alone.

CHAPTER XIV

On the Multiplication and Division of Angles.

234. Let z be any angle or arc of a unit circle. Let its sine be x, its cosine be y, and its tangent be t. Then we have $x^2 + y^2 = 1$, and $t = \frac{x}{y}$. As we have seen above, both the sine and cosine of the angles $z, 2z, 3z, 4z, 5z, \cdots$ form a recurrent series with a scale of relation given by $2y, -1$. First we consider the sine of these angles.

$$\begin{aligned}
\sin 0z &= 0\\
\sin 1z &= x\\
\sin 2z &= 2xy\\
\sin 3z &= 4xy^2 - x\\
\sin 4z &= 8xy^3 - 4xy\\
\sin 5z &= 16xy^4 - 12xy^2 + x\\
\sin 6z &= 32xy^5 - 32xy^3 + 6xy\\
\sin 7z &= 64xy^6 - 80xy^4 + 24xy^2 - x\\
\sin 8z &= 128xy^7 - 192xy^5 + 80xy^3 - 8xy.
\end{aligned}$$

We conclude that $\sin nz = x\left(2^{n-1}y^{n-1} - (n-2)2^{n-3}y^{n-3}\right)$

$$+ x\left(\frac{(n-3)}{1}\frac{(n-4)}{2}2^{n-5}y^{n-5} - \frac{(n-4)}{1}\frac{(n-5)}{2}\frac{(n-6)}{3}2^{n-7}y^{n-7}\right)$$

$$+ x\left(\frac{(n-5)}{1}\frac{(n-6)}{2}\frac{(n-7)}{3}\frac{(n-8)}{4}2^{n-9}y^{n-9} - \cdots\right)$$

235. If we let the arc $nz = s$, then

$\sin nz = \sin s = \sin(\pi - s) = \sin(2\pi + s) = \sin(3\pi - s)$ etc. These sines are all equal to each other. Now we obtain different values for x, such as $\sin\frac{s}{n}$, $\sin\frac{\pi - s}{n}$, $\sin\frac{2\pi + s}{n}$, $\sin\frac{3\pi - s}{n}$, $\sin\frac{4\pi + s}{n}$, etc., although each of the arcs is related to an arc in the above equation. There will be as many different values of x as there are units in the integer n. Each value of x will be a root of one of the equations given. We must be wary, lest we expect the same values from similar expressions. Since we know the roots of the equation after the fact, a comparison of the terms of the equation will give properties worthy of notice. In order to have an equation in which only x appears, we substitute $\sqrt{1 - x^2}$ for y, and we will have different operations depending on whether n is even or odd.

236. Let n be an odd number. Since the arcs $-z$, z, $3z$, $5z$ are different from the case $2z$, whose sine is $1 - 2x^2$, the scale of the relation in the sine progression will be $2 - 4x^2$ and -1. Hence we have

$$\sin(-z) = -x$$
$$\sin z = x$$
$$\sin 3z = 3x - 4x^3$$
$$\sin 5z = 5x - 20x^3 + 16x^5$$
$$\sin 7z = 7x - 56x^3 + 112x^5 - 64x^7$$
$$\sin 9z = 9x - 120x^3 + 432x^5 - 576x^7 + 256x^9.$$

It follows that

$$\sin nz = nx - \frac{n(n^2-1)}{1\cdot 2\cdot 3}x^3 + \frac{n(n^2-1)(n^2-9)}{1\cdot 2\cdot 3\cdot 4\cdot 5}x^5$$
$$- \frac{n(n^2-1)(n^2-9)(n^2-25)}{1\cdot 2\cdot 3\cdot 4\cdot 5\cdot 6\cdot 7}x^7 + \cdots,$$ provided n is an odd number.

The roots of this equation are $\sin z$, $\sin\left(\frac{2\pi}{n} + z\right)$, $\sin\left(\frac{4\pi}{n} + z\right)$,

$\sin\left(\frac{6\pi}{n} + z\right)$, $\sin\left(\frac{8\pi}{n} + z\right)$, etc., where there are n roots.

237. From the above equation we obtain $0 = 1 - \frac{nx}{\sin nz} + \frac{n(n^2 - 1)}{1\cdot 2\cdot 3 \sin nz}x^3$ $- \frac{n(n^2 - 1)(n^2 - 9)}{1\cdot 2\cdot 3\cdot 4\cdot 5 \sin nz}x^5 + \cdots \pm \frac{2^{n-1}}{\sin nz}x^n$, where the positive sign is used if n is one less than a multiple of four, otherwise, the negative sign is used. The factors are

$$\left(1 - \frac{x}{\sin z}\right)\left(1 - \frac{x}{\sin\left(\frac{2\pi}{n} + z\right)}\right)\left(1 - \frac{x}{\sin\left(\frac{4\pi}{n} + z\right)}\right)\cdots,$$

from which we conclude that

$$\frac{n}{\sin nz} = \frac{1}{\sin z} + \frac{1}{\sin\left(\frac{2\pi}{n} + z\right)} + \frac{1}{\sin\left(\frac{4\pi}{n} + z\right)}$$

$$+ \frac{1}{\sin\left(\frac{6\pi}{n} + z\right)} + \cdots,$$

until there are n terms. The product of all these terms is

$$\pm \frac{2^{n-1}}{\sin nz} = \frac{1}{\sin z \sin\left(\frac{2\pi}{n} + z\right)\sin\left(\frac{4\pi}{n} + z\right)\sin\left(\frac{6\pi}{n} + z\right)\cdots},$$

that is,

$$\sin nz = \pm 2^{n-1}\sin z \sin\left(\frac{2\pi}{n} + z\right)\sin\left(\frac{4\pi}{n} + z\right)\sin\left(\frac{6\pi}{n} + z\right)\cdots.$$

Since the penultimate term is lacking, we have

$$0 = \sin z + \sin\left(\frac{2\pi}{n} + z\right) + \sin\left(\frac{4\pi}{n} + z\right) + \sin\left(\frac{6\pi}{n} + z\right) + \cdots.$$

EXAMPLE I

If $n = 3$, then we have the equations

$$0 = \sin z + \sin\left(\frac{2\pi}{3} + z\right) + \sin\left(\frac{4\pi}{3} + z\right)$$

$$= \sin z + \sin\left(\frac{\pi}{3} - z\right) - \sin\left(\frac{\pi}{3} + z\right),$$

$$\frac{3}{\sin 3z} = \frac{1}{\sin z} + \frac{1}{\sin\left(\frac{2\pi}{3} + z\right)} + \frac{1}{\sin\left(\frac{4\pi}{3} + z\right)}$$

$$= \frac{1}{\sin z} + \frac{1}{\sin\left(\frac{\pi}{3} - z\right)} - \frac{1}{\sin\left(\frac{\pi}{3} + z\right)},$$

$$\sin 3z = -4 \sin z \sin\left(\frac{2\pi}{3} + z\right)\sin\left(\frac{4\pi}{3} + z\right)$$

$= 4 \sin z \sin\left(\frac{\pi}{3} - z\right)\sin\left(\frac{\pi}{3} + z\right)$. As we have noted above, we also have

$\sin\left(\frac{\pi}{3} + z\right) = \sin z + \sin\left(\frac{\pi}{3} - z\right)$ and

$$3 \csc 3z = \csc z + \csc\left(\frac{\pi}{3} - z\right) - \csc\left(\frac{\pi}{3} + z\right).$$

EXAMPLE II

We let $n = 5$, and we have the equations

$$0 = \sin z + \sin\left(\frac{2}{5}\pi + z\right) + \sin\left(\frac{4}{5}\pi + z\right)$$

$+ \sin\left(\frac{6}{5}\pi + z\right) + \sin\left(\frac{8}{5}\pi + z\right)$, or

$$0 = \sin z + \sin\left(\frac{2}{5}\pi + z\right) + \sin\left(\frac{1}{5}\pi - z\right)$$

$- \sin\left(\frac{1}{5}\pi + z\right) - \sin\left(\frac{2}{5}\pi - z\right),$ or

$$0 = \sin z + \sin\left(\frac{1}{5}\pi - z\right) - \sin\left(\frac{1}{5}\pi + z\right)$$

$+ \sin\left(\frac{2}{5}\pi + z\right) - \sin\left(\frac{2}{5}\pi - z\right)$. We also have

$$\frac{5}{\sin 5z} = \frac{1}{\sin z} + \frac{1}{\sin\left(\frac{1}{5}\pi - z\right)} - \frac{1}{\sin\left(\frac{1}{5}\pi + z\right)}$$

$- \frac{1}{\sin\left(\frac{2}{5}\pi - z\right)} + \frac{1}{\sin\left(\frac{2}{5}\pi + z\right)}$, and

$$\sin 5z = 16 \sin z \sin\left(\frac{1}{5}\pi - z\right)\sin\left(\frac{1}{5}\pi + z\right) \sin\left(\frac{2}{5}\pi - z\right)\sin\left(\frac{2}{5}\pi + z\right).$$

EXAMPLE III

If we let $n = 2m + 1$, then

$$0 = \sin z + \sin\left(\frac{\pi}{n} - z\right) - \sin\left(\frac{\pi}{n} + z\right) - \sin\left(\frac{2}{n}\pi - z\right)$$

$$+ \sin\left(\frac{2}{n}\pi + z\right) + \sin\left(\frac{3}{n}\pi - z\right) - \sin\left(\frac{3}{n}\pi + z\right) - \cdots$$

$\pm \sin\left(\frac{m}{n}\pi - z\right)$, where the upper sign is used when m is odd, and the lower sign when m is even. There is another equation:

$$\frac{n}{\sin nz} = \frac{1}{\sin z} + \frac{1}{\sin\left(\frac{\pi}{n} - z\right)} - \frac{1}{\sin\left(\frac{\pi}{n} + z\right)}$$

$$- \frac{1}{\sin\left(\frac{2}{n}\pi - z\right)} + \frac{1}{\sin\left(\frac{2}{n}\pi + z\right)} + \frac{1}{\sin\left(\frac{3}{n}\pi - z\right)}$$

$$- \frac{1}{\sin\left(\frac{3}{n}\pi + z\right)} - \cdots \pm \frac{1}{\sin\left(\frac{m}{n}\pi + z\right)},$$ and this equation is easily translated into cosecants. In the third place we have the product:

$$\sin nz = 2^{2m} \sin z \sin\left(\frac{\pi}{n} - z\right) \sin\left(\frac{\pi}{n} + z\right) \sin\left(\frac{2}{n}\pi - z\right) \sin\left(\frac{2}{n}\pi + z\right)$$

$$\sin\left(\frac{3}{n}\pi - z\right) \sin\left(\frac{3}{n}\pi + z\right) \cdots \sin\left(\frac{m}{n}\pi - z\right) \sin\left(\frac{m}{n}\pi + z\right).$$

238. Now suppose n is an even power. Since $y = \sqrt{1 - x^2}$ and $\cos 2z = 1 - 2x^2$, just as before, the scale of the relation is $2 - 4x^2$, and -1, so that

$$\sin 0z = 0$$

$$\sin 2z = 2x\sqrt{1 - x^2}$$

$$\sin 4z = (4x - 8x^3)\sqrt{1 - x^2}$$

$$\sin 6z = (6x - 32x^3 + 32x^5)\sqrt{1 - x^2}$$

$$\sin 8z = (8x - 80x^3 + 192x^5 - 128x^7)\sqrt{1 - x^2}$$

and in general

$$\sin nz = \left(nx - \frac{n(n^2 - 4)}{1 \cdot 2 \cdot 3}x^3 + \frac{n(n^2 - 4)(n^2 - 16)}{1 \cdot 2 \cdot 3 \cdot 4 \cdot 5}x^5\right)\sqrt{1 - x^2}$$

$$+ \left(- \frac{n(n^2 - 4)(n^2 - 16)(n^2 - 36)}{1 \cdot 2 \cdot 3 \cdot 4 \cdot 5 \cdot 6 \cdot 7}x^7 + \cdots \pm 2^{n-1}x^{n-1}\right)\sqrt{1 - x^2},$$

where n is any even number.

239. In order to remove the radical from the equation we square both sides of the equation to obtain $(\sin nz)^2 = n^2x^2 + Px^4 + Qx^6 + \cdots - 2^{2n-2}x^{2n}$ or $x^{2n} \cdots - \frac{n^2}{2^{2n-2}}x^2 + \frac{1}{2^{2n-2}}(\sin nz)^2 = 0$. The roots of this equation are both

negative and positive, that is,

$$\sin nz = \pm 2^{n-1}\sin z \sin\left(\frac{\pi}{n} - z\right)\sin\left(\frac{2}{n}\pi + z\right)\sin\left(\frac{3}{n}\pi - z\right) \cdots \text{, where}$$

either sign could be used, depending on the particular case.

EXAMPLE

When we substitute successively the numbers 2, 4, 6, etc. we obtain n different sines.

$$\sin 2z = 2 \sin z \sin\left(\frac{\pi}{2} - z\right)$$

$$\sin 4z = 8 \sin z \sin\left(\frac{\pi}{4} - z\right)\sin\left(\frac{\pi}{4} + z\right)\sin\left(\frac{\pi}{2} - z\right)$$

$$\sin 6z = 32 \sin z \sin\left(\frac{\pi}{6} - z\right)\sin\left(\frac{\pi}{6} + z\right)\sin\left(\frac{2}{6}\pi - z\right)$$

$$\sin\left(\frac{2}{6}\pi + z\right)\sin\left(\frac{3}{6}\pi - z\right)$$

$$\sin 8z = 128 \sin z \sin\left(\frac{\pi}{8} - z\right)\sin\left(\frac{\pi}{8} + z\right)\sin\left(\frac{2}{8}\pi - z\right)$$

$$\sin\left(\frac{2}{8}\pi + z\right)\sin\left(\frac{3}{8}\pi - z\right)\sin\left(\frac{3}{8}\pi + z\right)\sin\left(\frac{4}{8}\pi - z\right)$$

240. From the above it is clear that in general we have

$$\sin nz = 2^{n-1}\sin z \sin\left(\frac{\pi}{n} - z\right)\sin\left(\frac{\pi}{n} + z\right)\sin\left(\frac{2}{n}\pi - z\right)$$

$$\sin\left(\frac{2}{n}\pi + z\right)\sin\left(\frac{3}{n}\pi - z\right)\sin\left(\frac{3}{n}\pi + z\right) \cdots \sin\left(\frac{\pi}{2} - z\right)\text{, provided } n \text{ is even.}$$

If this result is compared with what we obtained above when n is odd, the results are so similar, that we can use one formula in either case. Whether n is odd or even, we have

$$\sin nz = 2^{n-1}\sin z \sin\left(\frac{\pi}{n} - z\right)\sin\left(\frac{\pi}{n} + z\right)\sin\left(\frac{2}{n}\pi - z\right)$$

$$\sin\left(\frac{2}{n}\pi + z\right)\sin\left(\frac{3}{n}\pi - z\right)\sin\left(\frac{3}{n}\pi + z\right)\cdots$$, until the number of factors is equal to n.

241. These formulas by which the sine of multiples of an angle are expressed as factors are useful, not only for finding the logarithms of sines of multiples of an angle, but also for other expressions for the sine by factors, similar to those given above in section 184. We now have

$$\sin z = 1 \sin z$$

$$\sin 2z = 2 \sin z \sin\left(\frac{\pi}{2} - z\right)$$

$$\sin 3z = 4 \sin z \sin\left(\frac{\pi}{3} - z\right)\sin\left(\frac{\pi}{3} + z\right)$$

$$\sin 4z = 8 \sin z \sin\left(\frac{\pi}{4} - z\right)\sin\left(\frac{\pi}{4} + z\right)\sin\left(\frac{2}{4}\pi - z\right)$$

$$\sin 5z = 16 \sin z \sin\left(\frac{\pi}{5} - z\right)\sin\left(\frac{\pi}{5} + z\right)\sin\left(\frac{2}{5}\pi - z\right)$$

$$\sin\left(\frac{2}{5}\pi + z\right)$$

$$\sin 6z = 32 \sin z \sin\left(\frac{\pi}{6} - z\right)\sin\left(\frac{\pi}{6} + z\right)\sin\left(\frac{2}{6}\pi - z\right)$$

$$\sin\left(\frac{2}{6}\pi + z\right)\sin\left(\frac{3}{6}\pi - z\right)$$

242. Since $\dfrac{\sin 2nz}{\sin nz} = 2 \cos nz$, we can also express the cosine of multiple angles by factors.

$$\cos z = 1 \sin\left(\frac{\pi}{2} - z\right)$$

$$\cos 2z = 2\sin\left(\frac{\pi}{4} - z\right)\sin\left(\frac{\pi}{4} + z\right)$$

$$\cos 3z = 4\sin\left(\frac{\pi}{6} - z\right)\sin\left(\frac{\pi}{6} + z\right)\sin\left(\frac{3}{6}\pi - z\right)$$

$$\cos 4z = 8\sin\left(\frac{\pi}{8} - z\right)\sin\left(\frac{\pi}{8} + z\right)\sin\left(\frac{3}{8}\pi - z\right)\sin\left(\frac{3}{8}\pi + z\right)$$

$$\cos 5z = 16\sin\left(\frac{\pi}{10} - z\right)\sin\left(\frac{\pi}{10} + z\right)\sin\left(\frac{3}{10}\pi - z\right)$$

$$\sin\left(\frac{3}{10}\pi + z\right)\sin\left(\frac{5}{10}\pi - z\right).$$

In general we have

$$\cos nz = 2^{n-1}\sin\left(\frac{\pi}{2n} - z\right)\sin\left(\frac{\pi}{2n} + z\right)\sin\left(\frac{3}{2n}\pi - z\right)$$

$$\sin\left(\frac{3}{2n}\pi + z\right)\sin\left(\frac{5}{2n}\pi - z\right)$$, where there are n factors.

243. The same formulas can be derived from a consideration of the cosine of multiple angles. If we let $\cos z = y$, then

$$\begin{aligned}
\cos 0z &= 1 \\
\cos 1z &= y \\
\cos 2z &= 2y^2 - 1 \\
\cos 3z &= 4y^3 - 3y \\
\cos 4z &= 8y^4 - 8y^2 + 1 \\
\cos 5z &= 16y^5 - 20y^3 + 5y \\
\cos 6z &= 32y^6 - 48y^4 + 18y^2 - 1 \\
\cos 7z &= 64y^7 - 112y^5 + 56y^3 - 7y
\end{aligned}$$

and in general,

$$\cos nz = 2^{n-1}y^n - \frac{n}{1}2^{n-3}y^{n-2} + \frac{n}{1}\frac{(n-3)}{2}2^{n-5}y^{n-4}$$

$$- \frac{n}{1}\frac{(n-4)}{2}\frac{(n-5)}{3}2^{n-7}y^{n-6}$$

$$+ \frac{n}{1}\frac{(n-5)}{2}\frac{(n-6)}{3}\frac{(n-7)}{4}2^{n-9}y^{n-8} - \cdots$$. Since

$\cos nz = \cos(2\pi - nz) = \cos(2\pi + nz) = \cos(4\pi \pm nz) = \cos(6\pi \pm nz)$ etc., the roots of the above equation are

$\cos z$, $\cos\left(\frac{2}{n}\pi \pm z\right)$, $\cos\left(\frac{4}{n}\pi \pm z\right)$, $\cos\left(\frac{6}{n}\pi \pm z\right)$, etc. These values will be different for as many expressions as there are roots, but there are n roots.

244. First we note that, with the exception when $n = 1$, since the second term is lacking, the sum of all these roots is equal to zero. That is

$$0 = \cos z + \cos\left(\frac{2}{n}\pi - z\right) + \cos\left(\frac{2}{n}\pi + z\right) + \cos\left(\frac{4}{n}\pi - z\right)$$

$$+ \cos\left(\frac{4}{n}\pi + z\right) + \cdots,$$ where there are n terms in the sum. This equality is immediate when n is even, since for each positive term there is a corresponding equal negative term. Hence we consider the case when n is odd, with the exception $n = 1$. Since $\cos v = -\cos(\pi - v)$, we have

$$0 = \cos z - \cos\left(\frac{\pi}{3} - z\right) - \cos\left(\frac{\pi}{3} + z\right)$$

$$0 = \cos z - \cos\left(\frac{\pi}{5} - z\right) - \cos\left(\frac{\pi}{5} + z\right) + \cos\left(\frac{2}{5}\pi - z\right) + \cos\left(\frac{2}{5}\pi + z\right)$$

$$0 = \cos z - \cos\left(\frac{\pi}{7} - z\right) - \cos\left(\frac{\pi}{7} + z\right) + \cos\left(\frac{2}{7}\pi - z\right)$$

$$+ \cos\left(\frac{2}{7}\pi + z\right) - \cos\left(\frac{3}{7}\pi - z\right) - \cos\left(\frac{3}{7}\pi + z\right).$$ In general we have, when n is any odd number,

$$0 = \cos z - \cos\left(\frac{\pi}{n} - z\right) - \cos\left(\frac{\pi}{n} + z\right) + \cos\left(\frac{2}{n}\pi - z\right)$$

$$+ \cos\left(\frac{2}{n}\pi + z\right) - \cos\left(\frac{3}{n}\pi - z\right) - \cos\left(\frac{3}{n}\pi + z\right)$$

$$+ \cos\left(\frac{4}{n}\pi - z\right) + \cos\left(\frac{4}{n}\pi + z\right) - \cdots,$$ with a total of n terms. As we have already remarked, n must be an odd number greater than 1.

245. As far as the product of these terms goes, the formulas depend on whether n is odd, oddly even, or evenly even, but they are all covered by the expressions derived in section 242, where each of the sines was transformed into a cosine. That is,

$$\cos z = 1 \cos z$$

$$\cos 2z = 2 \cos\left(\frac{\pi}{4} + z\right)\cos\left(\frac{\pi}{4} - z\right)$$

$$\cos 3z = 4 \cos\left(\frac{2}{6}\pi + z\right)\cos\left(\frac{2}{6}\pi - z\right)\cos z$$

$$\cos 4z = 8 \cos\left(\frac{3}{8}\pi + z\right)\cos\left(\frac{3}{8}\pi - z\right)\cos\left(\frac{\pi}{8} + z\right)\cos\left(\frac{\pi}{8} - z\right)$$

$$\cos 5z = 16 \cos\left(\frac{4}{8}\pi + z\right)\cos\left(\frac{4}{8}\pi - z\right)\cos\left(\frac{2}{8}\pi + z\right)$$

$$\cos\left(\frac{2}{8}\pi - z\right)\cos z$$

In general we have

$$\cos nz = 2^{n-1}\cos\left(\frac{n-1}{n}\pi + z\right)\cos\left(\frac{n-1}{n}\pi - z\right)$$

$$\cos\left(\frac{n-3}{n}\pi + z\right)\cos\left(\frac{n-3}{n}\pi - z\right)\cos\left(\frac{n-5}{n}\pi + z\right)$$

$$\cos\left(\frac{n-5}{n}\pi - z\right)\cos\left(\frac{n-7}{n}\pi + z\right)\cdots,$$ where there are n factors.

246. If n is odd and the equation begins with 1, then $0 = 1 \pm \dfrac{ny}{\cos nz} + \cdots$, where the negative sign is used when n is odd with the form $4m + 1$, while the positive is used when $n = 4m - 1$. Then we have

$$+\frac{1}{\cos z} = \frac{1}{\cos z}$$

$$-\frac{3}{\cos 3z} = \frac{1}{\cos z} - \frac{1}{\cos\left(\dfrac{\pi}{3} - z\right)} - \frac{1}{\cos\left(\dfrac{\pi}{3} + z\right)}$$

$$+\frac{5}{\cos 5z} = \frac{1}{\cos z} - \frac{1}{\cos\left(\dfrac{\pi}{5} - z\right)} - \frac{1}{\cos\left(\dfrac{\pi}{3} + z\right)}$$

$$+\frac{1}{\cos\left(\dfrac{2}{5}\pi - z\right)} + \frac{1}{\cos\left(\dfrac{2}{5}\pi + z\right)}.$$

In general, when $n = 2m + 1$, we have

$$\frac{n}{\cos nz} = \frac{2m + 1}{\cos(2m + 1)z} = \frac{1}{\cos\left(\dfrac{m}{n}\pi + z\right)} + \frac{1}{\cos\left(\dfrac{m}{n}\pi - z\right)}$$

$$- \frac{1}{\cos\left(\dfrac{m - 1}{n}\pi + z\right)} - \frac{1}{\cos\left(\dfrac{m - 1}{n}\pi - z\right)} + \frac{1}{\cos\left(\dfrac{m - 2}{n}\pi + z\right)}$$

$$+ \frac{1}{\cos\left(\dfrac{m - 2}{n}\pi - z\right)} - \frac{1}{\cos\left(\dfrac{m - 3}{n}\pi + z\right)} - \cdots,$$ where there are n terms.

247. Since $\dfrac{1}{\cos v} = \sec v$, we deduce the following important properties of secants.

$$\sec z = \sec z$$

$$3 \sec 3z = \sec\left(\frac{\pi}{3} + z\right) + \sec\left(\frac{\pi}{3} - z\right) - \sec\left(\frac{0}{3}\pi + z\right)$$

$$5 \sec 5z = \sec\left(\frac{2}{5}\pi + z\right) + \sec\left(\frac{2}{5}\pi - z\right) - \sec\left(\frac{\pi}{5} + z\right)$$

$$- \sec\left(\frac{\pi}{5} - z\right) + \sec\left(\frac{0}{5}\pi + z\right)$$

$$7 \sec 7z = \sec\left(\frac{3}{7}\pi + z\right) + \sec\left(\frac{3}{7}\pi - z\right) - \sec\left(\frac{2}{7}\pi + z\right)$$

$$- \sec\left(\frac{2}{7}\pi - z\right) + \sec\left(\frac{\pi}{7} + z\right) + \sec\left(\frac{\pi}{7} - z\right) - \sec\left(\frac{0}{7}\pi + z\right).$$

In general, when $n = 2m + 1$, then

$$n \sec nz = \sec\left(\frac{m}{n}\pi + z\right) + \sec\left(\frac{m}{n}\pi - z\right) - \sec\left(\frac{m-1}{n}\pi + z\right)$$

$$- \sec\left(\frac{m-1}{n}\pi - z\right) + \sec\left(\frac{m-2}{n}\pi + z\right) + \sec\left(\frac{m-2}{n}\pi - z\right)$$

$$- \sec\left(\frac{m-3}{n}\pi + z\right) - \sec\left(\frac{m-3}{n}\pi - z\right) + \sec\left(\frac{m-4}{n}\pi + z\right)$$

$$+ \cdots \pm \sec z.$$

248. From section 237, we have for cosecants

$$\csc z = \csc z$$

$$3 \csc 3z = \csc z + \csc\left(\frac{\pi}{3} - z\right) - \csc\left(\frac{\pi}{3} + z\right)$$

$$5 \csc 5z = \csc z + \csc\left(\frac{\pi}{5} - z\right) - \csc\left(\frac{\pi}{5} + z\right)$$

$$- \csc\left(\frac{2}{5}\pi - z\right) + \csc\left(\frac{2}{5}\pi + z\right)$$

$$7 \csc 7z = \csc z + \csc\left(\frac{\pi}{7} - z\right) - \csc\left(\frac{\pi}{7} + z\right)$$
$$- \csc\left(\frac{2}{7}\pi - z\right) + \csc\left(\frac{2}{7}\pi + z\right) + \csc\left(\frac{3}{7}\pi - z\right)$$
$$- \csc\left(\frac{3}{7}\pi + z\right).$$

In general, for $n = 2m + 1$, we have

$$n \csc nz = \csc z + \csc\left(\frac{\pi}{n} - z\right) - \csc\left(\frac{\pi}{n} + z\right)$$
$$- \csc\left(\frac{2}{n}\pi - z\right) + \csc\left(\frac{2}{n}\pi + z\right) + \csc\left(\frac{3}{n}\pi - z\right)$$
$$- \csc\left(\frac{3}{n}\pi + z\right) - \cdots \pm \csc\left(\frac{m}{n}\pi + z\right),$$ where the upper sign is used if m is even and the lower when m is odd.

249. As we have seen above,

$\cos nz \pm i \sin nz = (\cos z \pm i \sin z)^n$, so that

$$\cos nz = \frac{(\cos z + i \sin z)^n + (\cos z - i \sin z)^n}{2},$$ and

$$\sin nz = \frac{(\cos z + i \sin z)^n - (\cos z - i \sin z)^n}{2i}.$$ Furthermore

$$\tan nz = \frac{(\cos z + i \sin z)^n - (\cos z - i \sin z)^n}{(\cos z + i \sin z)^n i + (\cos z - i \sin z)^n i}.$$

We let $\tan z = \dfrac{\sin z}{\cos z} = t$, then

$$\tan nz = \frac{(1 + ti)^n - (1 - ti)^n}{(1 + ti)^n i + (1 - ti)^n i},$$ and we have the following tangents of multiple angles.

$$\tan z = t$$

$$\tan 2z = \frac{2t}{1 - t^2}$$

$$\tan 3z = \frac{3t - t^3}{1 - 3t^2}$$

$$\tan 4z = \frac{4t - 4t^3}{1 - 6t^2 + t^4}$$

$$\tan 5z = \frac{5t - 10t^3 + t^5}{1 - 10t^2 + 5t^4}.$$

In general we have

$$\tan nz = \frac{nt - \frac{n}{1}\frac{(n-1)}{2}\frac{(n-2)}{3}t^3 + \cdots}{1 - \frac{n}{1}\frac{(n-1)}{2}t^2 + \cdots}.$$

Since $\tan nz = \tan(\pi + nz) = \tan(2\pi + nz) = \tan(3\pi + nz) = \cdots$, the roots of the equation are

$\tan z$, $\tan\left(\frac{\pi}{n} + z\right)$, $\tan\left(\frac{2}{n}\pi + z\right)$, $\tan\left(\frac{3}{n}\pi + z\right)$, $\ldots$, where there are n roots.

250. If the equation begins with 1, it will be

$$0 = 1 - \frac{nt}{\tan nz} - \frac{n}{1}\frac{(n-1)}{2}t^2 + \frac{n}{1}\frac{(n-1)}{2}\frac{(n-2)}{3}\frac{t^3}{\tan nz} + \cdots.$$

When we compare the coefficients with the roots, we obtain

$$n \cot nz = \cot z + \cot\left(\frac{\pi}{n} + z\right) + \cot\left(\frac{2}{n}\pi + z\right) + \cot\left(\frac{3}{n}\pi + z\right) + \cot\left(\frac{4}{n}\pi + z\right) + \cdots + \cot\left(\frac{n-1}{n}\pi + z\right).$$

Furthermore, the sum of the squares of all of the cotangents is equal to $\frac{n^2}{(\sin nz)^2} - n$. In the same way higher powers can be defined. When we substitute specific numbers for n we obtain

$$\cot z = \cot z$$

$$2 \cot 2z = \cot z + \cot\left(\frac{\pi}{2} + z\right)$$

$$3 \cot 3z = \cot z + \cot\left(\frac{\pi}{3} + z\right) + \cot\left(\frac{2}{3}\pi + z\right)$$

$$4 \cot 4z = \cot z + \cot\left(\frac{\pi}{4} + z\right) + \cot\left(\frac{2}{4}\pi + z\right) + \cot\left(\frac{3}{4}\pi + z\right)$$

$$5 \cot 5z = \cot z + \cot\left(\frac{\pi}{5} + z\right) + \cot\left(\frac{2}{5}\pi + z\right)$$

$$+ \cot\left(\frac{3}{5}\pi + z\right) + \cot\left(\frac{4}{5}\pi + z\right).$$

251. Since $\cot v = -\cot(\pi - v)$, we have

$$\cot z = \cot z$$

$$2 \cot 2z = \cot z - \cot\left(\frac{\pi}{2} - z\right)$$

$$3 \cot 3z = \cot z - \cot\left(\frac{\pi}{3} - z\right) + \cot\left(\frac{\pi}{3} + z\right)$$

$$4 \cot 4z = \cot z - \cot\left(\frac{\pi}{4} - z\right) + \cot\left(\frac{\pi}{4} + z\right) - \cot\left(\frac{2}{4}\pi - z\right)$$

$$5 \cot 5z = \cot z - \cot\left(\frac{\pi}{5} - z\right) + \cot\left(\frac{\pi}{5} + z\right)$$

$$- \cot\left(\frac{2}{5}\pi - z\right) + \cot\left(\frac{2}{5}\pi + z\right)$$

In general we have $n \cot nz = \cot z - \cot\left(\frac{\pi}{n} - z\right) + \cot\left(\frac{\pi}{n} + z\right)$

$$- \cot\left(\frac{2}{n}\pi - z\right) + \cot\left(\frac{2}{n}\pi + z\right) - \cot\left(\frac{3}{n}\pi - z\right) + \cot\left(\frac{3}{n}\pi + z\right) - \text{etc.},$$

where there are n terms.

252. We consider the equation found from the highest power, and first we distinguish the cases where n is even and odd. Suppose n is odd, of the form $n = 2m + 1$, then

$$t - \tan z = 0$$
$$t^3 - 3t^2\tan 3z - 3t + \tan 3z = 0$$
$$t^5 - 5t^4\tan 5z - 10t^3 + 10t^2\tan 5z + 5t - \tan 5z = 0.$$

In general we have $t^n - nt^{n-1}\tan nz - \cdots \pm \tan nz = 0$ where the negative sign is used when m is even, and positive sign when m is odd. From the coefficient of the second term we have

$$\tan z = \tan z$$
$$3 \tan 3z = \tan z + \tan\left(\frac{\pi}{3} + z\right) + \tan\left(\frac{2}{3}\pi + z\right)$$
$$5 \tan 5z = \tan z + \tan\left(\frac{\pi}{5} + z\right) + \tan\left(\frac{2}{5}\pi + z\right)$$
$$+ \tan\left(\frac{3}{5}\pi + z\right) + \tan\left(\frac{4}{5}\pi + z\right)$$

etc.

253. Since $\tan v = -\tan(\pi - v)$, angles larger than a right angle can be reduced to an angle less than a right angle, with the result that

$$\tan z = \tan z$$
$$3 \tan 3z = \tan z - \tan\left(\frac{\pi}{3} - z\right) + \tan\left(\frac{\pi}{3} + z\right)$$
$$5 \tan 5z = \tan z - \tan\left(\frac{\pi}{5} - z\right) + \tan\left(\frac{\pi}{5} + z\right)$$
$$- \tan\left(\frac{2}{5}\pi - z\right) + \tan\left(\frac{2}{5}\pi + z\right)$$

$$7 \tan 7z = \tan z - \tan\left(\frac{\pi}{7} - z\right) + \tan\left(\frac{\pi}{7} + z\right) - \tan\left(\frac{2}{7}\pi - z\right)$$

$$+ \tan\left(\frac{2}{7}\pi + z\right) - \tan\left(\frac{3}{7}\pi - z\right) + \tan\left(\frac{3}{7}\pi + z\right).$$

In general we have, when $n = 2m + 1$,

$$n \tan nz = \tan z - \tan\left(\frac{\pi}{n} - z\right) + \tan\left(\frac{\pi}{n} + z\right)$$

$$- \tan\left(\frac{2}{n}\pi - z\right) + \tan\left(\frac{2}{n}\pi + z\right) - \cdots$$

$$- \tan\left(\frac{m}{n}\pi - z\right) + \tan\left(\frac{m}{n}\pi + z\right).$$

254. Next we note that the product of all of these tangents is equal to $\tan nz$ since, due to the alternation of the sign with the odd and even values of m, the ambiguity of the sign in the product is removed. Thus we have

$$\tan z = \tan z$$

$$\tan 3z = \tan z \tan\left(\frac{\pi}{3} - z\right)\tan\left(\frac{\pi}{3} + z\right)$$

$$\tan 5z = \tan z \tan\left(\frac{\pi}{5} - z\right)\tan\left(\frac{\pi}{5} + z\right)\tan\left(\frac{2}{5}\pi - z\right)\tan\left(\frac{2}{5}\pi + z\right),$$

in general we have, when $n = 2m + 1$,

$$\tan nz = \tan z \tan\left(\frac{\pi}{n} - z\right)\tan\left(\frac{\pi}{n} + z\right)\tan\left(\frac{2}{n}\pi - z\right)\tan\left(\frac{2}{n}\pi + z\right)$$

$$\tan\left(\frac{3}{n}\pi - z\right) \cdots \tan\left(\frac{m}{n}\pi - z\right)\tan\left(\frac{m}{n}\pi + z\right).$$

255. Now if n is even and we begin with the highest power, we have

$$t^2 + 2t \cot 2z - 1 = 0$$

$$t^4 + 4t^3 \cot 4z - 6t^2 - 4t \cot 4z + 1 = 0.$$

In general, if $n = 2m$, then

$t^n + nt^{n-1}\cot nz - \cdots \pm 1 = 0$ where the negative sign is used when m is odd, and the positive when m is even. When we compare the roots with the coefficient of the second term, we have

$$-2\cot 2z = \tan z + \tan\left(\frac{\pi}{2} + z\right)$$

$$-4\cot 4z = \tan z + \tan\left(\frac{\pi}{4} + z\right) + \tan\left(\frac{2}{4}\pi + z\right) + \tan\left(\frac{3}{4}\pi + z\right)$$

$$-6\cot 6z = \tan z + \tan\left(\frac{\pi}{6} + z\right) + \tan\left(\frac{2}{6}\pi + z\right)$$

$$+ \tan\left(\frac{3}{6}\pi + z\right) + \tan\left(\frac{4}{6}\pi + z\right) + \tan\left(\frac{5}{6}\pi + z\right),$$

etc.

256. Since $\tan v = -\tan(\pi - v)$, the following equations can be stated

$$2\cot 2z = -\tan z + \tan\left(\frac{\pi}{2} - z\right)$$

$$4\cot 4z = -\tan z + \tan\left(\frac{\pi}{4} - z\right) - \tan\left(\frac{\pi}{4} + z\right) + \tan\left(\frac{2}{4}\pi - z\right)$$

$$6\cot 6z = -\tan z + \tan\left(\frac{\pi}{6} - z\right) - \tan\left(\frac{\pi}{6} + z\right) + \tan\left(\frac{2}{6}\pi - z\right)$$

$$- \tan\left(\frac{2}{6}\pi + z\right) + \tan\left(\frac{3}{6}\pi - z\right).$$

In general, if $n = 2m$, then

$$n\cot nz = -\tan z + \tan\left(\frac{\pi}{n} - z\right) - \tan\left(\frac{\pi}{n} + z\right) + \tan\left(\frac{2}{n}\pi - z\right)$$

$$- \tan\left(\frac{2}{n}\pi + z\right) + \tan\left(\frac{3}{n}\pi - z\right) - \tan\left(\frac{3}{n}\pi + z\right) + \cdots$$

$$+ \tan\left(\frac{m}{n}\pi - z\right).$$

257. In the even case the ambiguity of the sign in the product of all of the roots is also removed, so that we have

$$1 = \tan z \tan\left(\frac{\pi}{2} - z\right)$$

$$1 = \tan z \tan\left(\frac{\pi}{4} - z\right)\tan\left(\frac{\pi}{4} + z\right)\tan\left(\frac{2}{4}\pi - z\right)$$

$$1 = \tan z \tan\left(\frac{\pi}{6} - z\right)\tan\left(\frac{\pi}{6} + z\right)\tan\left(\frac{2}{6}\pi - z\right)$$

$$\tan\left(\frac{2}{6}\pi + z\right)\tan\left(\frac{3}{6}\pi - z\right)$$

etc. An obvious reason for these equations is seen when it is noticed that these roots occur in pairs, each pair being cosines of complementary angles. Since the product of each pair is equal to 1, the product of all the pairs is equal to 1.

258. Since the sine and cosine of angles making an arithmetic progression provide a recurrent series, so that by the results of the previous chapter, we can find the sum of any such sines and cosines. Let the angles in the arithmetic progression be

$a, a + b, a + 2b, a + 3b, a + 4b, a + 5b, \cdots$, and first we find the sum of the infinite series of sines of these angles. We let

$s = \sin a + \sin(a + b) + \sin(a + 2b) + \sin(a + 3b) + \cdots$. Since this is a recurrent series, whose scale of relation is $2 \cos b, - 1$, this series arises from a rational function whose denominator is $1 - 2z \cos b + z^2$ when we let $z = 1$. The rational function is

$$\frac{\sin a + z(\sin(a + b) - 2 \sin a \cos b)}{1 - 2z \cos b + z^2}$$ which, when $z = 1$, becomes

$$s = \frac{\sin a + \sin(a + b) - 2 \sin a \cos b}{2 - 2 \cos b} = \frac{\sin a - \sin(a - b)}{2(1 - \cos b)},$$ because

$2 \sin a \cos b = \sin(a + b) + \sin(a - b)$. Since

$\sin f - \sin g = 2 \cos\left(\frac{f + g}{2}\right)\sin\left(\frac{f - g}{2}\right)$, we have

$\sin a - \sin(a - b) = 2 \cos\left(a - \frac{1}{2}b\right)\sin\left(\frac{1}{2}b\right)$. Furthermore,

$1 - \cos b = 2\left(\sin\left(\frac{1}{2}b\right)\right)^2$, so that $s = \dfrac{\cos\left(a - \frac{1}{2}b\right)}{2 \sin\left(\frac{1}{2}b\right)}$.

259. It follows that the sum of any series of sines whose arcs are in an arithmetic progression can be found. For example, let us find the sum of the following progression:

$\sin a + \sin(a + b) + \sin(a + 2b) + \sin(a + 3b) + \cdots + \sin(a + nb)$.

Since, if this progression were to be continued to infinity, the sum would be

$\dfrac{\cos\left(a - \frac{1}{2}b\right)}{2 \sin\left(\frac{1}{2}b\right)}$, we consider the last terms of this infinite series:

$\sin(a + (n + 1)b) + \sin(a + (n + 2)b) + \sin(a + (n + 3)b) + \cdots$.

Since the sum of this series is $\dfrac{\cos\left(a + \left(n + \frac{1}{2}\right)b\right)}{2 \sin\left(\frac{1}{2}b\right)}$, when this latter is subtracted from the former sum, we have the desired sum. That is, if

$s = \sin a + \sin(a + b) + \sin(a + 2b) + \cdots + \sin(a + nb)$, then

$$s = \frac{\cos\left(a - \frac{1}{2}b\right) - \cos\left(a + \left(n + \frac{1}{2}\right)b\right)}{2 \sin\left(\frac{1}{2}b\right)}$$

$$= \frac{\sin\left(a + \frac{1}{2}nb\right)\sin\frac{1}{2}(n+1)b}{\sin\left(\frac{1}{2}b\right)}.$$

260. In a similar way, if we consider the sum of the cosines, and let

$s = \cos a + \cos(a+b) + \cos(a+2b) + \cos(a+3b) + \cdots$. Then

$s = \dfrac{\cos a + z(\cos(a+b) - 2\cos a \cos b)}{1 - 2z\cos b + z^2}$ when $z = 1$. Since

$2\cos a \cos b = \cos(a-b) + \cos(a+b)$ we have

$s = \dfrac{\cos a - \cos(a-b)}{2(1-\cos b)}$. Since

$\cos f - \cos g = 2\sin\left(\dfrac{f+g}{2}\right)\sin\left(\dfrac{g-f}{2}\right)$ we have

$\cos a - \cos(a-b) = -2\sin\left(a - \dfrac{1}{2}b\right)\sin\left(\dfrac{1}{2}b\right)$. Since

$1 - \cos b = 2\left(\sin\left(\dfrac{1}{2}b\right)\right)^2$, we have $s = \dfrac{-\sin\left(a - \frac{1}{2}b\right)}{2\sin\left(\frac{1}{2}b\right)}$. Furthermore, since

$\cos(a + (n+1)b) + \cos(a + (n+2)b) + \cos(a + (n+3)b) + \cdots$ has

the sum equal to $-\dfrac{\sin\left(a + \left(n + \frac{1}{2}\right)b\right)}{2\sin\left(\frac{1}{2}b\right)}$, when this expression is subtracted

from the sum of the first series, we have

$s = \cos a + \cos(a+b) + \cos(a+2b) + \cos(a+3b)$

$+ \cdots + \cos(a+nb)$

$$= -\frac{\sin\left(-\frac{1}{2}b\right) + \sin\left(a + \left(n + \frac{1}{2}\right)b\right)}{2\sin\left(\frac{1}{2}b\right)}$$

$$= \frac{\cos\left(a + \frac{1}{2}nb\right)\sin\left(\frac{1}{2}(n+1)b\right)}{\sin\left(\frac{1}{2}b\right)}.$$

261. There are many other questions about sines and tangents which could be settled on the basis of the principles which we have discussed, such as the sums of squares or higher powers of sines and tangents. Since these are derived in a similar way from the coefficients of some of the equations given above, we will not delay here any longer. There is one thing, however, which we here note in this regard. Any power of a sine or cosine can be expressed in terms of the sine or cosine. For the sake of clarity we make a brief exposition.

262. In order that the explanation be more expeditious, we recall the following lemmma.

$$2\sin a \sin z = \cos(a - z) - \cos(a + z)$$
$$2\cos a \sin z = \sin(a + z) - \sin(a - z)$$
$$2\sin a \cos z = \sin(a + z) + \sin(a - z)$$
$$2\cos a \cos z = \cos(a - z) + \cos(a + z).$$

First we treat the powers of the sine.

$$\sin z = \sin z$$

$$2(\sin z)^2 = 1 - \cos 2z$$

$$4(\sin z)^3 = 3\sin z - \sin 3z$$

$$8(\sin z)^4 = 3 - 4\cos 2z + \cos 4z$$

$$16(\sin z)^5 = 10 \sin z - 5 \sin 3z + \sin 5z$$

$$32(\sin z)^6 = 10 - 15 \cos 2z + 6 \cos 4z - \cos 6z$$

$$64(\sin z)^7 = 35 \sin z - 21 \sin 3z + 7 \sin 5z - \sin 7z$$

$$128(\sin z)^8 = 35 - 56 \sin 2z + 28 \cos 4z - 8 \cos 6z + \cos 8z$$

$$256(\sin z)^9 = 126 \sin z - 84 \sin 3z + 36 \sin 5z$$

$$- 9 \sin 7z + \sin 9z$$

etc. The law by which the coefficients progress is given by the binomial coefficients, except for the absolute number in the even powers, which is provided by the binomial coefficient of the previous power.

263. In the same way the powers of cosines are given.

$$\cos z = \cos z$$

$$2(\cos z)^2 = 1 + \cos 2z$$

$$4(\cos z)^3 = 3 \cos z + \cos 3z$$

$$8(\cos z)^4 = 3 + 4 \cos 2z + \cos 4z$$

$$16(\cos z)^5 = 10 \cos z + 5 \cos 3z + \cos 5z$$

$$32(\cos z)^6 = 10 + 15 \cos 2z + 6 \cos 4z + \cos 6z$$

$$64(\cos z)^7 = 35 \cos z + 21 \cos 3z + 7 \cos 5z + \cos 7z$$

etc. As for the law of the progression, the same applies as in the case of the sines.

CHAPTER XV

On Series Which Arise From Products.

264. Let a given product of factors, whether a finite or infinite product, be $(1 + \alpha z)(1 + \beta z)(1 + \gamma z)(1 + \delta z)(1 + \epsilon z)(1 + \zeta z) \cdots$. When we actually multiply the factors we obtain a series

$1 + Az + Bz^2 + Cz^3 + Dz^4 + Ez^5 + Fz^6 + \cdots$. It is clear that the coefficients A, B, C, D, E, etc. are obtained from the numbers α, β, γ, δ, ϵ, ζ, etc. as follows:

$A = \alpha + \beta + \gamma + \delta + \epsilon + \zeta + \cdots$ which is the sum of the individual numbers; B is the sum of the products taken two at a time; C is the sum of the products taken three at a time; D is the sum of the products taken four at a time; E is the sum of the products taken five at a time; etc., until we arrive at the product of all of the numbers.

265. Now if we let $z = 1$, then the product

$(1 + \alpha)(1 + \beta)(1 + \gamma)(1 + \delta)(1 + \epsilon) \cdots$ is equal to 1 and the series of all of the numbers α, β, γ, δ, ϵ, etc. taken singly, then products two at a time, etc. Furthermore, if the same number can occur in two or more ways, it will occur twice or more times in this series.

266. If we let $z = -1$, the product

$(1 - \alpha)(1 - \beta)(1 - \gamma)(1 - \delta)(1 - \epsilon) \cdots$ is equal to the sum of 1 and the series of all the numbers which arise from $\alpha, \beta, \gamma, \delta, \epsilon, \zeta, \cdots$ taken singly, and products of two or more at a time, as before, but with this difference: those taken singly, or three at a time, five at a time or any odd number at a time have a negative sign, while those taken two at a time, four at a time six at a time, or any even number at a time have a positive sign.

267. For $\alpha, \beta, \gamma, \delta, \cdots$ let us use all of the prime numbers $2, 3, 5, 7, 11, 13, \cdots$, and let

$P = (1 + 2)(1 + 3)(1 + 5)(1 + 7)(1 + 11)(1 + 13) \cdots$ be the product. The series which arises has 1, all of the prime numbers themselves, and numbers which arise from products of different prime numbers. Hence $P = 1 + 2 + 3 + 5 + 6 + 7 + 10 + 11 + 13 + 14 + 15 + \cdots$. We note that this series contains all of the natural numbers except powers and numbers which are divisible by powers. The series lacks the numbers 4, 8, 9, 12, 16, 18 since they are either powers, as 4, 8, 9, 16, or divisible by powers, as 12, 18.

268. A similar thing happens if we substitute for $\alpha, \beta, \gamma, \delta, \cdots$ any power of the prime numbers. For example, if we let

$$P = \left(1 + \frac{1}{2^n}\right)\left(1 + \frac{1}{3^n}\right)\left(1 + \frac{1}{5^n}\right)\left(1 + \frac{1}{7^n}\right)\left(1 + \frac{1}{11^n}\right) \cdots,$$

then $P = 1 + \frac{1}{2^n} + \frac{1}{3^n} + \frac{1}{5^n} + \frac{1}{6^n} + \frac{1}{7^n} + \frac{1}{10^n} + \frac{1}{11^n} + \cdots$, in which fractions all numbers appear except powers or those which are divisible by powers. Since all integers are either prime or are products of primes, the only numbers excluded from the series are those in which the same prime occurs two or more times.

269. If the numbers α, β, γ, δ, $\cdots$ are taken as negative, as was done in section 266, and we let

$$P = \left(1 - \frac{1}{2^n}\right)\left(1 - \frac{1}{3^n}\right)\left(1 - \frac{1}{5^n}\right)\left(1 - \frac{1}{7^n}\right)\left(1 - \frac{1}{11^n}\right)\cdots, \text{ then}$$

$$P = 1 - \frac{1}{2^n} - \frac{1}{3^n} - \frac{1}{5^n} + \frac{1}{6^n} - \frac{1}{7^n} + \frac{1}{10^n} - \frac{1}{11^n}$$

$- \frac{1}{13^n} + \frac{1}{15^n} - \cdots$. We note that the terms with primes, or products of three different primes, or any product of an odd number of different primes, appear with a negative sign. Those terms in which the product of two, four, six, or any even numbers of different primes, appear with a positive sign. Thus in this series there occurs the term $\frac{1}{30^n}$. Since $30 = 2 \cdot 3 \cdot 5$, it contains no power of a prime as a factor. The term $\frac{1}{30^n}$ appears with a negative sign, since 30 is the product of three different primes.

270. Let us now consider the expression

$$\frac{1}{(1 - \alpha z)(1 - \beta z)(1 - \gamma z)(1 - \delta z)(1 - \epsilon z)\cdots}.$$

When the division is carried out, we obtain the series $1 + Az + Bz^2 + Cz^3 + Dz^4 + Ez^5 + Fz^6 + \cdots$. It is clear that the coefficients A, B, C, D, E, F, etc. depend on the numbers α, β, γ, δ, ϵ, etc. in the following way: A is the sum of the numbers taken singly; B is the sum of the products taken two at a time; C is the sum of the products taken three at a time; D is the sum of products taken four at a time; etc., where we do not exclude products of the same factor.

271. When we let $z = 1$ the expression

$$\frac{1}{(1-\alpha)(1-\beta)(1-\gamma)(1-\delta)(1-\epsilon)\cdots}$$ is equal to the sum of 1 and the series of all the numbers which arise from α, β, γ, δ, ϵ, ζ, etc. taken either singly or the product of two, or more at a time, where we do not exclude the possibility that some of these terms may be equal. This is different from what we considered in section 265, since there we allowed only different factors. Here we allow the factor to occur twice or even more times. Here all numbers occur which can be obtained as products of α, β, γ, δ, etc.

272. For this reason, the series always has an infinite number of terms, whether the product has an infinite or finite number of factors. For example

$$\frac{1}{1-\frac{1}{2}} = 1 + \frac{1}{2} + \frac{1}{4} + \frac{1}{8} + \frac{1}{16} + \frac{1}{32} + \cdots$$

where all numbers occur which are powers of $\frac{1}{2}$. Then we have

$$\frac{1}{\left(1-\frac{1}{2}\right)\left(1-\frac{1}{3}\right)} = 1 + \frac{1}{2} + \frac{1}{3} + \frac{1}{4} + \frac{1}{6}$$
$$+ \frac{1}{8} + \frac{1}{9} + \frac{1}{12} + \frac{1}{16} + \frac{1}{18} + \cdots,$$

where numbers do not occur unless they are products of powers of 2 and 3, that is, unless they are divisible by 2 or 3.

273. If for α, β, γ, δ, etc. we substitute the reciprocals of all of the primes, and let

$$P = \frac{1}{\left(1-\frac{1}{2}\right)\left(1-\frac{1}{3}\right)\left(1-\frac{1}{5}\right)\left(1-\frac{1}{7}\right)\left(1-\frac{1}{11}\right)\cdots}$$

then $P = 1 + \frac{1}{2} + \frac{1}{3} + \frac{1}{4} + \frac{1}{5} + \frac{1}{6} + \frac{1}{7} + \frac{1}{8} + \frac{1}{9} \cdots$, where all numbers, whether prime or products of primes, occur. Since every natural number is either a prime or a product of primes, it is clear that all integers will appear in denominators.

274. The same happens if any power of the primes is used. We let

$$P = \frac{1}{\left(1-\frac{1}{2^n}\right)\left(1-\frac{1}{3^n}\right)\left(1-\frac{1}{5^n}\right)\left(1-\frac{1}{7^n}\right)\cdots},$$

then $P = 1 + \frac{1}{2^n} + \frac{1}{3^n} + \frac{1}{4^n} + \frac{1}{5^n} + \frac{1}{6^n} + \frac{1}{7^n} + \frac{1}{8^n} + \cdots$, where all natural numbers occur with no exception. If in the factors we have everywhere the positive sign, so that

$$P = \frac{1}{\left(1+\frac{1}{2^n}\right)\left(1+\frac{1}{3^n}\right)\left(1+\frac{1}{5^n}\right)\left(1+\frac{1}{7^n}\right)\cdots},$$

then

$$P = 1 - \frac{1}{2^n} - \frac{1}{3^n} + \frac{1}{4^n} - \frac{1}{5^n} + \frac{1}{6^n} - \frac{1}{7^n} - \frac{1}{8^n} + \frac{1}{9^n} + \frac{1}{10^n} - \cdots,$$

where the primes have a negative sign, the products of two primes, either the same or different, have a positive sign. In general those numbers whose number of prime factors is even have a positive sign, while those with an odd number of prime factors have a negative sign. For example, since $240 = 2 \cdot 2 \cdot 2 \cdot 2 \cdot 3 \cdot 5$, the term $\frac{1}{240^n}$ has a positive sign. The reason for this law is stated in section 270, when $z = -1$.

275. When we compare these results with those obtained above, and consider two series whose product is equal to 1, we have the following.

$$P = \frac{1}{\left(1-\frac{1}{2^n}\right)\left(1-\frac{1}{3^n}\right)\left(1-\frac{1}{5^n}\right)\left(1-\frac{1}{7^n}\right)\cdots}$$

$$Q = \left(1 - \frac{1}{2^n}\right)\left(1 - \frac{1}{3^n}\right)\left(1 - \frac{1}{5^n}\right)\left(1 - \frac{1}{7^n}\right)\cdots$$

then $P = \frac{1}{2^n} + \frac{1}{3^n} + \frac{1}{4^n} + \frac{1}{5^n} + \frac{1}{6^n} + \frac{1}{7^n} + \frac{1}{8^n} + \cdots$

$$Q = 1 - \frac{1}{2^n} - \frac{1}{3^n} - \frac{1}{5^n} - \frac{1}{6^n} - \frac{1}{7^n} + \frac{1}{10^n} - \frac{1}{11^n} - \cdots$$

(see section 269). It is clear that $PQ = 1$.

276. If we let

$$P = \frac{1}{\left(1+\frac{1}{2^n}\right)\left(1+\frac{1}{3^n}\right)\left(1+\frac{1}{5^n}\right)\left(1+\frac{1}{7^n}\right)\cdots},$$

and $Q = \left(1 + \frac{1}{2^n}\right)\left(1 + \frac{1}{3^n}\right)\left(1 + \frac{1}{5^n}\right)\left(1 + \frac{1}{7^n}\right)\cdots,$

then $P = 1 - \frac{1}{2^n} - \frac{1}{3^n} + \frac{1}{4^n} - \frac{1}{5^n} + \frac{1}{6^n} - \frac{1}{7^n} - \frac{1}{8^n} + \frac{1}{9^n} + \cdots$

and $Q = 1 + \frac{1}{2^n} + \frac{1}{3^n} + \frac{1}{5^n} + \frac{1}{6^n} + \frac{1}{7^n} + \frac{1}{10^n} + \frac{1}{11^n} + \cdots$. We also have $PQ = 1$, so that when the sum of one series is know, we also have the sum of the other.

277. On the other hand, from a knowledge of the sums of these series we can determine the values of the infinite factors. For example, if

$M = 1 + \frac{1}{2^n} + \frac{1}{3^n} + \frac{1}{4^n} + \frac{1}{5^n} + \frac{1}{6^n} + \frac{1}{7^n} + \cdots$ and

$$N = 1 + \frac{1}{2^{2n}} + \frac{1}{3^{2n}} + \frac{1}{4^{2n}} + \frac{1}{5^{2n}} + \frac{1}{6^{2n}} + \frac{1}{7^{2n}} + \cdots \text{, then}$$

$$M = \frac{1}{\left(1-\frac{1}{2^n}\right)\left(1-\frac{1}{3^n}\right)\left(1-\frac{1}{5^n}\right)\left(1-\frac{1}{7^n}\right)\cdots}$$

and $N = \dfrac{1}{\left(1-\frac{1}{2^{2n}}\right)\left(1-\frac{1}{3^{2n}}\right)\left(1-\frac{1}{5^{2n}}\right)\left(1-\frac{1}{7^{2n}}\right)\cdots}$.

Then by division we have

$$\frac{M}{N} = \left(1 + \frac{1}{2^n}\right)\left(1 + \frac{1}{3^n}\right)\left(1 + \frac{1}{5^n}\right)\left(1 + \frac{1}{7^n}\right)\left(1 + \frac{1}{11^n}\right)\cdots,$$ so that

$$\frac{M^2}{N} = \frac{2^n + 1}{2^n - 1} \cdot \frac{3^n + 1}{3^n - 1} \cdot \frac{5^n + 1}{5^n - 1} \cdot \frac{7^n + 1}{7^n - 1} \cdot \frac{11^n + 1}{11^n - 1} \cdots.$$ From a

knowledge of M and N, besides the values of these products, we also have the sum of the following series:

$$\frac{1}{M} = 1 - \frac{1}{2^n} - \frac{1}{3^n} - \frac{1}{5^n} + \frac{1}{6^n} - \frac{1}{7^n} + \frac{1}{10^n} - \frac{1}{11^n} - \cdots$$

$$\frac{1}{N} = 1 - \frac{1}{2^{2n}} - \frac{1}{3^{2n}} - \frac{1}{5^{2n}} + \frac{1}{6^{2n}} - \frac{1}{7^{2n}} + \frac{1}{10^{2n}} - \frac{1}{11^{2n}} - \cdots$$

$$\frac{M}{N} = 1 + \frac{1}{2^n} + \frac{1}{3^n} + \frac{1}{5^n} + \frac{1}{6^n} + \frac{1}{7^n} + \frac{1}{10^n} + \frac{1}{11^n} + \cdots$$

$$\frac{N}{M} = 1 - \frac{1}{2^n} - \frac{1}{3^n} + \frac{1}{4^n} - \frac{1}{5^n} + \frac{1}{6^n} - \frac{1}{7^n} - \frac{1}{8^n} + \frac{1}{9^n} + \frac{1}{10^n} - \cdots.$$

From combinations of these, many others can be deduced.

EXAMPLE I

Let $n = 1$, and since we have already seen that $\log\left(\frac{1}{1-x}\right) = x + \frac{x^2}{2} + \frac{x^3}{3} + \frac{x^4}{4} + \frac{x^5}{5} + \frac{x^6}{6} + \cdots$, if we let $x = 1$, then $\log\left(\frac{1}{1-1}\right) = \log\infty = 1 + \frac{1}{2} + \frac{1}{3} + \frac{1}{4} + \frac{1}{5} + \cdots$. Since the logarithm of an infinitely large number is itself infinitely large, we have

$$M = 1 + \frac{1}{2} + \frac{1}{3} + \frac{1}{4} + \frac{1}{5} + \frac{1}{6} + \frac{1}{7} + \cdots = \infty, \text{ so that } \frac{1}{M} = \frac{1}{\infty} = 0$$

and

$$0 = 1 - \frac{1}{2} - \frac{1}{3} - \frac{1}{5} + \frac{1}{6} - \frac{1}{7} + \frac{1}{10} - \frac{1}{11} - \frac{1}{13} + \frac{1}{14} + \frac{1}{15} - \cdots .$$

In the products we have

$$M = \infty = \frac{1}{\left(1-\frac{1}{2}\right)\left(1-\frac{1}{3}\right)\left(1-\frac{1}{5}\right)\left(1-\frac{1}{7}\right)\left(1-\frac{1}{11}\right)\cdots},$$

so that $\infty = \frac{2}{1} \cdot \frac{3}{2} \cdot \frac{5}{4} \cdot \frac{7}{6} \cdot \frac{11}{10} \cdot \frac{13}{12} \cdot \frac{17}{16} \cdot \frac{19}{18} \cdots$ and

$0 = \frac{1}{2} \cdot \frac{2}{3} \cdot \frac{4}{5} \cdot \frac{6}{7} \cdot \frac{10}{11} \cdot \frac{12}{13} \cdot \frac{16}{17} \cdot \frac{18}{19} \cdots$. We have seen above that

$$N = 1 + \frac{1}{2^2} + \frac{1}{3^2} + \frac{1}{4^2} + \frac{1}{5^2} + \frac{1}{6^2} + \frac{1}{7^2} + \cdots = \frac{\pi^2}{6}.$$

From this we obtain these sums of the series:

$$\frac{6}{\pi^2} = 1 - \frac{1}{2^2} - \frac{1}{3^2} - \frac{1}{5^2} + \frac{1}{6^2} - \frac{1}{7^2} + \frac{1}{10^2} - \frac{1}{11^2} - \cdots$$

$$\infty = 1 + \frac{1}{2} + \frac{1}{3} + \frac{1}{5} + \frac{1}{6} + \frac{1}{7} + \frac{1}{10} + \frac{1}{11} + \cdots$$

$$0 = 1 - \frac{1}{2} - \frac{1}{3} + \frac{1}{4} - \frac{1}{5} + \frac{1}{6} - \frac{1}{7} - \frac{1}{8} + \frac{1}{9} + \frac{1}{10} - \frac{1}{11} - \cdots .$$

For the factors we have

$$\frac{\pi^2}{6} = \frac{2^2}{2^2-1} \cdot \frac{3^2}{3^2-1} \cdot \frac{5^2}{5^2-1} \cdot \frac{7^2}{7^2-1} \cdot \frac{11^2}{11^2-1} \cdots ,$$

$$\frac{\pi^2}{6} = \frac{4}{3} \cdot \frac{9}{8} \cdot \frac{25}{24} \cdot \frac{49}{48} \cdot \frac{121}{120} \cdot \frac{169}{168} \cdots .$$

Since $\frac{M}{N} = \infty$ or $\frac{N}{M} = 0$, we have

$$\infty = \frac{3}{2} \cdot \frac{4}{3} \cdot \frac{6}{5} \cdot \frac{8}{7} \cdot \frac{12}{11} \cdot \frac{14}{13} \cdot \frac{18}{17} \cdot \frac{20}{19} \cdots$$

$$0 = \frac{2}{3} \cdot \frac{3}{4} \cdot \frac{5}{6} \cdot \frac{7}{8} \cdot \frac{11}{12} \cdot \frac{13}{14} \cdot \frac{17}{18} \cdot \frac{19}{20} \cdots$$

$$\infty = \frac{3}{1} \cdot \frac{4}{2} \cdot \frac{6}{4} \cdot \frac{8}{6} \cdot \frac{12}{10} \cdot \frac{14}{12} \cdot \frac{18}{16} \cdot \frac{20}{18} \cdots$$

$0 = \frac{1}{3} \cdot \frac{1}{2} \cdot \frac{2}{3} \cdot \frac{3}{4} \cdot \frac{5}{6} \cdot \frac{6}{7} \cdot \frac{8}{9} \cdot \frac{9}{10} \cdots$. Note that in this last product, except for the first fraction, in each fraction the numerator is always 1 less than the denominator, and also the sum of the numerator and denominator of each fraction gives the sequence of prime numbers, 3, 5, 7, 11, 13, 17, 19, $\cdots$.

EXAMPLE II

Let $n = 2$, then from above we have

$$M = 1 + \frac{1}{2^2} + \frac{1}{3^2} + \frac{1}{4^2} + \frac{1}{5^2} + \frac{1}{6^2} + \frac{1}{7^2} + \cdots = \frac{\pi^2}{6}$$

$N = 1 + \frac{1}{2^4} + \frac{1}{3^4} + \frac{1}{4^4} + \frac{1}{5^4} + \frac{1}{6^4} + \frac{1}{7^4} + \cdots = \frac{\pi^4}{90}$. From these we obtain the sums of the following series:

$$\frac{6}{\pi^2} = 1 - \frac{1}{2^2} - \frac{1}{3^2} - \frac{1}{5^2} + \frac{1}{6^2} - \frac{1}{7^2} + \frac{1}{10^2} - \frac{1}{11^2} - \cdots$$

$$\frac{90}{\pi^4} = 1 - \frac{1}{2^4} - \frac{1}{3^4} - \frac{1}{5^4} + \frac{1}{6^4} - \frac{1}{7^4} + \frac{1}{10^4} - \frac{1}{11^4} - \cdots$$

$$\frac{15}{\pi^2} = 1 + \frac{1}{2^2} + \frac{1}{3^2} + \frac{1}{5^2} + \frac{1}{6^2} + \frac{1}{7^2} + \frac{1}{10^2} + \frac{1}{11^2} + \cdots$$

$$\frac{\pi^2}{15} = 1 - \frac{1}{2^2} - \frac{1}{3^2} + \frac{1}{4^2} - \frac{1}{5^2} + \frac{1}{6^2} - \frac{1}{7^2} - \frac{1}{8^2} + \frac{1}{9^2} + \frac{1}{10^2} - \cdots .$$

Now we note the values of the following products:

$$\frac{\pi^2}{6} = \frac{2^2}{2^2 - 1} \cdot \frac{3^2}{3^2 - 1} \cdot \frac{5^2}{5^2 - 1} \cdot \frac{7^2}{7^2 - 1} \cdot \frac{11^2}{11^2 - 1} \cdots$$

$$\frac{\pi^4}{90} = \frac{2^4}{2^4 - 1} \cdot \frac{3^4}{3^4 - 1} \cdot \frac{5^4}{5^4 - 1} \cdot \frac{7^4}{7^4 - 1} \cdot \frac{11^4}{11^4 - 1} \cdots$$

$$\frac{15}{\pi^2} = \frac{2^2+1}{2^2} \cdot \frac{3^2+1}{3^2} \cdot \frac{5^2+1}{5^2} \cdot \frac{7^2+1}{7^2} \cdot \frac{11^2+1}{11^2} \cdots$$ or

$$\frac{\pi^2}{15} = \frac{4}{5} \cdot \frac{9}{10} \cdot \frac{25}{26} \cdot \frac{49}{50} \cdot \frac{121}{122} \cdot \frac{169}{170} \cdots$$

$$\frac{5}{2} = \frac{2^2+1}{2^2-1} \cdot \frac{3^2+1}{3^2-1} \cdot \frac{5^2+1}{5^2-1} \cdot \frac{7^2+1}{7^2-1} \cdot \frac{11^2+1}{11^2-1} \cdots$$

$$\frac{5}{2} = \frac{5}{3} \cdot \frac{5}{4} \cdot \frac{13}{12} \cdot \frac{25}{24} \cdot \frac{61}{60} \cdot \frac{85}{84} \cdots \qquad \frac{3}{2} = \frac{5}{4} \cdot \frac{13}{12} \cdot \frac{25}{24} \cdot \frac{61}{60} \cdot \frac{85}{84} \cdots .$$

In this last product, each numerator exceeds the denominator by 1, while the sums of numerator and denominator give the sequence of squares of the primes, $3^2, 5^2, 7^2, 11^2, \cdots$.

EXAMPLE III

Since we know, from our previous work, the values of M only if n is even, we can let $n = 4$, so that

$$M = 1 + \frac{1}{2^4} + \frac{1}{3^4} + \frac{1}{4^4} + \frac{1}{5^4} + \frac{1}{6^4} + \cdots = \frac{\pi^4}{90}$$

$$N = 1 + \frac{1}{2^8} + \frac{1}{3^8} + \frac{1}{4^8} + \frac{1}{5^8} + \frac{1}{6^8} + \cdots = \frac{\pi^8}{9450}.$$

From these we obtain the following sums of series:

$$\frac{90}{\pi^4} = 1 - \frac{1}{2^4} - \frac{1}{3^4} - \frac{1}{5^4} + \frac{1}{6^4} - \frac{1}{7^4} + \frac{1}{10^4} - \frac{1}{11^4} - \cdots$$

$$\frac{9450}{\pi^8} = 1 - \frac{1}{2^8} - \frac{1}{3^8} - \frac{1}{5^8} + \frac{1}{6^8} - \frac{1}{7^8} + \frac{1}{10^8} - \frac{1}{11^8} - \cdots$$

$$\frac{105}{\pi^4} = 1 + \frac{1}{2^4} + \frac{1}{3^4} + \frac{1}{5^4} + \frac{1}{6^4} + \frac{1}{7^4} + \frac{1}{10^4} + \frac{1}{11^4} + \cdots$$

$$\frac{\pi^4}{105} = 1 - \frac{1}{2^4} - \frac{1}{3^4} + \frac{1}{4^4} - \frac{1}{5^4} + \frac{1}{6^4} - \frac{1}{7^4} - \frac{1}{8^4} + \frac{1}{9^4} + \cdots .$$

We also have the values of the following products:

$$\frac{\pi^4}{90} = \frac{2^4}{2^4-1} \cdot \frac{3^4}{3^4-1} \cdot \frac{5^4}{5^4-1} \cdot \frac{7^4}{7^4-1} \cdot \frac{11^4}{11^4-1} \cdots$$

$$\frac{\pi^8}{9450} = \frac{2^8}{2^8 - 1} \cdot \frac{3^8}{3^8 - 1} \cdot \frac{5^8}{5^8 - 1} \cdot \frac{7^8}{7^8 - 1} \cdot \frac{11^8}{11^8 - 1} \cdots$$

$$\frac{105}{\pi^4} = \frac{2^4 + 1}{2^4} \cdot \frac{3^4 + 1}{3^4} \cdot \frac{5^4 + 1}{5^4} \cdot \frac{7^4 + 1}{7^4} \cdot \frac{11^4 + 1}{11^4} \cdots$$

$$\frac{7}{6} = \frac{2^4 + 1}{2^4 - 1} \cdot \frac{3^4 + 1}{3^4 - 1} \cdot \frac{5^4 + 1}{5^4 - 1} \cdot \frac{7^4 + 1}{7^4 - 1} \cdot \frac{11^4 + 1}{11^4 - 1} \cdots$$

$\frac{35}{34} = \frac{41}{40} \cdot \frac{313}{312} \cdot \frac{1201}{1200} \cdot \frac{7321}{7320} \cdots$. In this last expression the numerator of each factor is one greater than the denominator and the sum of the numerator and denominator of each factor is the fourth power of the prime numbers, $3, 5, 7, 11, \cdots$.

278. Because we can express the sum of the series

$M = 1 + \frac{1}{2^n} + \frac{1}{3^n} + \frac{1}{4^n} + \frac{1}{5^n} + \frac{1}{6^n} + \cdots$ as a product of factors, it is convenient to use logarithms. Since

$$M = \frac{1}{\left(1 - \frac{1}{2^n}\right)\left(1 - \frac{1}{3^n}\right)\left(1 - \frac{1}{5^n}\right)\left(1 - \frac{1}{7^n}\right)\cdots},$$ we have

$$\log M = -\log\left(1 - \frac{1}{2^n}\right) - \log\left(1 - \frac{1}{3^n}\right) - \log\left(1 - \frac{1}{5^n}\right) - \log\left(1 - \frac{1}{7^n}\right) - \cdots .$$

If we use natural logarithms, then

$$\log M = 1\left(\frac{1}{2^n} + \frac{1}{3^n} + \frac{1}{5^n} + \frac{1}{7^n} + \frac{1}{11^n} + \cdots\right)$$

$$+ \frac{1}{2}\left(\frac{1}{2^{2n}} + \frac{1}{3^{2n}} + \frac{1}{5^{2n}} + \frac{1}{7^{2n}} + \frac{1}{11^{2n}} + \cdots\right)$$

$$+ \frac{1}{3}\left(\frac{1}{2^{3n}} + \frac{1}{3^{3n}} + \frac{1}{5^{3n}} + \frac{1}{7^{3n}} + \frac{1}{11^{3n}} + \cdots\right)$$

$$+ \frac{1}{4}\left(\frac{1}{2^{4n}} + \frac{1}{3^{4n}} + \frac{1}{5^{4n}} + \frac{1}{7^{4n}} + \frac{1}{11^{4n}} + \cdots\right)$$

etc. If we also let

$$N = 1 + \frac{1}{2^{2n}} + \frac{1}{3^{2n}} + \frac{1}{4^{2n}} + \frac{1}{5^{2n}} + \frac{1}{6^{2n}} + \cdots, \text{ then}$$

$$N = \frac{1}{\left(1-\frac{1}{2^{2n}}\right)\left(1-\frac{1}{3^{2n}}\right)\left(1-\frac{1}{5^{2n}}\right)\left(1-\frac{1}{7^{2n}}\right)\cdots}$$

and taking natural logarithms, we have

$$\log N = 1\left(\frac{1}{2^{2n}} + \frac{1}{3^{2n}} + \frac{1}{5^{2n}} + \frac{1}{7^{2n}} + \frac{1}{11^{2n}} + \cdots\right)$$

$$+ \frac{1}{2}\left(\frac{1}{2^{4n}} + \frac{1}{3^{4n}} + \frac{1}{5^{4n}} + \frac{1}{7^{4n}} + \frac{1}{11^{4n}} + \cdots\right)$$

$$+ \frac{1}{3}\left(\frac{1}{2^{6n}} + \frac{1}{3^{6n}} + \frac{1}{5^{6n}} + \frac{1}{7^{6n}} + \frac{1}{11^{6n}} + \cdots\right)$$

$$+ \frac{1}{4}\left(\frac{1}{2^{8n}} + \frac{1}{3^{8n}} + \frac{1}{5^{8n}} + \frac{1}{7^{8n}} + \frac{1}{11^{8n}} + \cdots\right)$$

etc. From the two expressions we obtain

$$\log M - \frac{1}{2}\log N = 1\left(\frac{1}{2^{n}} + \frac{1}{3^{n}} + \frac{1}{5^{n}} + \frac{1}{7^{n}} + \frac{1}{11^{n}} + \cdots\right)$$

$$+ \frac{1}{3}\left(\frac{1}{2^{3n}} + \frac{1}{3^{3n}} + \frac{1}{5^{3n}} + \frac{1}{7^{3n}} + \frac{1}{11^{3n}} + \cdots\right)$$

$$+ \frac{1}{5}\left(\frac{1}{2^{5n}} + \frac{1}{3^{5n}} + \frac{1}{5^{5n}} + \frac{1}{7^{5n}} + \frac{1}{11^{5n}} + \cdots\right)$$

$$+ \frac{1}{7}\left(\frac{1}{2^{7n}} + \frac{1}{3^{7n}} + \frac{1}{5^{7n}} + \frac{1}{7^{7n}} + \frac{1}{11^{7n}} + \cdots\right)$$

etc.

279. If $n = 1$, then $M = 1 + \frac{1}{2} + \frac{1}{3} + \frac{1}{4} + \cdots = \log \infty$ and $N = \frac{\pi^2}{6}$. Then

$$\log\log\infty - \frac{1}{2}\log\left(\frac{\pi^2}{6}\right) = 1\left(\frac{1}{2} + \frac{1}{3} + \frac{1}{5} + \frac{1}{7} + \frac{1}{11} + \cdots\right)$$

$$+ \frac{1}{3}\left(\frac{1}{2^3} + \frac{1}{3^3} + \frac{1}{5^3} + \frac{1}{7^3} + \frac{1}{11^3} + \cdots\right)$$

$$+ \frac{1}{5}\left(\frac{1}{2^5} + \frac{1}{3^5} + \frac{1}{5^5} + \frac{1}{7^5} + \frac{1}{11^5} + \cdots\right)$$

$$+ \frac{1}{7}\left(\frac{1}{2^7} + \frac{1}{3^7} + \frac{1}{5^7} + \frac{1}{7^7} + \frac{1}{11^7} + \cdots\right)$$

etc. But these series, except for the first ones, not only have finite sums, but the sum of all of them taken together is still finite, and reasonably small. It follows that the first series $\frac{1}{2} + \frac{1}{3} + \frac{1}{5} + \frac{1}{7} + \frac{1}{11} + \cdots$ has an infinite sum.

280. Let $n = 2$, then $M = \frac{\pi^2}{6}$ and $N = \frac{\pi^4}{90}$. It follows then that

$$2\log\pi - \log 6 = 1\left(\frac{1}{2^2} + \frac{1}{3^2} + \frac{1}{5^2} + \frac{1}{7^2} + \frac{1}{11^2} + \cdots\right)$$

$$+ \frac{1}{2}\left(\frac{1}{2^4} + \frac{1}{3^4} + \frac{1}{5^4} + \frac{1}{7^4} + \frac{1}{11^4} + \cdots\right)$$

$$+ \frac{1}{3}\left(\frac{1}{2^6} + \frac{1}{3^6} + \frac{1}{5^6} + \frac{1}{7^6} + \frac{1}{11^6} + \cdots\right)$$

etc., and

$$4\log\pi - \log 90 = 1\left(\frac{1}{2^4} + \frac{1}{3^4} + \frac{1}{5^4} + \frac{1}{7^4} + \frac{1}{11^4} + \cdots\right)$$

$$+ \frac{1}{2}\left(\frac{1}{2^8} + \frac{1}{3^8} + \frac{1}{5^8} + \frac{1}{7^8} + \frac{1}{11^8} + \cdots\right)$$

$$+ \frac{1}{3}\left(\frac{1}{2^{12}} + \frac{1}{3^{12}} + \frac{1}{5^{12}} + \frac{1}{7^{12}} + \frac{1}{11^{12}} + \cdots\right)$$

etc., and

$$\frac{1}{2}\log\frac{5}{2} = 1\left(\frac{1}{2^2} + \frac{1}{3^2} + \frac{1}{5^2} + \frac{1}{7^2} + \frac{1}{11^2} + \cdots\right)$$

$$+ \frac{1}{3}\left(\frac{1}{2^6} + \frac{1}{3^6} + \frac{1}{5^6} + \frac{1}{7^6} + \frac{1}{11^6} + \cdots\right)$$

$$+ \frac{1}{5}\left(\frac{1}{2^{10}} + \frac{1}{3^{10}} + \frac{1}{5^{10}} + \frac{1}{7^{10}} + \frac{1}{11^{10}} + \cdots\right)$$

etc.

281. Although the law by which the sequence of prime numbers proceeds is not known, still it is not difficult to assign an approximate value to the sum of the series of higher powers. Let the series

$$M = 1 + \frac{1}{2^n} + \frac{1}{3^n} + \frac{1}{4^n} + \frac{1}{5^n} + \frac{1}{6^n} + \frac{1}{7^n} + \cdots \text{ and}$$

$$S = \frac{1}{2^n} + \frac{1}{3^n} + \frac{1}{5^n} + \frac{1}{7^n} + \frac{1}{11^n} + \frac{1}{13^n} + \cdots, \text{ then}$$

$$S = M - 1 - \frac{1}{4^n} - \frac{1}{6^n} - \frac{1}{8^n} - \frac{1}{9^n} - \frac{1}{10^n} - \cdots \text{ and}$$

$$\frac{M}{2^n} = \frac{1}{2^n} + \frac{1}{4^n} + \frac{1}{6^n} + \frac{1}{8^n} + \frac{1}{10^n} + \frac{1}{12^n} + \cdots, \text{ so that}$$

$$S = M - \frac{M}{2^n} - 1 + \frac{1}{2^n} - \frac{1}{9^n} - \frac{1}{15^n} - \frac{1}{21^n} - \cdots \text{ or}$$

$$S = (M - 1)\left(1 - \frac{1}{2^n}\right) - \frac{1}{9^n} - \frac{1}{15^n} - \frac{1}{21^n} - \frac{1}{25^n} - \frac{1}{27^n} - \cdots .$$

Since

$$M\left(1 - \frac{1}{2^n}\right)\frac{1}{3^n} = \frac{1}{3^n} + \frac{1}{9^n} + \frac{1}{15^n} + \frac{1}{21^n} + \cdots, \text{ we have}$$

$$S = (M - 1)\left(1 - \frac{1}{2^n}\right)\left(1 - \frac{1}{3^n}\right) + \frac{1}{6^n} - \frac{1}{25^n} - \frac{1}{35^n} - \frac{1}{45^n} - \cdots .$$

Since we have the value of M, the value of S can conveniently be found, provided only that n is reasonably large.

282. Since we can find the sums S for larger values of n, we now consider the values for smaller values of n, which can be obtained from the formulas already found. In this way we find the following results for the series

$$\frac{1}{2^n} + \frac{1}{3^n} + \frac{1}{5^n} + \frac{1}{7^n} + \frac{1}{11^n} + \frac{1}{13^n} + \frac{1}{17^n} + \cdots .$$

If n equals	then the sum of the series equals
2	0.452247420041222
4	0.076993139764252
6	0.017070086850639
8	0.004061405366515
10	0.000993603573633
12	0.000246026470033
14	0.000061244396725
16	0.000015282026219
18	0.000003817278702
20	0.000000953961123
22	0.000000238450446
24	0.000000059608184
26	0.000000014901555
28	0.000000003725333
30	0.000000000931323
32	0.000000000232830
34	0.000000000058207
36	0.000000000014551

The remaining sums decrease by about one fourth at each step.

283. The series $1 + \frac{1}{2^n} + \frac{1}{3^n} + \frac{1}{4^n} + \cdots$ can be converted into a product directly as follows. Let

$$A = 1 + \frac{1}{2^n} + \frac{1}{3^n} + \frac{1}{4^n} + \frac{1}{5^n} + \frac{1}{6^n} + \frac{1}{7^n} + \frac{1}{8^n} + \cdots .$$

From A we subtract

$\frac{1}{2^n}A = \frac{1}{2^n} + \frac{1}{4^n} + \frac{1}{6^n} + \frac{1}{8^n} + \cdots$ to obtain

$B = \left(1 - \frac{1}{2^n}\right)A = 1 + \frac{1}{3^n} + \frac{1}{5^n} + \frac{1}{7^n} + \frac{1}{9^n} + \frac{1}{11^n} + \cdots$, in which there remain no terms divisible by 2. Then from B we subtract

$\frac{1}{3^n}B = \frac{1}{3^n} + \frac{1}{9^n} + \frac{1}{15^n} + \frac{1}{21^n} + \cdots$, to obtain

$C = \left(1 - \frac{1}{3^n}\right)B = 1 + \frac{1}{5^n} + \frac{1}{7^n} + \frac{1}{11^n} + \frac{1}{13^n} + \cdots$. In C there are no terms divisible by 3. From C we subtract

$\frac{1}{5^n}C = \frac{1}{5^n} + \frac{1}{25^n} + \frac{1}{35^n} + \frac{1}{55^n} + \cdots$ to obtain

$D = \left(1 - \frac{1}{5^n}\right)C = 1 + \frac{1}{7^n} + \frac{1}{11^n} + \frac{1}{13^n} + \frac{1}{17^n} + \cdots$ which contains no terms divisible by 5. In like manner we remove terms divisible by 7, 11 and all of the remaining prime numbers. It is clear that after all terms have been removed which are divisible by any prime, only 1 remains. Now for B, C, D, E, etc., we substitute their computed values to obtain

$1 = A\left(1 - \frac{1}{2^n}\right)\left(1 - \frac{1}{3^n}\right)\left(1 - \frac{1}{5^n}\right)\left(1 - \frac{1}{7^n}\right)\left(1 - \frac{1}{11^n}\right)\cdots$. It follows that the sum of the series

$$A = \frac{1}{\left(1-\frac{1}{2^n}\right)\left(1-\frac{1}{3^n}\right)\left(1-\frac{1}{5^n}\right)\left(1-\frac{1}{7^n}\right)\cdots}$$

or $A = \frac{2^n}{2^n - 1} \cdot \frac{3^n}{3^n - 1} \cdot \frac{5^n}{5^n - 1} \cdot \frac{7^n}{7^n - 1} \cdot \frac{11^n}{11^n - 1} \cdots .$

284. This method can conveniently be used to find infinite products for series for which we have already found sums. For instance in section 175 we found the sums of the series

$1 - \frac{1}{3^n} + \frac{1}{5^n} - \frac{1}{7^n} + \frac{1}{9^n} - \frac{1}{11^n} + \frac{1}{13^n} - \cdots$. If n is odd, the sum is equal to $N\pi^n$ where N has the various values we gave in that cited section. We note also that, since only odd numbers occur, when a number has the form $4m + 1$, we use a positive sign, while if it has the form $4m - 1$, the negative sign is used. Now we let

$A = 1 - \frac{1}{3^n} + \frac{1}{5^n} - \frac{1}{7^n} + \frac{1}{9^n} - \frac{1}{11^n} + \frac{1}{13^n} - \frac{1}{15^n} + \cdots$. We add

$\frac{1}{3^n}A = \frac{1}{3^n} - \frac{1}{9^n} + \frac{1}{15^n} - \frac{1}{21^n} + \frac{1}{27^n} - \cdots$ to obtain

$$B = \left(1 + \frac{1}{3^n}\right)A = 1 + \frac{1}{5^n} - \frac{1}{7^n} - \frac{1}{11^n} + \frac{1}{13^n} + \frac{1}{17^n} - \cdots .$$

Then we subtract $\frac{1}{5^n}B = \frac{1}{5^n} + \frac{1}{25^n} - \frac{1}{35^n} - \frac{1}{55^n} + \cdots$ to obtain

$C = \left(1 - \frac{1}{5^n}\right)B = 1 - \frac{1}{7^n} - \frac{1}{11^n} + \frac{1}{13^n} + \frac{1}{17^n} - \cdots$ where the numbers divisible by 3 or 5 have been removed. We add

$\frac{1}{7^n}C = \frac{1}{7^n} - \frac{1}{49^n} - \frac{1}{77^n} + \cdots$ to obtain

$D = \left(1 + \frac{1}{7^n}\right)C = 1 - \frac{1}{11^n} + \frac{1}{13^n} + \frac{1}{17^n} - \cdots$ which contains no numbers divisible by 7. We add $\frac{1}{11^n}D = \frac{1}{11^n} - \frac{1}{121^n} + \cdots$ to obtain $E = \left(1 + \frac{1}{11^n}\right)D = 1 + \frac{1}{13^n} + \frac{1}{17^n} - \cdots$, where numbers divisible by 11 have been removed. In this way we remove all of the remaining terms which are divisible by prime numbers, so that we finally have $1 = A\left(1 + \frac{1}{3^n}\right)\left(1 - \frac{1}{5^n}\right)\left(1 + \frac{1}{7^n}\right)\left(1 + \frac{1}{11^n}\right)\left(1 - \frac{1}{13^n}\right)\cdots$ and

$$A = \frac{3^n}{3^n + 1} \cdot \frac{5^n}{5^n - 1} \cdot \frac{7^n}{7^n + 1} \cdot \frac{11^n}{11^n + 1} \cdot \frac{13^n}{13^n - 1} \cdot \frac{17^n}{17^n - 1} \cdots,$$

where the the powers of all of the prime numbers in the numerators occur and the denominators are increased or decreased by 1 depending on whether the number has the form $4m - 1$ or $4m + 1$.

285. Let $n = 1$. Since $A = \frac{\pi}{4}$, we have $\frac{\pi}{4} = \frac{3}{4} \cdot \frac{5}{4} \cdot \frac{7}{8} \cdot \frac{11}{12} \cdot \frac{13}{12} \cdot \frac{17}{16} \cdot \frac{19}{20} \cdot \frac{23}{24} \cdots$. Previously we found that $\frac{\pi^2}{6} = \frac{4}{3} \cdot \frac{3^2}{2\cdot 4} \cdot \frac{5^2}{4\cdot 6} \cdot \frac{7^2}{6\cdot 8} \cdot \frac{11^2}{10\cdot 12} \cdot \frac{13^2}{12\cdot 14} \cdot \frac{17^2}{16\cdot 18} \cdot \frac{19^2}{18\cdot 20} \cdots$. If we divide the second expression by the first, we obtain

$$\frac{2\pi}{3} = \frac{4}{3} \cdot \frac{3}{2} \cdot \frac{5}{6} \cdot \frac{7}{6} \cdot \frac{11}{10} \cdot \frac{13}{14} \cdot \frac{17}{18} \cdot \frac{19}{18} \cdot \frac{23}{22} \cdots$$ or

$\frac{\pi}{2} = \frac{3}{2} \cdot \frac{5}{6} \cdot \frac{7}{6} \cdot \frac{11}{10} \cdot \frac{13}{14} \cdot \frac{17}{18} \cdot \frac{19}{18} \cdot \frac{23}{22} \cdots$ where the numerators are prime numbers and the denominators are oddly even, differing from the numerators by 1. If now we divide this last expression by the first expression for $\frac{\pi}{4}$, we obtain $2 = \frac{4}{2} \cdot \frac{4}{6} \cdot \frac{8}{6} \cdot \frac{12}{10} \cdot \frac{12}{14} \cdot \frac{16}{18} \cdot \frac{20}{18} \cdot \frac{24}{22} \cdots$ or

$2 = \frac{2}{1} \cdot \frac{2}{3} \cdot \frac{4}{3} \cdot \frac{6}{5} \cdot \frac{6}{7} \cdot \frac{8}{9} \cdot \frac{10}{9} \cdot \frac{12}{11} \cdots$. These fractions arise from the odd prime numbers, 3, 5, 7, 11, 13, 17, $\cdots$. For each of these primes we take the pair of numbers which differ from it by unity, in either direction. From each of the resulting pair of even numbers, when divided by 2, the odd number becomes the denominator and the even number the numerator.

286. We can compare these expressions with those of Wallis:

$$\frac{\pi}{2} = \frac{2\cdot2\cdot4\cdot4\cdot6\cdot6\cdot8\cdot8\cdot10\cdot10\cdot12}{1\cdot3\cdot3\cdot5\cdot5\cdot7\cdot7\cdot9\cdot9\cdot11\cdot11} \cdots$$

$$\frac{4}{\pi} = \frac{3\cdot3}{2\cdot4} \cdot \frac{5\cdot5}{4\cdot6} \cdot \frac{7\cdot7}{6\cdot8} \cdot \frac{9\cdot9}{8\cdot10} \cdot \frac{11\cdot11}{10\cdot12} \cdots$$

$$\frac{\pi^2}{8} = \frac{3\cdot3}{2\cdot4} \cdot \frac{5\cdot5}{4\cdot6} \cdot \frac{7\cdot7}{6\cdot8} \cdot \frac{11\cdot11}{10\cdot12} \cdot \frac{13\cdot13}{12\cdot14} \cdots .$$

If the previous expression is divided by the last one we have

$\frac{32}{\pi^3} = \frac{9\cdot9}{8\cdot10} \cdot \frac{15\cdot15}{14\cdot16} \cdot \frac{21\cdot21}{20\cdot22} \cdot \frac{25\cdot25}{24\cdot26} \cdots$, where all of the non-prime odd numbers appear in the numerators.

287. Let $n = 3$, then $A = \frac{\pi^3}{32}$ and

$$\frac{\pi^3}{32} = \frac{3^3}{3^3+1} \cdot \frac{5^3}{5^3-1} \cdot \frac{7^3}{7^3+1} \cdot \frac{11^3}{11^3+1} \cdot \frac{13^3}{13^3-1} \cdot \frac{17^3}{17^3-1} \cdots .$$

From the series $\frac{\pi^6}{945} = 1 + \frac{1}{2^6} + \frac{1}{3^6} + \frac{1}{4^6} + \frac{1}{5^6} + \cdots$ we have

$$\frac{\pi^6}{945} = \frac{2^6}{2^6-1} \cdot \frac{3^6}{3^6-1} \cdot \frac{5^6}{5^6-1} \cdot \frac{7^6}{7^6-1} \cdot \frac{11^6}{11^6-1} \cdot \frac{13^6}{13^6-1} \cdots$$

or $\frac{\pi^6}{960} = \frac{3^6}{3^6-1} \cdot \frac{5^6}{5^6-1} \cdot \frac{7^6}{7^6-1} \cdot \frac{11^6}{11^6-1} \cdot \frac{13^6}{13^6-1} \cdots$. When this last expression is divided by the first we obtain

$$\frac{\pi^3}{30} = \frac{3^3}{3^3-1} \cdot \frac{5^3}{5^3+1} \cdot \frac{7^3}{7^3-1} \cdot \frac{11^3}{11^3-1} \cdot \frac{13^3}{13^3+1} \cdot \frac{17^3}{17^3+1} \cdots .$$

When this expression is divided again by the first we have

$$\frac{16}{15} = \frac{3^3+1}{3^3-1} \cdot \frac{5^3-1}{5^3+1} \cdot \frac{7^3+1}{7^3-1} \cdot \frac{11^3+1}{11^3-1} \cdot \frac{13^3-1}{13^3+1} \cdot \frac{17^3-1}{17^3+1} \cdots$$

or $\frac{16}{15} = \frac{14}{13} \cdot \frac{62}{63} \cdot \frac{172}{171} \cdot \frac{666}{665} \cdot \frac{1098}{1099} \cdots$. These fractions arise from the cubes of odd prime numbers. After dividing by two the pair of even numbers which differ from the cube by unity, we take the even part for the numerator and the odd part for the denominator.

288. From these expressions we can derive new series in which the denominators consist of all the natural numbers. Since

$\frac{\pi}{4} = \frac{3}{3+1} \cdot \frac{5}{5-1} \cdot \frac{7}{7+1} \cdot \frac{11}{11+1} \cdot \frac{13}{13-1} \cdots$, we have

$$\frac{\pi}{6} = \frac{1}{\left(1+\frac{1}{2}\right)\left(1+\frac{1}{3}\right)\left(1-\frac{1}{5}\right)\left(1+\frac{1}{7}\right)\left(1+\frac{1}{11}\right)\cdots}.$$

It follows that the series is

$\frac{\pi}{6} = 1 - \frac{1}{2} - \frac{1}{3} + \frac{1}{4} + \frac{1}{5} + \frac{1}{6} - \frac{1}{7} - \frac{1}{8} + \frac{1}{9} - \frac{1}{10} - \cdots$, where the rule for the signs is as follows: for 2 the sign is negative; for the odd primes of the form $4m-1$, the sign is negative; for odd primes of the form $4m+1$, the sign is positive; for the composite numbers, the sign is the product of the signs of the prime factors. For example, the sign of the fraction $\frac{1}{60}$ is negative since $-60 = (-2)(-2)(-3)(+5)$. In a similar way we have

$$\frac{\pi}{2} = \frac{1}{\left(1-\frac{1}{2}\right)\left(1+\frac{1}{3}\right)\left(1-\frac{1}{5}\right)\left(1+\frac{1}{7}\right)\left(1+\frac{1}{11}\right)\cdots},$$

so that the series is

$\frac{\pi}{2} = 1 + \frac{1}{2} - \frac{1}{3} + \frac{1}{4} + \frac{1}{5} - \frac{1}{6} - \frac{1}{7} + \frac{1}{8} + \frac{1}{9} + \frac{1}{10} - \cdots$, where 2 has a positive sign; prime numbers of the form $4m - 1$ have a negative sign; prime numbers of the form $4m + 1$ have a positive sign; a composite number has the sign of the product of the signs of its factors, according to the rules for multiplication.

289. Since

$$\frac{\pi}{2} = \frac{1}{\left(1-\frac{1}{3}\right)\left(1+\frac{1}{5}\right)\left(1-\frac{1}{7}\right)\left(1-\frac{1}{11}\right)\cdots},$$

we have $\frac{\pi}{2} = 1 + \frac{1}{3} - \frac{1}{5} + \frac{1}{7} + \frac{1}{9} + \frac{1}{11} - \frac{1}{13} - \frac{1}{15} + \cdots$, where only the odd numbers occur and the signs follow this law: prime numbers of the form $4m - 1$ have a positive sign; prime numbers of the form $4m + 1$ have a negative sign; from the previous rules the sign of a composite number is defined. A second series can be formed, from this one, in which all the natural numbers occur, namely, $\pi = \frac{1}{\left(1-\frac{1}{2}\right)\left(1-\frac{1}{3}\right)\left(1+\frac{1}{5}\right)\left(1-\frac{1}{7}\right)\left(1-\frac{1}{11}\right)\cdots}$.

From this we have

$\pi = 1 + \frac{1}{2} + \frac{1}{3} + \frac{1}{4} - \frac{1}{5} + \frac{1}{6} + \frac{1}{7} + \frac{1}{8} + \frac{1}{9} - \frac{1}{10} + \cdots$, where 2 has a positive sign; prime numbers of the form $4m - 1$ have a positive sign; prime numbers of the form $4m + 1$ have a negative sign.

290. There are innumerable ways of assigning signs so that the series with $1, \frac{1}{2}, \frac{1}{3}, \frac{1}{4}, \frac{1}{5}, \frac{1}{6}, \frac{1}{7}, \frac{1}{8}, \cdots$ take some sought for sum. For example,

$$\frac{\pi}{2} = \frac{1}{\left(1-\frac{1}{2}\right)\left(1+\frac{1}{3}\right)\left(1-\frac{1}{5}\right)\left(1+\frac{1}{7}\right)\left(1+\frac{1}{11}\right)\cdots}.$$

If we multiply this expression by $\dfrac{\left(1+\frac{1}{3}\right)}{\left(1-\frac{1}{3}\right)} = 2$, then,

$$\pi = \frac{1}{\left(1-\frac{1}{2}\right)\left(1-\frac{1}{3}\right)\left(1-\frac{1}{5}\right)\left(1+\frac{1}{7}\right)\left(1+\frac{1}{11}\right)\cdots}$$ and

$$\pi = 1 + \frac{1}{2} + \frac{1}{3} + \frac{1}{4} + \frac{1}{5} + \frac{1}{6} - \frac{1}{7} + \frac{1}{8} + \frac{1}{9} + \frac{1}{10} - \frac{1}{11} + \cdots,$$

where 2 has a positive sign, 3 has a positive sign and the remaining prime numbers of the form $4m - 1$ have a negative sign; prime numbers of the form $4m + 1$ have a positive sign; the sign of the composite numbers is known from the signs of its factors. In the same way, since

$$\pi = \frac{1}{\left(1-\frac{1}{2}\right)\left(1-\frac{1}{3}\right)\left(1+\frac{1}{5}\right)\left(1-\frac{1}{7}\right)\left(1-\frac{1}{11}\right)\cdots},$$ when this expression is multiplied by $\dfrac{1+\frac{1}{5}}{1-\frac{1}{5}} = \dfrac{3}{2}$, we have

$$\frac{3\pi}{2} = \frac{1}{\left(1-\frac{1}{2}\right)\left(1-\frac{1}{3}\right)\left(1-\frac{1}{5}\right)\left(1-\frac{1}{7}\right)\left(1-\frac{1}{11}\right)\cdots}$$

and $\dfrac{3\pi}{2} = 1 + \dfrac{1}{2} + \dfrac{1}{3} + \dfrac{1}{4} + \dfrac{1}{5} + \dfrac{1}{6} + \dfrac{1}{7} + \dfrac{1}{8} + \dfrac{1}{9} + \cdots$, where 2 has a positive sign; prime numbers of the form $4m - 1$ have a positive sign; prime numbers of the form $4m + 1$, except 5, have a negative sign.

291. There are innumerable ways of constructing a series with a sum equal to zero. Since $0 = \frac{2}{3} \cdot \frac{3}{4} \cdot \frac{5}{6} \cdot \frac{7}{8} \cdot \frac{11}{12} \cdot \frac{13}{14} \cdot \frac{17}{18} \cdots$, we have

$$0 = \frac{1}{\left(1+\frac{1}{2}\right)\left(1+\frac{1}{3}\right)\left(1+\frac{1}{5}\right)\left(1+\frac{1}{7}\right)\left(1+\frac{1}{11}\right)\cdots},$$

so that $0 = 1 - \frac{1}{2} - \frac{1}{3} + \frac{1}{4} - \frac{1}{5} + \frac{1}{6} - \frac{1}{7} - \frac{1}{8} + \frac{1}{9} + \frac{1}{10} - \cdots$,

where all prime numbers have a negative sign and composite numbers have the sign given by the rules of multiplication. If we multiply the given product expression by

$$\frac{1 + \frac{1}{2}}{1 - \frac{1}{2}} = 3, \text{ then}$$

$$0 = \frac{1}{\left(1-\frac{1}{2}\right)\left(1+\frac{1}{3}\right)\left(1+\frac{1}{5}\right)\left(1+\frac{1}{7}\right)\left(1+\frac{1}{11}\right)\cdots},$$

so that $0 = 1 + \frac{1}{2} - \frac{1}{3} + \frac{1}{4} - \frac{1}{5} - \frac{1}{6} - \frac{1}{7} + \frac{1}{8} + \frac{1}{9} - \frac{1}{10} - \cdots$,

where 2 has a positive sign; the other prime numbers have a negative sign. Likewise,

$$0 = \frac{1}{\left(1+\frac{1}{2}\right)\left(1-\frac{1}{3}\right)\left(1-\frac{1}{5}\right)\left(1+\frac{1}{7}\right)\left(1+\frac{1}{11}\right)\cdots},$$

so that

$0 = 1 - \frac{1}{2} + \frac{1}{3} + \frac{1}{4} + \frac{1}{5} - \frac{1}{6} - \frac{1}{7} - \frac{1}{8} + \frac{1}{9} - \frac{1}{10} - \cdots$, where all primes except 3 and 5 have a negative sign. In general, as long as all prime numbers, except for a finite collection, have negative signs, then the sum of the series will be equal to zero. On the other hand, if all but a finite collection of primes have a positive sign, then the sum of the series will be infinitely large.

292. In section 176 we found sums for the series

$A = 1 - \frac{1}{2^n} + \frac{1}{4^n} - \frac{1}{5^n} + \frac{1}{7^n} - \frac{1}{8^n} + \frac{1}{10^n} - \frac{1}{11^n} + \frac{1}{13^n} - \cdots$. If n is an odd number, then

$\frac{1}{2^n}A = \frac{1}{2^n} - \frac{1}{4^n} + \frac{1}{8^n} - \frac{1}{10^n} + \frac{1}{14^n} - \cdots$. When these two series are added we obtain

$$B = \left(1 + \frac{1}{2^n}\right)A = 1 - \frac{1}{5^n} + \frac{1}{7^n} - \frac{1}{11^n} + \frac{1}{13^n} - \frac{1}{17^n} + \frac{1}{19^n} - \frac{1}{23^n} + \frac{1}{25^n} - \cdots$$

and $\frac{1}{5^n}B = \frac{1}{5^n} - \frac{1}{25^n} + \frac{1}{35^n} - \frac{1}{55^n} + \cdots$. When these two series are added we obtain

$$C = \left(1 + \frac{1}{5^n}\right)B = 1 + \frac{1}{7^n} - \frac{1}{11^n} + \frac{1}{13^n} - \frac{1}{17^n} + \frac{1}{19^n} - \frac{1}{25^n} + \cdots$$

and $\frac{1}{7^n}C = \frac{1}{7^n} + \frac{1}{49^n} - \frac{1}{77^n} + \cdots$. When the last is subtracted from the other series we obtain

$$D = \left(1 - \frac{1}{7^n}\right)C = 1 - \frac{1}{11^n} + \frac{1}{13^n} - \frac{1}{17^n} + \frac{1}{19^n} - \cdots .$$

From all of these we obtain

$A\left(1 + \frac{1}{2^n}\right)\left(1 + \frac{1}{5^n}\right)\left(1 - \frac{1}{7^n}\right)\left(1 + \frac{1}{11^n}\right)\left(1 - \frac{1}{13^n}\right)\cdots = 1$ where the prime numbers which are 1 greater than a multiple of 6 have a negative sign and those which are 1 less than a multiple of 6 have a positive sign. It follows that

$$A = \frac{2^n}{2^n + 1} \cdot \frac{5^n}{5^n + 1} \cdot \frac{7^n}{7^n - 1} \cdot \frac{11^n}{11^n + 1} \cdot \frac{13^n}{13^n - 1} \cdots .$$

293. Let us consider the case when $n = 1$. Since $A = \dfrac{\pi}{3\sqrt{3}}$, we have

$\dfrac{\pi}{3\sqrt{3}} = \dfrac{2}{3} \cdot \dfrac{5}{6} \cdot \dfrac{7}{6} \cdot \dfrac{11}{12} \cdot \dfrac{13}{12} \cdot \dfrac{17}{18} \cdot \dfrac{19}{18} \cdots$, where all of the prime numbers except 3 occur in the numerators, and the denominators are all divisible by 6 and differ from the numerator by 1. Since

$\dfrac{\pi^2}{6} = \dfrac{4}{3} \cdot \dfrac{9}{8} \cdot \dfrac{5 \cdot 5}{4 \cdot 6} \cdot \dfrac{7 \cdot 7}{6 \cdot 8} \cdot \dfrac{11 \cdot 11}{10 \cdot 12} \cdot \dfrac{13 \cdot 13}{12 \cdot 14} \cdots$, when this expression is divided by the previous expression, we obtain

$\dfrac{\pi\sqrt{3}}{2} = \dfrac{9}{4} \cdot \dfrac{5}{4} \cdot \dfrac{7}{8} \cdot \dfrac{11}{10} \cdot \dfrac{13}{14} \cdot \dfrac{17}{16} \cdot \dfrac{19}{20} \cdots$, where the denominators are not divisible by 6. We also have

$$\frac{\pi}{2\sqrt{3}} = \frac{5}{6} \cdot \frac{7}{6} \cdot \frac{11}{12} \cdot \frac{13}{12} \cdot \frac{17}{18} \cdot \frac{19}{18} \cdot \frac{23}{24} \cdots$$

$$\frac{2\pi}{3\sqrt{3}} = \frac{5}{4} \cdot \frac{7}{8} \cdot \frac{11}{10} \cdot \frac{13}{14} \cdot \frac{17}{16} \cdot \frac{19}{20} \cdot \frac{23}{22} \cdots .$$

When the last expression is divided by the previous one, we obtain

$\dfrac{4}{3} = \dfrac{6}{4} \cdot \dfrac{6}{8} \cdot \dfrac{12}{10} \cdot \dfrac{12}{14} \cdot \dfrac{18}{16} \cdot \dfrac{18}{20} \cdot \dfrac{24}{22} \cdots$ or

$\dfrac{4}{3} = \dfrac{3}{2} \cdot \dfrac{3}{4} \cdot \dfrac{6}{5} \cdot \dfrac{6}{7} \cdot \dfrac{9}{8} \cdot \dfrac{9}{10} \cdot \dfrac{12}{11} \cdots$. In this expression each fraction arises from a prime number 5, 7, 11, $\cdots$ by taking the pair of even numbers which differ from the prime by unity, and by using the number divisible by 3 for the numerator.

294. Since we saw above that $\dfrac{\pi}{4} = \dfrac{3}{4} \cdot \dfrac{5}{4} \cdot \dfrac{7}{8} \cdot \dfrac{11}{12} \cdot \dfrac{13}{12} \cdot \dfrac{17}{16} \cdots$ or

$\dfrac{\pi}{3} = \dfrac{5}{4} \cdot \dfrac{7}{8} \cdot \dfrac{11}{12} \cdot \dfrac{13}{12} \cdot \dfrac{17}{16} \cdot \dfrac{19}{20} \cdots$, when the expressions above for

$\frac{\pi}{2\sqrt{3}}$ and $\frac{2\pi}{3\sqrt{3}}$ are divided by this latest expression, we obtain

$\frac{\sqrt{3}}{2} = \frac{2}{3} \cdot \frac{4}{3} \cdot \frac{8}{9} \cdot \frac{10}{9} \cdot \frac{14}{15} \cdot \frac{16}{15} \cdots$ and

$\frac{2}{\sqrt{3}} = \frac{6}{5} \cdot \frac{6}{7} \cdot \frac{12}{11} \cdot \frac{18}{19} \cdot \frac{24}{23} \cdot \frac{30}{29} \cdots$. In the first of these two expressions, the fractions arise from primes of the form $12m + 6 \pm 1$, while in the second they arise from primes of the form $12m \pm 1$. In both cases, the pair of numbers, differing from the prime by unity, form the numerator and denominator is such a way that, when reduced, the numerator is even.

295. Now we consider the series discussed in section 179 which is expressed by

$A = \frac{\pi}{2\sqrt{2}} = 1 + \frac{1}{3} - \frac{1}{5} - \frac{1}{7} + \frac{1}{9} + \frac{1}{11} - \frac{1}{13} - \frac{1}{15} + \cdots$. Then

$\frac{1}{3}A = \frac{1}{3} + \frac{1}{9} - \frac{1}{15} - \frac{1}{21} + \frac{1}{27} + \frac{1}{33} - \cdots$.

When this is subtracted from the previous expression we have

$B = \left(1 - \frac{1}{3}\right)A = 1 - \frac{1}{5} - \frac{1}{7} + \frac{1}{11} - \frac{1}{13} + \frac{1}{17} + \frac{1}{19} + \cdots$. Then

$\frac{1}{5}B = \frac{1}{5} - \frac{1}{25} - \frac{1}{35} + \frac{1}{55} - \cdots$, which when added to the previous produces $C = \left(1 + \frac{1}{5}\right)B = 1 - \frac{1}{7} + \frac{1}{11} - \frac{1}{13} + \frac{1}{17} + \cdots$. If we proceed in the same way we finally arrive at the expression

$$1 = \frac{\pi}{2\sqrt{2}}\left(1 - \frac{1}{3}\right)\left(1 + \frac{1}{5}\right)\left(1 + \frac{1}{7}\right)\left(1 - \frac{1}{11}\right)\left(1 + \frac{1}{13}\right)\left(1 - \frac{1}{17}\right)\cdots,$$

where the signs are assigned as follows. If the prime has the form $8m + 1$ or $8m + 3$, then the negative sign is used, while if the form is $8m + 5$ or $8m + 7$, the sign is positive. It follows that

$$\frac{\pi}{2\sqrt{2}} = \frac{3}{2} \cdot \frac{5}{6} \cdot \frac{7}{8} \cdot \frac{11}{10} \cdot \frac{13}{14} \cdot \frac{17}{16} \cdot \frac{19}{18} \cdot \frac{23}{24} \cdots ,$$

where all denominators are divisible by 8 or are oddly even. Since we have

$$\frac{\pi}{4} = \frac{3}{4} \cdot \frac{5}{4} \cdot \frac{7}{8} \cdot \frac{11}{12} \cdot \frac{13}{12} \cdot \frac{17}{16} \cdot \frac{19}{20} \cdot \frac{23}{24} \cdots$$

and

$$\frac{\pi}{2} = \frac{3}{2} \cdot \frac{5}{6} \cdot \frac{7}{6} \cdot \frac{11}{10} \cdot \frac{13}{14} \cdot \frac{17}{18} \cdot \frac{19}{18} \cdot \frac{23}{22} \cdots ,$$

when these two expressions are multiplied we have

$$\frac{\pi^2}{8} = \frac{3\cdot 3}{2\cdot 4} \cdot \frac{5\cdot 5}{4\cdot 6} \cdot \frac{7\cdot 7}{6\cdot 8} \cdot \frac{11\cdot 11}{10\cdot 12} \cdot \frac{13\cdot 13}{12\cdot 14} \cdots .$$

When this expression is divided by the expression for $\frac{\pi}{2\sqrt{2}}$ we obtain

$$\frac{\pi}{2\sqrt{2}} = \frac{3}{4} \cdot \frac{5}{4} \cdot \frac{7}{6} \cdot \frac{11}{12} \cdot \frac{13}{12} \cdot \frac{17}{18} \cdot \frac{19}{20} \cdot \frac{23}{22} \cdots ,$$

where no denominator is divisible by 8, but those divisible by 4 are present whenever the numerator differs by unity. When the first expression for $\frac{\pi}{2\sqrt{2}}$ is divided by this second expression, we obtain

$$1 = \frac{2}{1} \cdot \frac{2}{3} \cdot \frac{3}{4} \cdot \frac{6}{5} \cdot \frac{6}{7} \cdot \frac{9}{8} \cdot \frac{10}{9} \cdot \frac{11}{12} \cdots ,$$

where each fraction arises from a prime number. For each prime the pair of numbers which differ from the prime by unity are divided by 2 and then the even number becomes the numerator unless it is divisible by 4.

296. In a like manner the other series, which express circular arcs, found in sections 179 and following, could be expressed as products dependent on the prime numbers. In this way we could develop important properties of both

infinite series and infinite products, but since we have discussed the principal results, we will not delay any longer to develop more. We will now proceed to a related topic. In this chapter we have considered the expression of numbers by products; in the next we move to the generation of numbers through addition.

CHAPTER XVI

On the Partition of Numbers

297. Let the following expression be given:

$(1 + x^\alpha z)(1 + x^\beta z)(1 + x^\gamma z)(1 + x^\delta z)(1 + x^\epsilon z) \cdots$. We ask about the form if the factors are actually multiplied. We suppose that it has the form $1 + Pz + Qz^2 + Rz^3 + Sz^4 + \cdots$, where it is clear that P is equal to the sum of the powers $x^\alpha + x^\beta + x^\gamma + x^\epsilon + \cdots$. Then Q is the sum of the products of the powers taken two at a time, that is Q is the sum of the different powers of x whose exponents are the sum of two of the different terms in the sequence $\alpha, \beta, \gamma, \delta, \epsilon, \zeta, \eta$, etc. In like manner R is the sum of powers of x whose exponents are the sum of three of the different terms. Further, S is the sum of powers of x whose exponents are the sum of four of the different terms in that same sequence $\alpha, \beta, \gamma, \delta, \epsilon$, etc., and so forth.

298. The individual powers of x which constitute the values of the letters P, Q, R, S, etc. have a coefficient of 1 if their exponents can be formed in only one way from $\alpha, \beta, \gamma, \delta$, etc. If the same exponent of a power of x can be obtained in several ways as the sum of two, three, or more terms of the sequence $\alpha, \beta, \gamma, \delta, \epsilon$, etc., then that power has a coefficient equal to the number of ways the exponent can be obtained. Thus, if in the value of Q there is found Nx^n, this is because n has N different ways of being expressed as a sum of two terms

from the sequence α, β, γ, etc. Further, if in the expression of the given product the term $Nx^n z^m$ occurs, this is because there are N different ways in which n can be a sum of m terms of the sequence α, β, γ, δ, ϵ, ζ, etc.

299. If the given product $(1 + x^\alpha z)(1 + x^\beta z)(1 + x^\gamma z)(1 + x^\delta z) \cdots$ is actually multiplied, then from the resulting expression it becomes immediately apparent how many different ways a given number can be the sum of any desired number of terms from the sequence α, β, γ, δ, ϵ, ζ, etc. For example, if it is desired to know how many different ways the number n can be the sum of m terms of the given sequence, then we find the term $x^n z^m$, and its coefficient is the desired number.

300. In order that this may become clearer, let us consider the following infinite product, $(1 + xz)(1 + x^2z)(1 + x^3z)(1 + x^4z)(1 + x^5z) \cdots$, which by actual multiplication gives

$$\begin{aligned} 1 &+ z(x + x^2 + x^3 + x^4 + x^5 + x^6 + x^7 + x^8 + x^9 + \cdots) \\ &+ z^2(x^3 + x^4 + 2x^5 + 2x^6 + 3x^7 + 3x^8 + 4x^9 + 4x^{10} \\ &\qquad + 5x^{11} + \cdots) \\ &+ z^3(x^6 + x^7 + 2x^8 + 3x^9 + 4x^{10} + 5x^{11} + 7x^{12} + 8x^{13} \\ &\qquad + 10x^{14} + \cdots) \\ &+ z^4(x^{10} + x^{11} + 2x^{12} + 3x^{13} + 5x^{14} + 6x^{15} + 9x^{16} \\ &\qquad + 11x^{17} + 15x^{18} + \cdots) \\ &+ z^5(x^{15} + x^{16} + 2x^{17} + 3x^{18} + 5x^{19} + 7x^{20} + 10x^{21} \\ &\qquad + 13x^{22} + 18x^{23} + \cdots) \\ &+ z^6(x^{21} + x^{22} + 2x^{23} + 3x^{24} + 5x^{25} + 7x^{26} + 11x^{27} \end{aligned}$$

$$+ 14x^{28} + 20x^{29} + \cdots)$$

$$+ z^7(x^{28} + x^{29} + 2x^{30} + 3x^{31} + 5x^{32} + 7x^{33} + 11x^{34}$$

$$+ 15x^{35} + 21x^{36} + \cdots)$$

$$+ z^8(x^{36} + x^{37} + 2x^{38} + 3x^{39} + 5x^{40} + 7x^{41} + 11x^{42}$$

$$+ 15x^{43} + 22x^{44} + \cdots)$$

etc. From this series we can immediately determine how many ways a given number can arise as the sum of a given number of terms in the sequence 1, 2, 3, 4, 5, 6, 7, 8, $\cdots$. For example, if we want to know how many ways the number 35 can be the sum of seven different terms in the sequence 1, 2, 3, 4, 5, 6, 7, $\cdots$, we look in the series for z^7 multiplying the power x^{35}. The coefficient, 15, is the number of ways in which 35 can be the sum of seven different terms in the sequence 1, 2, 3, 4, 5, 6, 7, $\cdots$.

301. If we let $z = 1$ and put all of the same powers of x in one term, or, what comes to the same thing, if we express the following infinite product $(1 + x)(1 + x^2)(1 + x^3)(1 + x^4)(1 + x^5)(1 + x^6) \cdots$, we obtain the series $1 + x + x^2 + 2x^3 + 2x^4 + 3x^5 + 4x^6 + 5x^7 + 6x^8 + \cdots$, where each coefficient indicates in how many different ways the exponent of x can be expressed as the sum of different terms of the sequence 1, 2, 3, 4, 5, 6, 7, $\cdots$. For example, 8 can be produced six ways as a sum of different numbers, as follows:

$$8 = 8 \qquad 8 = 4 + 3 + 1 \qquad 8 = 7 + 1$$
$$8 = 5 + 2 + 1 \qquad 8 = 6 + 2 \qquad 8 = 5 + 3.$$

We note that the proposed number itself should be counted since the number of terms is not specified with the result that a single one is not excluded.

302. We know now how many ways a number can be expressed as the sum of different numbers. The condition that the numbers be different is omitted if the product is put into the denominator. Let us consider the following expression, $\frac{1}{(1 - x^\alpha z)(1 - x^\beta z)(1 - x^\gamma z)(1 - x^\delta z)(1 - x^\epsilon z) \cdots}$. By actual division we obtain an infinite series $1 + Pz + Qz^2 + Rz^3 + Sz^4 + \cdots$, where it is clear that P is the sum of the powers of x whose exponents are contained in the sequence $\alpha, \beta, \gamma, \delta, \epsilon, \zeta, \eta \cdots$. Then Q is the sum of the powers of x whose exponents are sums of the two terms of the sequence, whether the same or different. Further, R is the sum of powers of x whose exponents are the sum of three terms of the sequence; S is the sum of powers whose exponents are the sums of four terms in the sequence; and so forth.

303. If the total expression is written out and similar terms added, then we will know how many ways a given number n can be the sum of m terms, whether the same or different, of the sequence $\alpha, \beta, \gamma, \delta, \epsilon, \zeta, \cdots$. Indeed, we look for the coefficient N of the term with $x^n z^m$, where the total term will be $Nx^n z^m$, and N is the number of different ways in which n can be expressed as the sum of m terms in the sequence $\alpha, \beta, \gamma, \delta, \epsilon, \cdots$. After having settled the previous question we resolve the present one in a similar way.

304. We consider a particular case which is especially worth noting. If the given expression is $\frac{1}{(1 - xz)(1 - x^2 z)(1 - x^3 z)(1 - x^4 z)(1 - x^5 z) \cdots}$, when the actual division is carried out we have

$$1 + z(x + x^2 + x^3 + x^4 + x^5 + x^6 + x^7 + x^8 + x^9 + \cdots)$$
$$+ z^2(x^2 + x^3 + 2x^4 + 2x^5 + 3x^6 + 3x^7 + 4x^8 + 4x^9$$
$$+ 5x^{10} + \cdots)$$
$$+ z^3(x^3 + x^4 + 2x^5 + 3x^6 + 4x^7 + 5x^8 + 7x^9 + 8x^{10}$$
$$+ 10x^{11} + \cdots)$$
$$+ z^4(x^4 + x^5 + 2x^6 + 3x^7 + 5x^8 + 6x^9 + 9x^{10} + 11x^{11}$$
$$+ 15x^{12} + \cdots)$$
$$+ z^5(x^5 + x^6 + 2x^7 + 3x^8 + 5x^9 + 7x^{10} + 10x^{11} + 13x^{12}$$
$$+ 18x^{13} + \cdots)$$
$$+ z^6(x^6 + x^7 + 2x^8 + 3x^9 + 5x^{10} + 7x^{11} + 11x^{12} + 14x^{13}$$
$$+ 20x^{14} + \cdots)$$
$$+ z^7(x^7 + x^8 + 2x^9 + 3x^{10} + 5x^{11} + 7x^{12} + 11x^{13} + 15x^{14}$$
$$+ 21x^{15} + \cdots)$$
$$+ z^8(x^8 + x^9 + 2x^{10} + 3x^{11} + 5x^{12} + 7x^{13} + 11x^{14} + 15x^{15}$$
$$+ 22x^{16} + \cdots)$$

From these series we can immediately state how many different ways a given number can be written as a sum of a given number of terms from the sequence 1, 2, 3, 4, 5, 6, 7 $\cdots$. For example, if we desire to know how many ways 13 can be written as the sum of 5 whole numbers, we look for $x^{13}z^5$ and notice that 18 is the coefficient of that term. This gives the answer that 13 can be written in 18 ways as the sum of 5 numbers.

305. If we let $z = 1$ and express together all similar powers of x, the expression

$$\frac{1}{(1 - x)(1 - x^2)(1 - x^3)(1 - x^4)(1 - x^5)(1 - x^6)\cdots}$$

gives rise to the series

$$1 + x + 2x^2 + 3x^3 + 5x^4 + 7x^5 + 11x^6 + 15x^7 + 22x^8 + \cdots .$$

In this series the coefficient indicates how many different ways the exponent of the power can be expressed as the sum of the whole numbers, whether they are equal or unequal. For example, from the term $11x^6$ we know that the number 6 can be expressed in 11 different ways as the sum of whole numbers. These eleven ways are:

$6 = 6$ $\quad 6 = 3 + 1 + 1 + 1$ $\quad 6 = 5 + 1$ $\quad 6 = 2 + 2 + 2$

$6 = 4 + 2$ $\quad 6 = 2 + 2 + 1 + 1$

$6 = 4 + 1 + 1$ $\quad 6 = 2 + 1 + 1 + 1 + 1$

$6 = 3 + 3$ $\quad 6 = 1 + 1 + 1 + 1 + 1 + 1$

$6 = 3 + 2 + 1$.

It is worth noting that the number 6 itself, since it is a term in the sequence $1, 2, 3, 4, 5, 6 \cdots$, is one of the ways.

306. After having given this general view, we now, with some diligence, investigate how we might find this information. In the first place we consider sums of whole numbers in which we allow only different addends, as we did previously. To this purpose we consider the expression $Z = (1 + xz)(1 + x^2z)(1 + x^3z)(1 + x^4z)(1 + x^5z) \cdots$. This gives rise to the series according to powers of z,

$$Z = 1 + Pz + Qz^2 + Rz^3 + Sz^4 + Tz^5 + \cdots,$$

where we desire a reasonable method for finding the functions of x: P, Q, R, S, T, etc. This is the most convenient way of answering the question posed above.

307. It is clear that when xz is substituted for z, we obtain $(1 + x^2z)(1 + x^3z)(1 + x^4z)(1 + x^5z) = \frac{Z}{1 + xz}$. It follows that when xz is substituted for z, we have $\frac{Z}{1 + xz}$ for Z. It follows that since

$Z = 1 + Pz + Qz^2 + Rz^3 + Sz^4 + \cdots$, we have

$\dfrac{Z}{1 + xz} = 1 + Pxz + Qx^2z^2 + Rx^3z^3 + Sx^4z^4 + \cdots$. When we multiply

this expression by $1 + xz$ we obtain

$$Z = 1 + Pxz + Qx^2z^2 + Rx^3z^3 + Sx^4z^4 + \cdots$$
$$+ xz + Px^2z^2 + Qx^3z^3 + Rx^4z^4 + \cdots .$$

When this is compared with the original definition of Z we obtain

$$P = \frac{x}{1 - x}$$
$$Q = \frac{x^3}{(1 - x)(1 - x^2)}$$
$$R = \frac{x^6}{(1 - x)(1 - x^2)(1 - x^3)}$$
$$S = \frac{x^{10}}{(1 - x)(1 - x^2)(1 - x^3)(1 - x^4)}$$
$$T = \frac{x^{15}}{(1 - x)(1 - x^2)(1 - x^3)(1 - x^4)}$$

etc.

308. In this way we can give each of these series by itself, from which it is possible to say in how many different ways a given number can be expressed as the sum of a given number of integers. It is clear that each of these series is recurrent, since it arises from a rational function of x. For example, the first expression, $P = \dfrac{x}{1 - x}$, gives the geometric series

$x + x^2 + x^3 + x^4 + x^5 + x^6 + x^7 + \cdots$. From this it is clear that each number occurs once in the sequence of whole numbers.

309. The second expression $\dfrac{x^3}{(1 - x)(1 - x^2)}$ gives the series

$$x^3 + x^4 + 2x^5 + 2x^6 + 3x^7 + 3x^8 + 4x^9 + 4x^{10} + \cdots ,$$

in which the coefficient of each term indicates in how many ways the exponent of x can be expressed as the sum of two different numbers. For example, $4x^9$ indicates that the number 9 can be cut into two unequal parts in four ways. If this series is divided by x^3, we obtain the series which arises from the rational function $\dfrac{1}{(1-x)(1-x^2)}$, namely,

$$1 + x + 2x^2 + 2x^3 + 3x^4 + 3x^5 + 4x^6 + 4x^7 + \cdots,$$

whose general term is Nx^n. From the nature of this series we understand the coefficient N to be the number of different ways in which the exponent n can be expressed as the sum of the numbers 1 and 2. Since the general term of the previous series is Nx^{n+3}, we can state the following theorem.

The number of different ways in which the number n can be expressed as the sum of the numbers 1 and 2 is the same as the number of different ways that $n + 3$ can be expressed as the sum of two different numbers.

310. The third expression in the series, $\dfrac{x^6}{(1-x)(1-x^2)(1-x^3)}$, when developed as a series gives

$x^6 + x^7 + 2x^8 + 3x^9 + 4x^{10} + 5x^{11} + 7x^{12} + 8x^{13} + \cdots$. In this series the coefficient of each term indicates the number of different ways the exponent of the power of x can be written as the sum of three different numbers. If the rational function $\dfrac{1}{(1-x)(1-x^2)(1-x^3)}$ is developed in a series, we have

$$1 + x + 2x^2 + 3x^3 + 4x^4 + 5x^5 + 7x^6 + 8x^7 + \cdots.$$

If we let Nx^n be the general term, then the coefficient N indicates the number of different ways in which the number n can be expressed as the sum of the numbers 1, 2, and 3. Since the general term of the preceding series is Nx^{n+6}, we have the following theorem.

The number of different ways in which n can be expressed as the sums of the numbers 1, 2, and 3 is the same as the number of different ways that the number $n + 6$ can be expressed as the sum of three different numbers.

311. The fourth expression in the series,

$$\frac{x^{10}}{(1 - x)(1 - x^2)(1 - x^3)(1 - x^4)},$$

when developed in a series gives

$$x^{10} + x^{11} + 2x^{12} + 3x^{13} + 5x^{14} + 6x^{15} + 9x^{16} + \cdots .$$

In this series the coefficient of any term indicates the number of different ways in which the exponent of the power of x can be expressed as the sum of four unequal numbers. If the expression

$\dfrac{1}{(1 - x)(1 - x^2)(1 - x^3)(1 - x^4)}$ is developed in a series, we obtain the above series after division by x^{10}, namely,

$$1 + x + 2x^2 + 3x^3 + 5x^4 + 6x^5 + 9x^6 + 11x^7 + \cdots .$$

The general term of this series is Nx^n, and the coefficient, N, indicates the number of different ways in which n can be expressed as the sums of numbers 1, 2, 3, 4. Since the previous series has a general term Nx^{n+10}, we deduce the following theorem.

The number of different ways in which the number n can be expressed as the sums of the numbers 1, 2, 3, and 4 is the same as the number of different ways in which the number $n + 10$ can be written as the sum of four different numbers.

312. In general, if the expression $\dfrac{1}{(1-x)(1-x^2)(1-x^3)\cdots(1-x^m)}$ is developed in a series with general term Nx^n, then the coefficient N indicates the number of different ways in which the number n can be expressed as sums of the numbers 1, 2, 3, 4, $\cdots$,m. On the other hand, if the expression

$$\frac{x^{\frac{m(m+1)}{2}}}{(1-x)(1-x^2)(1-x^3)\cdots(1-x^m)}$$

is developed in a series with general term $Nx^{n+\frac{m(m+1)}{2}}$, then the coefficient N indicates the number of different ways in which the number $n+\dfrac{m(m+1)}{1\cdot 2}$ can be expressed as the sum of m different numbers. We then have the following theorem.

The number of different ways in which n can be expressed as the sums of the numbers 1, 2, 3, 4, $\cdots$,m, *is the same as the number of different ways in which* $n+\dfrac{m(m+1)}{1\cdot 2}$ *can be expressed as the sum of m different numbers.*

313. After considering the partition of numbers into unequal parts, we now turn to those in which equality of parts is not excluded. This kind of partition has its origin in this expression:

$$Z = \frac{1}{(1-xz)(1-x^2z)(1-x^3z)(1-x^4z)(1-x^5z)\cdots}.$$

We let the resulting series be

$$Z = 1 + Pz + Qz^2 + Rz^3 + Sz^4 + Tz^5 + \cdots .$$

It is clear that when we substitute xz for z we obtain

$$\frac{1}{(1-x^2z)(1-x^3z)(1-x^4z)(1-x^5z)\cdots} = (1-xz)Z.$$

If we make the same substitution in the series, we have

$$(1-xz)Z = 1 + Pxz + Qx^2z^2 + Rx^3z^3 + Sx^4z^4 + \cdots .$$

If we multiply the original series by $(1-xz)$, the result is

$$\begin{aligned}(1-xz)Z &= 1 + Pz + Qz^2 + Rz^3 + Sz^4 + \cdots \\ &- xz - Pxz^2 - Qxz^3 - Rxz^4 - \cdots .\end{aligned}$$

When we compare the two expressions, we have

$P = \dfrac{x}{1-x}$, $Q = \dfrac{Px}{1-x^2}$, $R = \dfrac{Qx}{1-x^3}$, $S = \dfrac{Rx}{1-x^4}$, etc. It follows that for P, Q, R, S, etc., we have the following values.

$$\begin{aligned}P &= \frac{x}{1-x} \\ Q &= \frac{x^2}{(1-x)(1-x^2)} \\ R &= \frac{x^3}{(1-x)(1-x^2)(1-x^3)} \\ S &= \frac{x^4}{(1-x)(1-x^2)(1-x^3)(1-x^4)}\end{aligned}$$

etc.

314. These expressions differ from the earlier expressions only in that the exponents of the powers in the numerator are less. For this reason the coefficients of the resulting series correspond to those found in sections 300 and 304. It follows that we have the following theorems.

The number of different ways in which the number n can be expressed as the sums of the numbers 1 and 2, is the same as the number of different ways in which the number $n+2$ can be expressed by sums of two numbers.

The number of different ways in which the number n can be expressed as the sums of the numbers 1, 2, and 3, is the same as the number of different ways in which the number $n + 3$ can be expressed as the sum of three numbers.

The number of different ways in which the number n can be expressed as the sums of the numbers 1, 2, 3, and 4, is the same as the number of different ways in which the number $n + 4$ can be expressed as the sum of four numbers.

In general we have this theorem.

The number of different ways in which the number n can be expressed as sums of the numbers 1, 2, 3, . . . , m is the same as the number of ways the number $n + m$ can be expressed as the sum of m numbers.

315. It follows that if we want to know in how many different ways a given number can be expressed either with m unequal numbers, or with m numbers where we allow equality, then both questions can be answered if we know the number of ways in which each number can be expressed as sums of the numbers 1, 2, 3, 4, $\cdots$, m. The answers are given by the following theorems, which are derived from the previous theorems.

The number of different ways in which the number n can be expressed as the sum of m different numbers is the same as the number of different ways in which $n - \frac{m(m+1)}{2}$ can be expressed as sums of the numbers 1, 2, 3, 4, $\cdots$, m.

The number of different ways in which the number n can be expressed as the sum of m numbers, either equal or unequal, is the same as the number of different ways in which $n - m$ can be expressed as sums of the numbers 1, 2, 3, $\cdots$, m.

From these there arise the following theorems.

The number of different ways in which the number n can be expressed as the sum of m unequal numbers is the same as the number of different ways in which $n - \frac{m(m-1)}{2}$ *can be expressed as the sum of m numbers, either equal or unequal.*

The number of different ways in which the number n can be expressed as the sum of m numbers, either equal or unequal, is the same as the number of different ways in which the number $n + \frac{m(m-1)}{2}$ *can be expressed as the sum of m unequal numbers.*

316. We can discover the number of different ways in which the number n can be written as sums of the numbers $1, 2, 3, \cdots, m$, through the formation of a recurrent series. For this reason we develop the rational function $\frac{1}{(1-x)(1-x^2)(1-x^3)\cdots(1-x^m)}$ in a recurrent series up to the general term Nx^n. The coefficient N of this general term indicates the number of different ways in which the number n can be expressed as sums of the numbers $1, 2, 3, 4, \cdots, m$. However, this method of finding N has not a few difficulties when the numbers m and n are reasonably large. Since the denominator of the rational function has many terms, it is quite tedious to find a large number of terms of the series.

317. This business will be less burdensome if we clarify the simpler cases at first, since from these it will be easy to move on to the more complicated cases. Let the series which arises from the rational function

$\dfrac{1}{(1-x)(1-x^2)(1-x^3)\cdots(1-x^m)}$ have the general term Nx^n, and let the general term of the series which arises from

$\dfrac{x^m}{(1-x)(1-x^2)(1-x^3)\cdots(1-x^m)}$ be Mx^n, where the coefficient M indicates the number of different ways in which the number $n-m$ can be expressed as sums of the numbers $1, 2, 3, \cdots, m$. We subtract the second expression from the first, so that what remains is $\dfrac{1}{(1-x)(1-x^2)(1-x^3)\cdots(1-x^{m-1})}$. It is clear that the series which arises from this rational function has the general term $(N-M)x^n$. It follows that the coefficient $N-M$ indicates the number of different ways in which the number n can be expressed as sums of the numbers $1, 2, 3, \cdots, (m-1)$.

318. In this way we arrive at the following rule. Let L be the number of different ways in which the number n can be expressed as sums of the numbers $1, 2, 3, \cdots, (m-1)$.

Let M be the number of different ways in which the number $n-m$ can be expressed as sums of the numbers $1, 2, 3, \cdots, m$.

Let N be the number of different ways in which the number n can be expressed as sums of the numbers $1, 2, 3, \cdots, m$.

With these hypotheses, as we have seen, $L = N - M$ or $N = L + M$. It follows that if we have already found how many different ways n can be expressed as sums of the number $1, 2, 3, \cdots, (m-1)$, and how many different ways in which $n-m$ can be expressed as sums of the numbers $1, 2, 3, \cdots, m$, then by addition we know the number of different ways in which the number n can be expressed as sums of the numbers 1, 2, 3, $\cdots$, m. With the aid of this theorem, we use the results of the easier cases, which offer no difficulties, to find a solution to the more complicated cases. This is the way in which the appended table was computed, and this table can be used for further results. (See page 280.)

If we want to know the number of different ways in which the number 50 can be expressed as the sum of 7 unequal numbers, then we find in the column on the left the number $50 - \dfrac{7\cdot 8}{2} = 22$, and in this row under the heading VII we find the answer to be 522.

Suppose we want to know the number of different ways in which the number 50 can be expressed as the sum of 7 numbers, either equal or unequal. We find in the first column the number $50 - 7 = 43$, and in that row under the seventh column the answer is 8946.

319. The vertical columns of this table are the coefficients of a series. Although this series is recurrent, the coefficients have an intimate connection with the natural numbers, the triangular numbers, the tetrahedral numbers, etc. An explanation of this will not require much work. Since from the rational function $\dfrac{1}{(1 - x)(1 - x^2)}$ we have the series

$1 + x + 2x^2 + 2x^3 + 3x^4 + 3x^5 + \cdots$, it follows that

$\dfrac{x}{(1 - x)(1 - x^2)} = x + x^2 + 2x^3 + 2x^4 + 3x^5 + 3x^6 + \cdots$. If we add these two series we have $1 + 2x + 3x^2 + 4x^3 + 5x^4 + 6x^5 + 7x^6 + \cdots$, which arises by division from the rational function

$\frac{1 + x}{(1 - x)(1 - x^2)} = \frac{1}{(1 - x)^2}$. From this it is clear that the last series has the natural numbers for its coefficients. Thus in the series for the second column, if we let $x = 1$ and add two successive terms we obtain the series of the natural numbers,

$$1 + 1 + 2 + 2 + 3 + 3 + 4 + 4 + 5 + 5 + 6 + 6 + \cdots$$

$$1 + 2 + 3 + 4 + 5 + 6 + 7 + 8 + 9 + 10 + 11 + 12 + \cdots .$$

Conversely, if we begin with the series of natural numbers as in the lower series, then we obtain the upper series by subtracting each term of the upper series already found from the following term in the lower series.

320. The series for the third column arises from the rational function $\frac{1}{(1 - x)(1 - x^2)(1 - x^3)}$. Since $\frac{1}{(1 - x)^3} = \frac{(1 + x)(1 + x + x^2)}{(1 - x)(1 - x^2)(1 - x^3)}$, it is clear that if at first the terms of this series are added three at a time, and then the terms of the new series are added two at a time, we should obtain a series of the triangular numbers. This is what appears in the following scheme.

$$1 + 1 + 2 + 3 + 4 + 5 + 7 + 8 + 10 + 12 + 14 + 16 + \cdots$$

$$1 + 2 + 4 + 6 + 9 + 12 + 16 + 20 + 25 + 30 + 36 + 42 + \cdots$$

$$1 + 3 + 6 + 10 + 15 + 21 + 28 + 36 + 45 + 55 + 66 + 78 + \cdots .$$

From this scheme it should be clear how to obtain the original series from the triangular numbers.

321. In a similar way the series of the fourth column arises from the rational function $\frac{1}{(1 - x)(1 - x^2)(1 - x^3)(1 - x^4)}$, and

$\frac{(1 + x)(1 + x + x^2)(1 + x + x^2 + x^3)}{(1 - x)(1 - x^2)(1 - x^3)(1 - x^4)} = \frac{1}{(1 - x)^4}$. If in the given series the terms are added four at a time to give a second series, then the third series is obtained by adding the terms of the second series three at a time. Finally the series of tetrahedral numbers is obtained by adding the terms of the third series two at a time. This should be clear from the following calculations.

$$1 + 2 + 3 + 5 + 6 + 9 + 11 + 15 + 18 + 23 + 27 + \cdots$$

$$1 + 2 + 4 + 7 + 11 + 16 + 23 + 31 + 41 + 53 + 67 + \cdots$$

$$1 + 3 + 7 + 13 + 22 + 34 + 50 + 70 + 95 + 125 + 161 + \cdots$$

$$1 + 4 + 10 + 20 + 35 + 56 + 84 + 120 + 165 + 220 + 286 + \cdots$$

In a similar way the fifth series can be calculated from the series of tetrahedral numbers of the second order, the sixth series from the tetrahedral numbers of the third order, and so forth.

322. On the other hand, from the series of geometrical numbers, the series of numbers in the table can be calculated by the operations which should be obvious from the following schemes.

II $\quad 1 + 2 + 3 + 4 + 5 + 6 + 7 + 8 + 9 + 10 + \cdots$

$\quad 1 + 1 + 2 + 2 + 3 + 3 + 4 + 4 + 5 + 5 + \cdots$

III $\quad 1 + 3 + 6 + 10 + 15 + 21 + 28 + 36 + 45 + 55 + \cdots$

$\quad 1 + 2 + 4 + 6 + 9 + 12 + 16 + 20 + 25 + 30 + \cdots$

$\quad 1 + 1 + 2 + 3 + 4 + 5 + 7 + 8 + 10 + 12 + \cdots$

IV $\quad 1 + 4 + 10 + 20 + 35 + 56 + 84 + 120 + 165 + \cdots$

$\quad 1 + 3 + 7 + 13 + 22 + 34 + 50 + 70 + 95 + \cdots$

$$1 + 2 + 4 + 7 + 11 + 16 + 23 + 31 + 41 + \cdots$$

$$1 + 1 + 2 + 3 + 5 + 6 + 9 + 11 + 15 + \cdots$$

$$V \quad 1 + 5 + 15 + 35 + 70 + 126 + 210 + 330 + 495 + \cdots$$

$$1 + 4 + 11 + 24 + 46 + 80 + 130 + 200 + 195 + \cdots$$

$$1 + 3 + 7 + 14 + 25 + 41 + 64 + 95 + 136 + \cdots$$

$$1 + 2 + 4 + 7 + 12 + 18 + 27 + 38 + 53 + \cdots$$

$$1 + 1 + 2 + 3 + 5 + 7 + 10 + 13 + 18 + \cdots .$$

In each of these schemes the first series is a series of geometrical numbers. The second series is formed by subtracting a term of the second series from the following term of the first series. The third series is formed by subtracting the sum of two terms of the third series from the following term of the second series. In this way we proceed, subtracting the sum of three terms, then the sum of four terms, and so forth, from the following term of the preceding series, until we obtain the desired series, which begins with $1 + 1 + 2 + \cdots$. This is how we obtain the series exhibited in the columns of the table.

323. The columns of the table all have the same beginnings and continue to have some terms in common. From this we understand that if there were an infinite number of columns, the series would agree completely. These series arise from the function

$$\frac{1}{(1-x)(1-x^2)(1-x^3)(1-x^4)(1-x^5)(1-x^6)(1-x^7)\cdots}.$$ Because the series is recurrent, first we consider the denominator, since from it we obtain the

scale of the relation. If the factors of the denominator are multiplied one by one, we obtain

$$1 - x - x^2 + x^5 + x^7 - x^{12} - x^{15} + x^{22} + x^{26} - x^{35} - x^{40} + x^{51} + \cdots .$$

If we consider this sequence with some attention we will note that the only exponents which appear are of the form $\frac{3n^2 \pm n}{2}$ and that the sign of the corresponding term is negative when n is odd, and the sign is positive when n is even.

324. Since the scale of the relation is

$$+1, +1, 0, 0, -1, 0, -1, 0, 0, 0, 0, +1, 0, 0, +1, 0, 0, \cdots ,$$

the recurrent series which arises from the rational function

$$\frac{1}{(1-x)(1-x^2)(1-x^3)(1-x^4)(1-x^5)(1-x^6)(1-x^7)\cdots} \text{ is}$$

$$1 + x + 2x^2 + 3x^3 + 5x^4 + 7x^5 + 11x^6 + 15x^7 + 22x^8 + 30x^9 + 42x^{10} + 56x^{11} + 77x^{12} + 101x^{13} + 135x^{14} + 176x^{15} + 231x^{16} + 297x^{17} + 385x^{18} + 490x^{19} + 627x^{20} + 792x^{21} + 1002x^{22} + 1250x^{23} + 1570x^{24} + \cdots .$$

In this series each coefficient indicates the number of different ways in which the exponent itself can be expressed as the sum of whole numbers. For example, the number 7 can be expressed in the fifteen ways shown below:

$7=7$	$7=4+2+1$	$7=3+1+1+1+1$
$7=6+1$	$7=4+1+1+1$	$7=2+2+2+1$
$7=5+2$	$7=3+3+1$	$7=2+2+1+1+1$
$7=5+1+1$	$7=3+2+2$	$7=2+1+1+1+1+1$
$7=4+3$	$7=3+2+1+1$	$7=1+1+1+1+1+1+1$

325. If the product

$(1 + x)(1 + x^2)(1 + x^3)(1 + x^5)(1 + x^6) \cdots$ is developed, we have the series

$$1 + x + x^2 + 2x^3 + 2x^4 + 3x^5 + 4x^6 + 5x^7 + 6x^8 + 8x^9 + 10x^{10} + \cdots ,$$

in which each coefficient indicates the number of different ways in which the exponent can be expressed as the sum of different numbers. For example, the number 9 can be expressed in the following eight ways as the sum of different numbers.

$9 = 9$	$9 = 6 + 2 + 1$	$9 = 8 + 1$	$9 = 5 + 4$
$9 = 7 + 2$	$9 = 5 + 3 + 1$	$9 = 6 + 3$	$9 = 4 + 3 + 2$

326. In order that we may compare these two forms, we let

$$P = (1 - x)(1 - x^2)(1 - x^3)(1 - x^4)(1 - x^5)(1 - x^6) \cdots$$
$$Q = (1 + x)(1 + x^2)(1 + x^3)(1 + x^4)(1 + x^5)(1 + x^6) \cdots$$
$$PQ = (1 - x^2)(1 - x^4)(1 - x^6)(1 - x^8)(1 - x^{10})(1 - x^{12}) \cdots .$$

Since all of the factors of PQ are contained in P, when P is divided by PQ we obtain

$\dfrac{1}{Q} = (1 - x)(1 - x^3)(1 - x^5)(1 - x^7)(1 - x^9) \cdots$. It follows that

$Q = \dfrac{1}{(1 - x)(1 - x^3)(1 - x^5)(1 - x^7)(1 - x^9) \cdots}$. If this rational function is developed in an infinite series, the coefficient of each term will indicate the number of different ways in which the exponent can be expressed as the sum of odd numbers. Since this series is the one which we considered in the preceding section, we have the following theorem.

The number of different ways a given number can be expressed as the sum of different whole numbers is the same as the number of ways in which that same

number can be expressed as the sum of odd numbers, whether the same or different.

327. As we have already seen,

$$P = 1 - x - x^2 + x^5 + x^7 - x^{12} - x^{15} + x^{22} + x^{26} - x^{35} - x^{40} + \cdots .$$

If we substitute x^2 for x, we have

$$PQ = 1 - x^2 - x^4 + x^{10} + x^{14} - x^{24} - x^{30} + x^{44} + x^{52} - \cdots .$$

When PQ is divided by P we obtain

$$Q = \frac{1 - x^2 - x^4 + x^{10} + x^{14} - x^{24} - x^{30} + \cdots}{1 - x - x^2 + x^5 + x^7 - x^{12} - x^{15} + x^{22} + x^{26} - \cdots}.$$

It follows that Q is also a recurrent series, and that it arises from the series $\frac{1}{P}$ multiplied by $1 - x^2 - x^4 + x^{10} + x^{14} - x^{24} - \cdots$.

That is, from section 324 we know that

$$\frac{1}{P} = 1 + x + 2x^2 + 3x^3 + 5x^4 + 7x^5 + 11x^6 + 15x^7 + 22x^8 + 30x^9 + \cdots .$$

If this is multiplied by $1 - x^2 - x^4 + x^{10} + x^{14} - \cdots$, we obtain

$$1 + x + 2x^2 + 3x^3 + 5x^4 + 7x^5 + 11x^6 + 15x^7 + 22x^8 + 30x^9 + \cdots$$

$$- x^2 - x^3 - 2x^4 - 3x^5 - 5x^6 - 7x^7 - 11x^8 - 15x^9 - \cdots$$

$$- x^4 - x^5 - 2x^6 - 3x^7 - 5x^8 - 7x^9 - \cdots, \text{ or}$$

$$Q = 1 + x + x^2 + 2x^3 + 2x^4 + 3x^5 + 4x^6 + 5x^7 + 6x^8 + 8x^9 + \cdots .$$

It follows that if the formation of numbers by addition of numbers, either different or the same, is known, then we can deduce the formation of numbers by addition of unequal numbers and thus also the formation of numbers by addition of only odd numbers.

328. There remain a few problems of this type which are worth notice, and which also have some use for the understanding of the nature of numbers. Let us consider the expression

$(1 + x)(1 + x^2)(1 + x^4)(1 + x^8)(1 + x^{16})(1 + x^{32}) \cdots$, in which each exponent in the sequence is double its predecessor. The series which arises from this product is $1 + x^2 + x^3 + x^4 + x^5 + x^6 + x^7 + x^8 + \cdots$. There could remain some doubt as to whether this series is actually a geometric progression, and for this reason we investigate the series. We let

$P = (1 + x)(1 + x^2)(1 + x^4)(1 + x^8)(1 + x^{16}) \cdots$, and suppose

$P = 1 + \alpha x + \beta x^2 + \gamma x^3 + \delta x^4 + \epsilon x^5 + \zeta x^6 + \eta x^7 + \theta x^8 + \cdots$. It is clear that if we substitute x^2 for x, then the product becomes

$(1 + x^2)(1 + x^4)(1 + x^8)(1 + x^{16})(1 + x^{32}) \cdots$, which is equal to $\frac{P}{1 + x}$. It follows that

$$\frac{P}{1 + x} = 1 + \alpha x^2 + \beta x^4 + \gamma x^6 + \delta x^8 + \epsilon x^{10} + \zeta x^{12} + \cdots .$$

When we multiply by $1 + x$ we obtain

$$P = 1 + x + \alpha x^2 + \alpha x^3 + \beta x^4 + \beta x^5 + \gamma x^6 + \gamma x^7 \quad + \delta x^8 + \delta x^9 + \cdots .$$

If we compare this expression for P with the original expression we see that $\alpha = 1$, $\beta = \alpha$, $\gamma = \alpha$, $\delta = \beta$, $\epsilon = \beta$, $\zeta = \gamma$, $\eta = \gamma$, etc., so that all of the coefficients are equal to 1. It follows that the given product P is indeed equal to the geometric series $1 + x + x^2 + x^3 + x^4 + x^5 + x^6 + x^7 + \cdots$.

329. Since each whole number occurs exactly once as an exponent in the series which arises from the product $(1 + x)(1 + x^2)(1 + x^4) \cdots$, it follows that every whole number can be expressed as the sum of terms in the geometric

progression with common ratio equal to 2: 1, 2, 4, 8, 16, 32, $\cdots$, and this sum is unique. It can be noted that this property is used in weighing. If one has a set of weights weighing 1, 2, 4, 8, 16, 32, etc. pounds, then with only these weights, anything can be weighed, unless parts of pounds are required. That is, with only the ten weights, 1 lb., 2 lb., 4 lb., 8 lb., 16 lb., 32 lb., 64 lb., 128 lb., 256 lb., 512 lb., anything up to 1024 lb. can be weighed. Indeed, if an eleventh weight weighing 1024 lb. is added to the set, then anything up to 2048 lb. can be weighed.

330. It is frequently shown that in weighing one can use fewer weights, namely those in a geometric progression with common ratio equal to 3, that is, 1, 3, 9, 27, 81, $\cdots$. With these anything can be weighed, unless fractional parts of pounds are required. In this case, however, two pans of the balance, rather than just one, must be available for placing weights, when necessary. The procedure is based on the fact that, from the terms of the geometric progression with common ratio equal to 3, any number can be formed when both addition and subtraction are allowed. For example

$1 = 1$ $\qquad 5 = 9 - 3 - 1 \qquad 9 = 9$

$2 = 3 - 1 \qquad 6 = 9 - 3 \qquad 10 = 9 + 1$

$3 = 3 \qquad 7 = 9 - 3 + 1 \qquad 11 = 9 + 3 - 1$

$4 = 3 + 1 \qquad 8 = 9 - 1 \qquad 12 = 9 + 3$

etc.

331. To prove this fact we consider the following infinite product

$$P = (x^{-1} + 1 + x^1)(x^{-3} + 1 + x^3)(x^{-9} + 1 + x^9)\,(x^{-27} + 1 + x^{27}) \cdots .$$

When the series is developed, only exponents will appear which can be expressed from the numbers 1, 3, 9, 27, 81, $\cdots$, either by adding or subtracting. Whether each and every whole number appears as an exponent of x in this series is the question. Let $P = \cdots + cx^{-3} + bx^{-2} + ax^{-1} + 1 + \alpha x^1 + \beta x^2 + \gamma x^3 + \delta x^4 + \epsilon x^5 + \cdots$.

It is clear that when x^3 is substituted for x we obtain

$$\frac{P}{x^{-1} + 1 + x^1} = \cdots + bx^{-6} + ax^{-3} + 1 + \alpha x^3 + \beta x^6 + \gamma x^9 + \cdots .$$

It follows that $P = \cdots + ax^{-4} + ax^{-3} + ax^{-2} + x^{-1} + 1 + x + \alpha x^2 + \alpha x^3 + \alpha x^4 + \beta x^5 + \beta x^6 + \beta x^7 + \cdots$. When this expression for P is compared with the original expression for P, we have $\alpha = 1$, $\beta = \alpha$, $\gamma = \alpha$, $\delta = \alpha$, $\epsilon = \beta$, $\zeta = \beta$, etc. and $a = 1$, $b = a$, $c = a$, $d = a$, $e = b$, etc. It follows that

$$P = 1 + x + x^2 + x^3 + x^4 + x^5 + x^6 + x^7 + \cdots$$
$$+ x^{-1} + x^{-2} + x^{-3} + x^{-4} + x^{-5} + x^{-6} + x^{-7} + \cdots ,$$

so that we see that every power of x, both negative and positive, does occur, and hence every whole number can be expressed from the terms of the geometric progression with common ratio equal to 3, by either adding or subtracting, and this is done in a unique way.

TABLE

	I	II	III	IV	V	VI	VII	VIII	IX	X	XI
1	1	1	1	1	1	1	1	1	1	1	1
2	1	2	2	2	2	2	2	2	2	2	2
3	1	2	3	3	3	3	3	3	3	3	3
4	1	3	4	5	5	5	5	5	5	5	5
5	1	3	5	6	7	7	7	7	7	7	7
6	1	4	7	9	10	11	11	11	11	11	11
7	1	4	8	11	13	14	15	15	15	15	15
8	1	5	10	15	18	20	21	22	22	22	22
9	1	5	12	18	23	26	28	29	30	30	30
10	1	6	14	23	30	35	38	40	41	42	42
11	1	6	16	27	37	44	49	52	54	55	56
12	1	7	19	34	47	58	65	70	73	75	76
13	1	7	21	39	57	71	82	89	94	97	99
14	1	8	24	47	70	90	105	116	123	128	131
15	1	8	27	54	84	110	131	146	157	164	169
16	1	9	30	64	101	136	164	186	201	212	219
17	1	9	33	72	119	163	201	230	252	267	278
18	1	10	37	84	141	199	248	288	318	340	355
19	1	10	40	94	164	235	300	352	393	423	445
20	1	11	44	108	192	282	364	434	488	530	560
21	1	11	48	120	221	331	436	525	598	653	695
22	1	12	52	136	255	391	522	638	732	807	863
23	1	12	56	150	291	454	618	764	887	984	1060

T A B L E

	I	II	III	IV	V	VI	VII	VIII	IX	X	XI
24	1	13	61	169	333	532	733	919	1076	1204	1303
25	1	13	65	185	377	612	860	1090	1291	1455	1586
26	1	14	70	206	427	709	1009	1297	1549	1761	1930
27	1	14	75	225	480	811	1175	1527	1845	2112	2331
28	1	15	80	249	540	931	1367	1801	2194	2534	2812
29	1	15	85	270	603	1057	1579	2104	2592	3015	3370
30	1	16	91	297	674	1206	1824	2462	3060	3590	4035
31	1	16	96	321	748	1360	2093	2857	3589	4242	4802
32	1	17	102	351	831	1540	2400	3319	4206	5013	5788
33	1	17	108	378	918	1729	2738	3828	4904	5888	6751
34	1	18	114	411	1014	1945	3120	4417	5708	6912	7972
35	1	18	120	441	1115	2172	3539	5066	6615	8070	9373
36	1	19	127	478	1226	2432	4011	5812	7657	9418	11004
37	1	19	133	511	1342	2702	4526	6630	8824	10936	12866
38	1	20	140	551	1469	3009	5102	7564	10156	12690	15021
39	1	20	147	588	1602	3331	5731	8588	11648	14663	17475
40	1	21	154	632	1747	3692	6430	9749	13338	16928	20298
41	1	21	161	672	1898	4070	7190	11018	15224	19466	23501
42	1	22	169	720	2062	4494	8033	12450	17354	22367	27169
43	1	22	176	764	2233	4935	8946	14012	19720	25608	31316
44	1	23	184	816	2418	5427	9953	15765	22380	29292	36043
45	1	23	192	864	2611	5942	11044	17674	25331	33401	41373
46	1	24	200	920	2818	6510	12241	19805	28629	38047	47420

TABLE

	I	II	III	IV	V	VI	VII	VIII	IX	X	XI
47	1	24	208	972	3034	7104	13534	22122	32278	43214	54218
48	1	25	217	1033	3266	7760	14950	24699	36347	49037	61903
49	1	25	225	1089	3507	8442	16475	27493	40831	55494	70515
50	1	26	234	1154	3765	9192	18138	30588	45812	62740	80215
51	1	26	243	1215	4033	9975	19928	33940	51294	70760	91058
52	1	27	252	1285	4319	10829	21873	37638	57358	79725	103226
53	1	27	261	1350	4616	11720	23961	41635	64015	89623	116792
54	1	28	271	1425	4932	12692	26226	46031	71362	100654	131970
55	1	28	280	1495	5260	13702	28652	50774	79403	112804	148847
56	1	29	290	1575	5608	14800	31275	55974	88252	126299	167672
57	1	29	300	1650	5969	15944	34082	61575	97922	141136	188556
58	1	30	310	1735	6351	17580	37108	67696	108527	157564	211782
59	1	30	320	1815	6747	18467	40340	74280	120092	175586	237489
60	1	31	331	1906	7166	19858	43819	81457	132751	195491	266006
61	1	31	341	1991	7599	21301	47527	89162	146520	217280	297495
62	1	32	352	2087	8056	22856	51508	97539	161554	241279	332337
63	1	32	363	2178	8529	24473	55748	106522	177884	267507	370733
64	1	33	374	2280	9027	26207	60289	116263	195666	296320	413112
65	1	33	385	2376	9542	28009	65117	126692	214944	327748	459718
66	1	34	397	2484	10083	29941	70281	137977	235899	362198	511045
67	1	34	408	2586	10642	31943	75762	150042	258569	399705	567377
68	1	35	420	2700	11229	34085	81612	163069	283161	440725	629281
69	1	35	432	2808	11835	36308	87816	176978	309729	485315	697097

CHAPTER XVII

Using Recurrent Series to Find Roots of Equations.

332. The celebrated Daniel Bernoulli made significant use of recurrent series in his investigation of the roots of equations of any degree, in Volume III of the *Commentaries of the St. Petersburg Academy.* There he showed that the values of the roots of any algebraic equation can be accurately approximated by the use of recurrent series. Since this discovery is frequently very useful, we now will explain the process with special care, so that it may be clear in what cases it can be used. From time to time the unexpected does occur, and this method does not yield a root for a given equation. For this reason we consider from among the properties of recurrent series, that very basic one upon which the procedure rests, so that the force of this method is perfectly clear.

333. Since every recurrent series arises from some rational function, let such a function be $\frac{a + bz + cz^2 + dz^3 + ez^4 + \cdots}{1 - \alpha z - \beta z^2 - \gamma z^3 - \delta z^4 - \cdots}$. From this function there arises the series $A + Bz + Cz^2 + Dz^3 + Ez^4 + Fz^5 + \cdots$, whose coefficients A, B, C, D, E, F, etc. are determined by

$$A = a$$
$$B = \alpha A + b$$
$$C = \alpha B + \beta A + c$$
$$D = \alpha C + \beta B + \gamma A + d$$
$$E = \alpha D + \beta C + \gamma B + \delta A + e$$

etc. The general term, or the coefficient of the power z^n, is found by expressing the given rational function by partial fractions, whose denominators are factors of $1 - \alpha z - \beta z^2 - \gamma z^3 - \cdots$. This has been shown in Chapter XIII.

334. The form of the general term depends especially on the nature of the linear factors of the denominator, whether they are real or complex, and whether they are all different or two or more are equal. These different cases will be treated as we proceed. First we consider the case in which all linear factors are real and distinct. Let all of the linear factors of the denominator be $(1 - pz)(1 - qz)(1 - rz)(1 - sz) \cdots$. From these we obtain the following partial fractions, $\frac{U}{1 - pz} + \frac{V}{1 - qz} + \frac{W}{1 - rz} + \frac{X}{1 - sz} + \cdots$. When these are known, the general term of the recurrent series is $z^n(Up^n + Vq^n + Wr^n + Xs^n + \cdots)$ which we designate Px^n. That is, we let P be the coefficient of z^n, and we let $Q, R, \cdots$ be the coefficients of the following terms. Then the recurrent series is

$$A + Bz + Cz^2 + Dz^3 + \cdots + Pz^n + Qz^{n+1} + Rz^{n+2} + \cdots .$$

335. Now we let n be very large, that is, we continue the recurrent series to very many terms. If one number is larger than another, the powers of the larger will be even larger than the powers of the smaller. There will be a great difference between the powers Up^n, Vq^n, Wr^n, etc., so that, that which arises

from the largest of the numbers p, q, r, etc. will be much greater than the others. In comparison, the smaller ones are almost insignificant when n is an infinitely large number. Since the numbers p, q, r, etc. are all different, we suppose that p is the largest. When n is infinite, then $P = Ap^n$, or at least when n is very large P is approximately equal to Ap^n. Similarly, $Q = Ap^{n+1}$, so that $\frac{Q}{P} = p$. From this it is clear that when a series is continued for very many terms the coefficient of any term when divided by its predecessor leaves a quotient which approximates the largest number p.

336. If in the given rational function

$\frac{a + bz + cz^2 + dz^3 + \cdots}{1 - \alpha z - \beta z^2 - \gamma z^3 - \delta z^4 - \cdots}$, the denominator has factors which are all real and distinct, then from the recurrent series we can find one simple factor, namely $1 - pz$, in which p has the largest value. In this business the coefficients of the numerators, a, b, c, d, etc., play no part. Indeed, whatever values be given to these coefficients, the same value of the greatest of the numbers, p, will be found. The true value of p is found when the series is continued to infinity, while an approximate value is found when very many terms are found. The more terms we find the better is the approximation; the approximation is better also to the extent that p is greater than the other values q, r, s, etc. Finally, it makes no difference whether p is positive or negative, since in either case the powers increase.

337. By now it should be clear how we can use this investigation to find the roots of any algebraic equation. From a knowledge of the factors of the denomi-

nator $1 - \alpha z - \beta z^2 - \gamma z^3 - \delta z^4 - \cdots$, it is easy to find the roots of the equation $1 - \alpha z - \beta z^2 - \gamma z^3 - \delta z^4 - \cdots = 0$. That is, if $1 - pz$ is a factor, then $z = \frac{1}{p}$ will be a root. Since we know the largest value p from the recurrent series, we immediately know the smallest root of the equation $1 - \alpha z - \beta z^2 - \gamma z^3 - \cdots = 0$. If we let $z = \frac{1}{x}$ in this equation, we obtain $x^m - \alpha x^{m-1} - \beta x^{m-2} - \gamma x^{m-3} - \cdots = 0$. By means of the same method we can find the largest root of this equation, $x = p$.

338. If the given equation is $x^m - \alpha x^{m-1} - \beta x^{m-2} - \gamma x^{m-3} - \cdots = 0$ and if the factors are all linear and distinct, then the largest of the roots can be found in the following way. We form from the coefficients of the equation the following rational function $\frac{a + bz + cz^2 + dz^3 + \cdots}{1 - \alpha z - \beta z^2 - \gamma z^3 - \delta z^4 - \cdots}$. We then form the recurrent series, where we are free to choose any numerator of the rational function, which means that we can choose arbitrarily the first terms of the series. Let the series be $A + Bz + Cz^2 + Dz^3 + \cdots + Pz^n + Qz^{n+1} + \cdots$, then the fraction $\frac{Q}{P}$ gives the value of the largest root x for the original equation and the approximation will be closer, the larger we choose the number n.

EXAMPLE I

Let the given equation be $x^2 - 3x - 1 = 0$, and we want to find the largest root.

We form the function $\frac{a + bz}{1 - 3z - z^2}$, then choose the first two terms to be 1 and 2. The resulting recurrent series has the following coefficients:

1, 2, 7, 23, 76, 251, 829, 2738, $\cdots$. It follows that the quotient $\frac{2738}{829}$ is approximately equal to the largest root of the proposed equation. The decimal approximation of this fraction is 3.3027744, while the true value of the greatest root is $\frac{3 + \sqrt{13}}{2}$, whose decimal approximation is 3.3027756. The difference is only one in a million. We further note that the fractions $\frac{Q}{P}$ are alternately greater and less than the true root.

EXAMPLE II

The proposed equation is $3x - 4x^3 = \frac{1}{2}$, *whose roots are the sines of three arcs such that the sine of three times the arc is equal to* $\frac{1}{2}$.

The equation is expressed in the form $1 - 6x + 8x^3 = 0$ so that the coefficients may all be integers. We seek the minimum root, so that it is not necessary to substitute $\frac{1}{z}$ for x. We form the rational function $\frac{a + bx + cx^2}{1 - 6x + 8x^3}$ and freely choose the first three coefficients to be 0, 0, and 1. We make this choice to make the calculations easier for the remaining coefficients in the recurrent series. We omit the powers of x and give only the sequence of coefficients : 0, 0, 1, 6, 36, 208, 1200, 6912, 39808, 229248. The quotient $\frac{39808}{229248} = \frac{311}{1791} = 0.1736515$ is an approximation of the minimum root, which is actually the sine of 10 degrees. From the tables we find that the sine of 10 degrees is equal to 0.1736482. The calculated value exceeds the tabular value by $\frac{33}{10000000}$. This root could more easily be found by letting $x = \frac{1}{2}y$ to obtain the equation $1 - 3y + y^3$. In a similar way we obtain the sequence of coefficients 0, 0, 1, 3, 9, 26, 75, 216, 622, 1791, 5157, $\cdots$. Then the approximate root of this equation is $y = \frac{1791}{5157} = \frac{199}{573} = 0.3472949$, so that $x = \frac{1}{2y} = 0.1736479$. This approximate value differs by one tenth the difference given by the previous approximation.

EXAMPLE III

Suppose the largest root of the equation $1 - 6x + 8x^3$ *is desired.*

We let $x = \frac{y}{2}$, then $y^3 - 3y + 1 = 0$. We obtain the largest root of this equation from the recurrent series with scale of relation 0, 3, -1. When we take the first three terms arbitrarily we obtain

1, 1, 1, 2, 2, 5, 4, 13, 7, 35, 8, 98, $-$ 11, $\cdots$.

In this series, since we obtain a negative coefficient, it is indicated that the root is negative and $x = - 0.9396926$, which is the negative of the sine of 70 degrees. If we change the initial terms we obtain

1, $-$ 2, 4, $-$ 7, 14, $-$ 25, 49, $-$ 89, 172, $-$ 316, 605, $-$ $\cdots$

and $y = -\frac{605}{316}$ or $x = -\frac{605}{632} = - 0.957$. This misses the true value significantly.

339. The main reason for this discrepancy is that the roots of the proposed equation are the sine of 10 degrees, the sine of 50 degrees, and the negative of the sine of 70 degrees. Of these roots the two largest are not very different, so that the powers of these roots to which we continued are not different enough from each others that the second root, the sine of 50 degrees, does not vanish in

comparison with the largest. This deviation is also a result of the fact that the calculated values are alternately too large and too small. That is, when we take $y = -\frac{316}{172}$, $x = -\frac{158}{172} = -\frac{79}{86} = -0.918$. Since the powers of the largest root are alternately positive and negative, so also the powers of the second root are alternately added and subtracted. For this reason, in order that the discrepancy become negligible it is necessary to take very many terms in the series.

340. There is another remedy for this problem. We can transform the equation by a suitable substitution so that the roots are no longer so close to each other. For example, in the equation $1 - 6x + 8x^3 = 0$, whose roots are the negative sine of 70 degrees, the sine of 50 degrees, and the sine of 10 degrees, we let $x = y - 1$. The resulting equation is $8y^3 - 24y^2 + 18y - 1 = 0$, with roots 1 minus the sine of 70 degrees, 1 plus the sine of 50 degrees, and 1 plus the sine of 10 degrees. In this case the smallest root is 1 minus the sine of 70 degrees, although the sine of 70 degrees is the largest root of the previous equation. Now 1 plus the sine of 50 degrees is the largest root, while before the sine of 50 degrees was the middle root. Furthermore, in this way any root can be transformed into the minimum or maximum root of a new equation by substitution. It follows that any root can be found by this method. In addition, since the root 1 minus the sine of 70 degrees is much less than the other two, it can easily be calculated from the recurrent series.

EXAMPLE IV

Find the smallest root of the equation $8y^3 - 24y^2 + 18y - 1 = 0$, *which when subtracted from 1 gives the value of the sine of 70 degrees.*

We let $y = \frac{1}{2}z$, so that $z^3 - 6z^2 + 9z - 1 = 0$, and we find the smallest root of this equation by a recurrent series whose scale of relation is 9, -6, 1. In order to find the largest root, the scale of relation must be 6, -9, 1. For the smallest root we form the series with coefficients
1, 1, 1, 4, 31, 256, 2122, 17593, 145861, $\cdots$. The approximate value for z is $\frac{17593}{145861} = 0.12061483$. Then $y = 0.06030741$ and the sine of 70 degrees is equal to $1 - y = 0.93969258$ which hardly differs from the true value even in the last digit. From this example it should be clear how useful an appropriate transformation of the equation by substitution can be in order to find a root of the equation. It also shows that once this is done, not only the smallest and largest roots can be found, but all roots can be found.

341. If we already know an approximate value of any root of a given equation, so that, for instance, the number k is very close to the root, then we let $x - k = y$ or $x = y + k$. In this way we produce an equation whose smallest root will be $x - k$, whose value can be found from a recurrent series very easily, since this root is much smaller than any others. Then when k is added to the this root we have the root of the proposed equation. This trick has such a wide application that it can be used even when the equation has complex roots.

342. In particular, without this trick, it is impossible to find a root of an equation if the equation has another root which is equal to the first but with opposite sign. For example, if an equation has p for its largest root and $-p$ is also a root, then although the recurrent series might be continued to infinity, still the root p would never be found. For the sake of an example, consider the equation $x^3 - x^2 - 5x + 5 = 0$. The largest root of this equation is $\sqrt{5}$, but there is also a root equal to $-\sqrt{5}$. If we follow the method we have been using to find this largest root, and we form the recurrent series from the scale of relation 1, 5, - 5, we obtain the sequence of coefficients 1, 2, 3, 8, 13, 38, 63, 188, 313, 938, 1563, $\cdots$, but there is no constant ratio in this sequence. We note however, that when we consider the quotient of alternate terms we do have a number which is approximated and this number is the square of the largest root. That is, $\frac{1563}{313}, \frac{938}{188}, \frac{313}{63}$ are all approximations of 5. The fact is that whenever only the alternate term quotients approximate a constant, then that approximation is of the square of the desired root. The root itself $x = \sqrt{5}$ can be found by making the substitution $x = y + 2$ to obtain $1 - 3y - 5y^2 - y^3 = 0$. The smallest root is found from the series with coefficients 1, 1, 1, 9, 33, 145, 609, 2585, 10945, $\cdots$. The approximate value is $\frac{2585}{10945} = 0.2361$. But 2.2361 is approximately equal to $\sqrt{5}$, which is the largest root of the original equation.

343. Although the numerator of the rational function from which the recurrent series arises can be chosen freely, nevertheless a judicious choice is an important factor in arriving quickly at a suitable approximation. Since we have

assumed, as in section 334, that the general term of the recurrent series is $z^n(Up^n + Vq^n + Wr^n + \cdots)$, these coefficients U, V, W, etc. are determined by the numerator of the function. It follows that if the value of U is large or small, the approximate value of the largest root, p, will be obtained quickly or slowly respectively. Indeed a numerator could be chosen in such a way that U would vanish, and in that case, even though the series be indefinitely continued, the largest root, p, would never be found. This would happen if a numerator were chosen which had precisely the factor $1 - pz$, since in that case the factor would be removed from the quotient. For example, let the given equation be $x^3 - 6x^2 + 10x - 3 = 0$, whose largest root is equal to 3. If we consider $\frac{1 - 3z}{1 - 6z + 10z^2 - 3z^3}$, the recurrent series has a scale of relation 6, -10, 3 and the sequence of coefficients is 1, 3, 8, 21, 55, 144, 377, $\cdots$. However, the sequence of ratios does not converge to 3. The same series arises from the rational function $\frac{1}{1 - 3z + z^2}$, so that the sequence approximates the largest root of the equation $z^2 - 3z + 1 = 0$.

344. Indeed, the numerator can be chosen so that any desired root of the equation may appear. This is done by letting the numerator be the product of all the factors of the denominator except the one corresponding to the desired root. For instance, in the example just discussed, if we choose $1 - 3z + z^2$ as numerator, the rational function $\frac{1 - 3z + z^2}{1 - 6z + 10z^2 - 3z^3}$ gives the sequence 1, 3, 9, 27, 81, 243, $\cdots$ which is geometric and immediately shows the root to be equal to 3. The given rational function is equal to $\frac{1}{1 - 3z}$. It is clear that if

the initial terms, which can be freely chosen, are chosen in such a way that they form a geometric progression equal to powers of one of the roots of the equation, then the whole recurrent series will be geometric and show the root itself, even though the root may neither be the smallest nor the largest root.

345. We must beware, lest in seeking the largest or smallest root, unexpectedly we obtain a different root from the recurrent series. Care must be taken to choose a numerator which has no roots in common with the denominator. This can be accomplished by choosing the numerator to be simply equal to 1 so that the remaining terms are completely determined by the scale of relation. In this way we are certain to obtain the largest or smallest root from the proposed method. For example, given the equation $y^3 - 3y + 1 = 0$, we want to find the largest root. From the scale of relation 0, 3, -1 we begin with 1 and obtain the terms of the recurrent series 1, 0, 3 $-$ 1, 9, $-$ 6, 28, $-$ 27, 90, $-$ 109, 297, $-$ 517, 1000, $-$ 1848, 3517, $-$ 6544, + $\cdots$. This clearly converges to a constant ratio and indicates that the largest root is negative. The approximate value of the root is $y = -\dfrac{6544}{3517} = -1.860676$, while the root should be -1.86793852. We have already discussed why the approximation converges so slowly, namely, since there is a different root which is not much smaller than the largest root and is a positive root.

346. We have carefully considered the great usefulness of this method for finding the roots of equations, we have advised about the general method and also special cases where difficulties may arise. We have also shown some artifices whereby the process may produce results more quickly. It remains only to

discuss the cases when the equation has multiple roots or the roots are complex. We suppose that the denominator of the function

$\dfrac{a + bz + cz^2 + dz^3 + \cdots}{1 - \alpha z - \beta z^2 - \gamma z^3 - \delta z^4 - \cdots}$ has the factor $(1 - pz)^2$ and that the other factors are $1 - qz$, $1 - rz$, etc. The recurrent series which arises has a general term of the form

$$z^n((n + 1)\mathrm{A}p^n + \mathrm{B}p^n + \mathrm{C}q^n + \cdots)$$

We suppose here that n is very large and distinguish two cases. The first case is when p is larger than q, r, etc., while the second case is when p is not the largest root. In the first case, where p is the largest root, due to the coefficient $n + 1$, the remaining terms $\mathrm{B}p^n + \mathrm{C}q^n$ etc. do not vanish so quickly, in comparison, as before. Indeed if q were greater than p, then the term $(n + 1)\mathrm{A}p^n$ vanishes slowly in comparison to $\mathrm{B}p^n$. It turns out that the investigation of the root involves a great deal of work.

EXAMPLE I

Let the given equation be $x^3 - 3x^2 + 4 = 0$, whose largest root is double.

We seek the largest root with the method proposed earlier. Consider the rational function $\dfrac{1}{1 - 3z + 4z^3}$, whose recurrent series has the terms

1, 3, 9, 23, 57, 135, 313, 711, 1593, $\cdots$. We note that the quotient of any term by its predecessor is always greater than 2. The reason for this is easily seen from the general term. When we omit the terms $\mathrm{B}p^n$, $\mathrm{C}q^n$, etc., the coefficient of z^n is equal to $(n + 1)\mathrm{A}p^n + \mathrm{B}p^n$, the next coefficient is equal to $(n + 2)\mathrm{A}p^{n+1} + \mathrm{B}p^{n+1}$. The quotient is $\dfrac{(n + 2)A + B}{(n + 1)A + B}p$ which is always greater then p unless n has already increased to infinity.

EXAMPLE II

Let the equation be $x^3 - x^2 - 5x - 3 = 0$, *whose largest root is 3 and whose other two roots are equal to -1.*

We want to find the largest root by means of a recurrent series, whose scale of relation is 1, 5, 3. We obtain 1, 1, 6, 14, 47, 135, 412, 1228, $\cdots$, which approximates the value 3 with reasonable rapidity. This is because the powers of the smaller root, -1, even when multiplied by $n + 1$, vanish quickly when compared to the powers of 3.

EXAMPLE III

If the given equation is $x^3 + x^2 - 8x - 12 = 0$, whose roots are 3, -2, -2, the largest root emerges much more slowly. The sequence of coefficients for the series is 1, $-$ 1, 9, $-$ 5, 65, 3, 457, 347, 3345, 4915, $\cdots$ and this must be continued much further before it becomes clear that the root approximated is equal to 3.

347. In a similar way, if three factors are equal, so that one of the factors of the denominator is $(1 - pz)^3$ and the others are $1 - qz, 1 - rz$, etc., then the recurrent series has a general term of the form

$$z^n\left(\frac{(n+1)(n+2)}{1\cdot 2}Ap^n + (n+1)Bp^n + Cp^n + Dq^n + Er^n + \cdots\right).$$ If p is the largest root and if n is large enough so that the powers q^n, r^n, etc. vanish in comparison to p^n, then from the recurrent series we obtain the root approximated by $$\frac{\frac{1}{2}(n+2)(n+3)A + (n+2)B + C}{\frac{1}{2}(n+1)(n+2)A + (n+1)B + C}p.$$ Unless n is extremely large,

that is, almost infinite, the quotient will be larger than p. Indeed, the quotient is equal to $p + \dfrac{(n+2)A + B}{\frac{1}{2}(n+1)(n+2)A + (n+1)B + C}p$. If it happens that p is not the largest root, then the calculation of the root is much more difficult. It follows that equations which contain repeated roots are much more difficult to solve by this method of recurrent series than those equations in which the roots are all different.

348. We now consider how the recurrent series is to be arranged when the denominator of the rational function has complex factors. Let the function be $\dfrac{a + bz + cz^2 + dz^3 + \cdots}{1 - \alpha z - \beta z^2 - \gamma z^3 - \delta z^4 - \cdots}$ with real factors $1 - qz$, $1 - rz$, etc. of the denominator and also a trinomial factor $1 - 2pz \cos \phi + p^2z^2$, which contains two linear complex factors. If the recurrent series arising from this function is $A + Bz + Cz^2 + Dz^3 + \cdots + Pz^n + Qz^{n+1}$, then, according to our previous discussions, the coefficient of P is equal to $\dfrac{A \sin(n+1)\phi + B \sin n\phi}{\sin \phi}p^n + Cq^n + Dr^n + \cdots$. If the number p is less than one of the others, q, r, etc., so that the largest root of the equation $x^m - \alpha x^{m-1} - \beta x^{m-2} - \gamma x^{m-3} - \cdots = 0$ is real, then that largest root will be approximated by the recurrent series just as if there were no complex roots.

349. The method of finding the largest real root is not disturbed by the presence of complex roots as long as the product of the two conjugate complex roots in smaller than the square of the largest real root. On the other hand, if

the product of the conjugate complex roots is greater than or equal to the square of the largest real root, then the procedure just investigated will not produce the desired result. The reason is that the power p^n, when compared to the same power of the largest real root, will not vanish even if the series is continued to infinity. In order to illustrate these facts we give the following example.

EXAMPLE I

Let the given equation be $x^3 - 2x - 4 = 0$, *whose largest root we would like to investigate.*

This equation has the two factors $(x - 2)(x^2 + 2x + 2)$. From these we see that there is one real root, 2, and two complex roots whose product is equal to 2, which is less than the square of the real root. It follows that the real root can be found by the method we have given. We form the recurrent series from the scale of relation 0, 2, 4 and find the sequence

1, 0, 2, 4, 4, 16, 24, 48, 112, 192, 416, 832, $\cdots$. From this it is sufficiently clear that the root is 2.

EXAMPLE II

Let the given equation be $x^3 - 4x^2 + 8x - 8 = 0$, *with one real root equal to 2 and two complex roots whose product is 4, which is equal to the square of the real root.*

We try to find the real root by a recurrent series. In order to simplify the process, we let $x = 2y$. We then have $y^3 - 2y^2 + 2y - 1 = 0$, from which we obtain the coefficients of the recurrent series,

1, 2, 2, 1, 0, 0, 1, 2, 2, 1, 0, 0, 1, 2, 2, 1, $\cdots$. Since the same terms continue to repeat, we can conclude nothing except that the largest root is either not real or that there are complex roots whose product is equal to or greater than the square of the real root.

EXAMPLE III

Let the given equation be $x^3 - 3x^2 + 4x - 2 = 0$, whose real root is 1, the product of the two complex roots is equal to 2.

From the relational scale 3, - 4, 2 we form the series with coefficients $1, 3, 5, 5, 1, -7, -15, -15, 1, 33, 65, 65, 1, \cdots$. In this sequence some terms are positive, some are negative and the real root 1 can in no way be recognized.

350. Suppose that the product of two complex roots is p^2 and that this is greater than the square of any real root. It follows that compared to p^n, the other powers q^n, r^n, etc. vanish as n goes to infinity. In this case we let

$$P = \frac{A \sin(n+1)\phi + B \sin n\phi}{\sin \phi} p^n \text{ and}$$

$$Q = \frac{A \sin(n+2)\phi + B \sin(n+1)\phi}{\sin \phi} p^{n+1}, \text{ so that}$$

$$\frac{Q}{P} = \frac{A \sin(n+2)\phi + B \sin(n+1)\phi}{A \sin(n+1)\phi + B \sin n\phi} p.$$

This expression never takes a constant value, even when n goes to infinity, since the sine always oscillates between negative and positive values.

351. In the meantime, if the fractions $\frac{R}{Q}, \frac{S}{R}$ are also calculated, then the constant A and B can be eliminated, and the number n also leaves the expression to give $Pp^2 + R = 2Qp \cos \phi$. From this expression we find

$\cos \phi = \dfrac{Pp^2 + R}{2Qp}$. Likewise we find $\cos \phi = \dfrac{Qp^2 + S}{2Rp}$. From these two expressions we find $p = \sqrt{(r^2 - QS)/(Q^2 - PR)}$ and

$\cos \phi = \dfrac{QR - PS}{2\sqrt{(Q^2 - PR)(R^2 - QS)}}$. It follows that if the recurrent series is continued far enough that in comparison to p^n the powers of the other roots vanish, then we can find the trinomial factor $1 - 2pz \cos \phi + p^2z^2$.

352. Since the calculations which produce the results stated in the previous section may give some difficulty, we here set it out in detail. From the value for $\dfrac{Q}{P}$ we find

$$\mathrm{A}Pp \sin(n + 2)\phi + \mathrm{B}Pp \sin(n + 1)\phi = \mathrm{A}Q \sin(n + 1)\phi + \mathrm{B}Q \sin n\phi,$$

so that $\dfrac{\mathrm{A}}{\mathrm{B}} = \dfrac{Q \sin n\phi - Pp \sin(n + 1)\phi}{Pp \sin(n + 2)\phi - Q \sin(n + 1)\phi}$. Likewise

$$\frac{\mathrm{A}}{\mathrm{B}} = \frac{R \sin(n + 1)\phi - Qp \sin(n + 2)\phi}{Qp \sin(n + 3)\phi - R \sin(n + 2)\phi}.$$

When we equate these two expression we obtain

$$0 = Q^2p \sin n\phi \sin(n + 3)\phi - QR \sin n\phi \sin(n + 2)\phi$$

$$- PQp^2\sin(n + 1)\phi \sin(n + 3)\phi - q^2p \sin(n + 1)\phi \sin(n + 2)\phi$$

$$+ QR \sin(n + 1)\phi \sin(n + 1)\phi + PQp^2\sin(n + 1)\phi \sin(n + 2)\phi .$$

We recall that $\sin a \sin b = \frac{1}{2}\cos(a - b) - \frac{1}{2}\cos(a + b)$, so that

$$0 = \frac{1}{2}q^2p(\cos 3\phi - \cos \phi) + \frac{1}{2}QR(1 - \cos 2\phi) + \frac{1}{2}PQp^2(1 - \cos 2\phi).$$

When this expression is divided by $\dfrac{1}{2Q}$, we obtain

$(Pp^2 + R)(1 - \cos 2\phi) = Qp(\cos \phi - \cos 3\phi)$. We now recall that

$\cos \phi = \cos 2\phi \cos \phi + \sin 2\phi \sin \phi$ and

$\cos 3\phi = \cos 2\phi \cos \phi - \sin 2\phi \sin \phi$,

so that $\cos \phi - \cos 3\phi = 2 \sin 2\phi \sin \phi = 4(\sin \phi)^2 \cos \phi$. Recall also that $1 - \cos 2\phi = 2(\sin \phi)^2$, so that $Pp^2 + R = 2Qp \cos \phi$ and $\cos \phi = \dfrac{Qp^2 + S}{2Rp}$.

From these expressions we obtain the results stated above, that is, $p = \sqrt{(R^2 - QS)/(Q^2 - PR)}$ and $\cos \phi = \dfrac{QR - PS}{2\sqrt{(Q^2 - PR)(R^2 - QS)}}$.

353. If the denominator of the rational function from which the recurrent series is found has several trinomial factors which are equal, then when the form of the general term, which was given above, is considered, it becomes clear that the discovery of roots becomes much more precarious. On the other hand, if any real root is already approximately known, then by a transformation of the equation a much closer approximation can be found. We let x be equal to the sum of the approximate value of the root and y, then in the new equation we look for the smallest root for y. This value is added to the previous approximation for the true value of x.

EXAMPLE

Let the given equation be $x^3 - 3x^2 + 5x - 4 = 0$, *which has a root near 1, as is clear when we let* $x = 1$ *to produce* $x^3 - 3x^2 + 5x - 4 = -1$.

We let $x = 1 + y$ and obtain $1 - 2y - y^3 = 0$. In order to find the smallest root we form a recurrent series with relational scale 2, 0, 1. The sequence of terms is 1, 2, 4, 9, 20, 44, 97, 214, 472, 1041, 2296, $\cdots$, and the smallest root for y is approximately $\dfrac{1041}{2296} = 0.453397$. The approximate value for x is 1.453397, which is very close and there is hardly any other method which would find the root so easily.

354. If any recurrent series finally approximates a geometric progression, then from the law of the progression this fact is easily detected, that is, some root of the equation is the quotient of one term divided by the preceding term. Let $P, Q, R, S, T, \cdots$ be terms of the recurrent series far removed from the beginning, so that it might be taken for a geometric progression. Let $T = \alpha S + \beta R + \gamma Q + \delta P$, that is the relational scale is $\alpha, \beta, \gamma, \delta$. We let $\frac{P}{Q} = x$, then $\frac{R}{P} = x^2$, $\frac{S}{P} = x^3$, and $\frac{T}{P} = x^4$. When these values are substituted in the above equation we obtain $x^4 = \alpha x^3 + \beta x^2 + \gamma x + \delta$, so that it is clear that $\frac{Q}{P}$ is indeed a root of the discovered equation. This, with the previous method, informs us that the quotient $\frac{Q}{P}$ gives the largest root of the equation.

355. This method for finding roots can frequently be of help even if the equation is infinite. As an illustration consider the equation $\frac{1}{2} = z - \frac{z^3}{6} + \frac{z^5}{120} - \frac{z^7}{5040} + \cdots$ whose smallest root z gives the arc of 30 degrees , or one sixth of a semicircle. We write the equation in the form $0 = 1 - 2z + \frac{z^3}{3} - \frac{z^5}{60} + \frac{z^7}{2520} - \cdots$. From this we form a recurrent series with infinite relational scale 2, 0, $-\frac{1}{3}$, 0, $\frac{1}{60}$, 0, $-\frac{1}{2520}$, 0 + $\cdots$, so that the terms of the series are 1, 2, 4, $\frac{23}{3}$, $\frac{44}{3}$, $\frac{1681}{60}$, $\frac{2408}{45}$, $\cdots$. The approximate value for $z = \frac{1681 \cdot 45}{2408 \cdot 60} = \frac{1681 \cdot 3}{2408 \cdot 4} = \frac{5043}{9632} = 0.52356$. From our knowledge of the ratio of the circumference to the diameter the value should be

0.523598 which is an error of only $\frac{3}{100000}$. The method was useful in this case since all of the roots are real and all of the other roots are sufficiently different from the smallest root. Since this is not often the case in infinite equations, the method is seldom useful for their solution.

CHAPTER XVIII

On Continued Fractions.

356. Since we have discussed many facts about both infinite series and infinite products in the preceding chapters, it does not seem to be incongruous that we add something about a third kind of infinite expression, namely, about continued fractions or continuous division. Although this area has been little cultivated up until this time, we have no doubt that it will be much more widely used in the analysis of the infinite in the times to come. From time to time I have already given some examples of this kind, which make this prediction no less probable. It is especially the applications to the branches of arithmetic and common algebra, where I expect this topic to be of considerable aid, and that I briefly indicate and develop in this chapter.

357. By a continued fraction I mean a fraction of such a kind that the denominator consists of the sum of an integer and a fraction whose denominator again is the sum of an integer and a fraction of the same kind. This process can continue indefinitely or can stop at some point. A continued fraction has the following expression: $a + \cfrac{1}{b + \cfrac{1}{c + \cfrac{1}{d + \cfrac{1}{e + \cfrac{1}{f + \cdots}}}}}$

or

$$a + \cfrac{\alpha}{b + \cfrac{\beta}{c + \cfrac{\gamma}{d + \cfrac{\delta}{e + \cfrac{\epsilon}{f + \cdots}}}}}.$$

In the first form the numerators of all fractions are 1, and this is the form we consider first, while the second form can have any number for a numerator.

358. Now that we have given the form of a continued fraction, we next look for an equivalent expression in the usual way of expressing fractions. In order that we may find this more easily, we proceed step by step. We break the continued fraction first into a first fraction, then in a second, next in a third fraction, and so forth. We proceed as follows:

$$a = a$$

$$a + \frac{1}{b} = \frac{ab + 1}{b}$$

$$a + \cfrac{1}{b + \cfrac{1}{c}} = \frac{abc + a + c}{bc + 1}$$

$$a + \cfrac{1}{b + \cfrac{1}{c + \cfrac{1}{d}}} = \frac{abcd + ab + ad + cd + 1}{bcd + b + d}$$

$$a + \cfrac{1}{b + \cfrac{1}{c + \cfrac{1}{d + \cfrac{1}{e}}}}$$

$$= \frac{abcde + abe + ade + cde + abc + a + c + e}{bcde + be + de + bc + 1}.$$

359. Although in the ordinary expression of these fractions it is not easy to find a law of formation according to which the numerator and denominator are

to be composed of the letters a, b, c, d, etc., nevertheless with a bit of attention it becomes clear how one can form a given fraction from its predecessors. We note that any numerator is the sum of the numerator of its predecessor multiplied by the new letter and the numerator of its second predecessor. The same law holds for denominators. When we write in order the letters $a, b, c, d, \cdots$, the fractions formed from them are easily found as follows:

$$\begin{array}{cccccc} & a & b & c & d & e \\ \frac{1}{0} & \frac{a}{1} & \frac{ab+1}{b} & \frac{abc+a+c}{bc+1} & \frac{abcd+ab+ad+cd+1}{bcd+b+d} & \end{array}$$

where each numerator is found if the preceding numerator is multiplied by its superscript and added to the numerator of its second predecessor. The same procedure applies for the denominators. In order that this law may apply from the beginning we have prefixed the fraction $\frac{1}{0}$ which is not a part of the continued fraction, but makes the law of formation clearer. Any of these fractions gives the value of the continued fraction up to that letter which is above its predecessor.

360. In a similar way, the second form

$$a + \cfrac{\alpha}{b + \cfrac{\beta}{c + \cfrac{\gamma}{d + \cfrac{\delta}{e + \cfrac{\epsilon}{f + \cdots}}}}},$$

when broken up gives the following values

$$a = a$$

$$a + \frac{\alpha}{b} = \frac{ab + \alpha}{b}$$

$$a + \cfrac{\alpha}{b + \cfrac{\beta}{c}} = \frac{abc + \beta a + \alpha c}{bc + \beta}$$

$$a + \cfrac{\alpha}{b + \cfrac{\beta}{c + \cfrac{\gamma}{d}}} = \frac{abcd + \beta ad + \alpha cd + \gamma ab + \alpha\gamma}{bcd + \beta d + \gamma b}$$

etc. in which each fraction is found from its two predecessors in the following way.

$$\begin{array}{ccccc} a & b & c & d & e \\ \dfrac{1}{0} & \dfrac{a}{1} & \dfrac{ab + \alpha}{b} & \dfrac{abc + \beta a + \alpha c}{bc + \beta} & \dfrac{abcd + \beta ab + \alpha cd + \gamma ab + \alpha\gamma}{bcd + \beta d + \gamma b} \\ \alpha & \beta & \gamma & \delta & \epsilon \end{array}$$

361. In the above scheme each fraction has a superscript and a subscript. Again the first fraction is $\frac{1}{0}$, and the second is $\frac{a}{1}$. Thereafter, any fraction is formed by multiplying the numerator of the preceding fraction by the superscript and multiplying the second predecessor by the subscript. The new numerator is the sum of these two products. The new denominator likewise is the sum of the product of the denominator of the predecessor by the superscript and the product of the denominator of the second predecessor by its subscript. Any fraction found in this way gives the value of the continued fraction up to and including the denominator which is a superscript of the preceding fraction.

362. If these fractions are continued for as long as superscripts and subscripts are given, then the last fraction gives the true value of the continued fraction. The preceding fractions approximate the true value more and more closely. For this reason they provide a suitable approximation. We suppose that x is the true value of

$$a + \cfrac{\alpha}{b + \cfrac{\beta}{c + \cfrac{\gamma}{d + \cfrac{\delta}{e + \cdots}}}}.$$

It is obvious that the first fraction, $\frac{1}{0}$, is greater then x. The second fraction, $\frac{a}{1}$ is less than x, while the third, $a + \frac{\alpha}{b}$, is greater than x, and the fourth is again less, and so forth. The fractions are alternately greater and less than x. We also note that each fraction is closer to the true value than any of its predecessors. In this way we very quickly and easily obtain an approximate value for x. Although the continued fraction might continue indefinitely, still, provided the numerators α, β, γ, δ, $\cdots$ do not become too large, we obtain a good approximation. If all of the numerators are equal to 1, then there is no problem in the approximation.

363. In order that we may more easily see that the fractions really approximate the true value, we consider the differences in the fractions we have found. To begin with, we pass over the first fraction $\frac{1}{0}$, and we consider the difference beteen the second and the third. This difference is $\frac{\alpha}{b}$. When we subtract the fourth from the third we obtain $\frac{\alpha\beta}{b(bc + \beta)}$. The result of subtracting the fourth from the fifth is $\frac{\alpha\beta\gamma}{(bc + \beta)(bcd + \beta d + \gamma)}$, etc. It follows that we can express the value of the continued fraction by a series whose terms are formed as follows: $x = a + \frac{\alpha}{b} - \frac{\alpha\beta}{b(bc + \beta)} + \frac{\alpha\beta\gamma}{(bc + \beta)(bcd + \beta d + \gamma b)} - \cdots$.

If the continued fraction does not continue indefinitely, then the series breaks off after a finite number of terms.

364. Now we find an expression as alternating series for a continued fraction in which the first letter a vanishes. If

$$x = \frac{\alpha}{b} + \frac{\beta}{c} + \cfrac{\gamma}{d + \cfrac{\delta}{e + \cfrac{\epsilon}{f + \cdots}}},$$

then, by what we have already seen,

$$x = \frac{\alpha}{b} - \frac{\alpha\beta}{b(bc + \beta)} + \frac{\alpha\beta\gamma}{(bc + \beta)(bcd + \beta d + \gamma b)}$$

$$- \frac{\alpha\beta\gamma\delta}{(bcd + \beta d + \gamma b)(bcde + \beta de + \gamma be + \delta bc + \beta\delta)} + \cdots .$$ It follows that if $\alpha, \beta, \gamma, \delta \cdots$ are not increasing, for instance, all equal to 1, and if the denominators $a, b, c, d, \cdots$ are all positive integers, then the series of terms converges very quickly to the value of the continued fraction.

365. We can now consider the converse problem. Given an alternating series, find a continued fraction such that the series representing the value of the continued fraction is the given series. Let the given series be $x = A - B + C - D + E - F + \cdots$. By comparing terms of the given series and the series representing the continued fraction we obtain

$$A = \frac{\alpha}{b} \qquad \text{so that} \quad \alpha = Ab,$$

$$\frac{B}{A} = \frac{\beta}{bc + \beta} \qquad \text{so that} \quad \beta = \frac{Bbc}{A - B},$$

$$\frac{C}{B} = \frac{\gamma b}{bcd + \beta d + \gamma b} \qquad \text{and} \quad \gamma = \frac{Cd(bc + \beta)}{b(B - C)},$$

$$\frac{D}{C} = \frac{\delta(bc + \beta)}{bcde + \beta de + \gamma be + \delta bc + \beta\delta} \qquad \text{and} \quad \delta = \frac{De(bcd + \beta d + \gamma b)}{(bc + \beta)(C - D)},$$

etc. Since $\beta = \dfrac{Bbc}{A - B}$, we have $bc + \beta = \dfrac{Abc}{A - B}$. Then

$\gamma = \dfrac{ACcd}{(A - B)(B - C)}$. Since $bcd + \beta d + \gamma b = (bc + \beta)d + \gamma b$

$= \dfrac{Abcd}{A - B} + \dfrac{ACbcd}{(A - B)(B - C)} = \dfrac{ABbcd}{(A - B)(B - C)}$, we have

$\dfrac{bcd + \beta d + \gamma b}{bc + \beta} = \dfrac{Bd}{B - C}$ and $\delta = \dfrac{BDde}{(B - C)(C - D)}$.

In a similar way we find $\epsilon = \dfrac{CEef}{(C - D)(D - E)}$, and so forth.

366. In order that the law of formation may become clearer we let

$$\begin{aligned}
P &= b \\
Q &= bc + \beta \\
R &= bcd + \beta d + \gamma b \\
S &= bcde + \beta de + \gamma be + \delta bc + \beta\delta \\
T &= bcdef + \cdots \\
V &= bcdefg + \cdots .
\end{aligned}$$

From the law of formation for these expressions we have

$$\begin{aligned}
Q &= Pc + \beta \\
R &= Qd + \gamma P \\
S &= \mathrm{Re} + \delta Q \\
T &= Sf + \epsilon R \\
V &= Tg + \zeta S
\end{aligned}$$

etc.

When we use these letters we have

$$x = \frac{\alpha}{P} - \frac{\alpha\beta}{PQ} + \frac{\alpha\beta\gamma}{QR} - \frac{\alpha\beta\gamma\delta}{RS} + \frac{\alpha\beta\gamma\delta\epsilon}{ST} - \cdots .$$

367. Since we have let

$x = A - B + C - D + E - F + \cdots$, we have $A = \dfrac{\alpha}{P}$ so that $\alpha = AP$;

$\frac{B}{A} = \frac{\beta}{Q}$, so that $\beta = \frac{BQ}{A}$; $\frac{C}{B} = \frac{\gamma P}{R}$, so that $\gamma = \frac{CR}{BP}$; $\frac{D}{C} = \frac{\delta Q}{S}$, so that $\delta = \frac{DS}{CQ}$; $\frac{E}{D} = \frac{\epsilon R}{T}$ so that $\epsilon = \frac{ET}{DR}$; etc. Further, when we take the differences, we have

$$\begin{aligned}
A - B &= \frac{\alpha(Q - \beta)}{PQ} = \frac{\alpha c}{Q} = \frac{APc}{Q} \\
B - C &= \frac{\alpha\beta(R - \gamma P)}{PQR} = \frac{\alpha\beta d}{PR} = \frac{BQd}{R} \\
C - D &= \frac{\alpha\beta\gamma(S - \delta Q)}{QRS} = \frac{\alpha\beta\gamma e}{QS} = \frac{CRe}{S} \\
D - E &= \frac{\alpha\beta\gamma\delta(T - \epsilon R)}{RST} = \frac{\alpha\beta\gamma\delta f}{RT} = \frac{DSf}{T}
\end{aligned}$$

etc. If we consider products of differences we have

$$(A - B)(B - C) = \frac{ABcdP}{R} \text{ and so } \frac{R}{P} = \frac{ABcd}{(A - B)(B - C)};$$

$$(B - C)(C - D) = \frac{BCdeQ}{S} \text{ and so } \frac{S}{Q} = \frac{BCed}{(B - C)(C - D)};$$

$$(C - D)(D - E) = \frac{CDefR}{T} \text{ and so } \frac{T}{R} = \frac{CDef}{(C - D)(D - E)}.$$

Since $P = b$, $Q = \frac{ac}{A - B} = \frac{Abc}{A - B}$, we have

$$\alpha = Ab; \quad \beta = \frac{Bbc}{A - B}; \quad \gamma = \frac{ACcd}{(A - B)(B - C)}; \quad \delta = \frac{BDde}{(B - C)(C - D)};$$

$$\epsilon = \frac{CEef}{(C - D)(D - E)}; \text{ etc.}$$

368. Now that we have found values for the numerators $\alpha, \beta, \gamma, \delta, \cdots$, the denominators $b, c, d, e, \cdots$ can be arbitrarily chosen. It is convenient to choose them in such a way that the values for α, β, γ, δ, etc. will be integers. This of course depends on whether A, B, C, etc. are integers or fractions. If we suppose that these numbers are integers and make the following choices $b = 1$, then $\alpha = A$; $c = A - B$, then $\beta = B$; $d = B - C$, then $\gamma = AC$;

$e = C - D$, then $\delta = BD$; $f = D - E$, then $\epsilon = CE$; etc. It follows that if $x = A - B + C - D + E - F + \cdots$, then the value of x can be expressed as the continued fraction

$$x = \cfrac{A}{1+\cfrac{B}{A-B+\cfrac{AC}{B-C+\cfrac{BD}{C-D+\cfrac{CE}{D-E+\cdots}}}}}.$$

369. If all of the terms of the series are fractions, as for example, $x = \frac{1}{A} - \frac{1}{B} + \frac{1}{C} - \frac{1}{D} + \frac{1}{E} - \cdots$, then we have for $\alpha, \beta, \gamma, \delta, \cdots$ the following values, $\alpha = \frac{b}{A}$, $\beta = \frac{Abc}{B - A}$, $\gamma = \frac{B^2cd}{(B - A)(C - D)}$, $\delta = \frac{C^2de}{(C - B)(D - C)}$, $\epsilon = \frac{D^2ef}{(D - C)(E - D)}$, etc. We choose the arbitrary values as follows:

$b = A, \alpha = 1, \ c = B - A, \beta = A^2,$

$d = C - B, \gamma = B^2, \ e = D - C, \delta = C^2,$

etc. and the continued fraction will be

$$x = \cfrac{1}{A + \cfrac{A^2}{B - A + \cfrac{B^2}{C - B + \cfrac{C^2}{D - C + \cdots}}}}.$$

EXAMPLE I

Let the infinite series $1 - \frac{1}{2} + \frac{1}{3} - \frac{1}{4} + \frac{1}{5} - \cdots$ *be transformed into a continued fraction.*

We let $A = 1, B = 2, C = 3, D = 4, \cdots$. Since the value of the proposed series is log 2, it follows that

$$\log 2 = \cfrac{1}{1 + \cfrac{1}{1 + \cfrac{4}{1 + \cfrac{9}{1 + \cfrac{16}{1 + \cfrac{25}{1 + \cdots}}}}}}.$$

EXAMPLE II

Let the infinite series $\frac{\pi}{4} = 1 - \frac{1}{3} + \frac{1}{5} - \frac{1}{7} + \frac{1}{9} - \cdots$, *where* π *denotes the circumference of a circle with diameter equal to 1, be transformed into a continued fraction.*

When we substitute for $A, B, C, D, \cdots$, the numbers $1, 3, 5, 7, \cdots$,

we obtain $$\frac{\pi}{4} = \cfrac{1}{1 + \cfrac{1}{2 + \cfrac{9}{2 + \cfrac{25}{2 + \cfrac{49}{2 + \cdots}}}}}.$$

If we invert the fraction we obtain $$\frac{4}{\pi} = 1 + \cfrac{1}{2 + \cfrac{9}{2 + \cfrac{25}{2 + \cfrac{49}{2 + \cdots}}}}.$$ This is the expression first found by BROUNCKER as a quadrature of the circle.

EXAMPLE III

Let the given infinite series be

$$x = \frac{1}{m} - \frac{1}{m + n} + \frac{1}{m + 2n} - \frac{1}{m + 3n} + \cdots .$$ Since

$A = m, B = m + n, C = m + 2n, \cdots$, we have the continued fraction

$$x = \cfrac{1}{m + \cfrac{m^2}{n + \cfrac{(m+n)^2}{n + \cfrac{(m+2n)^2}{n + \cfrac{(m+3n)^2}{n + \cdots}}}}}.$$

When this fraction is inverted we obtain

$$\frac{1}{x} - m = \cfrac{m^2}{n + \cfrac{(m+n)^2}{n + \cfrac{(m+2n)^2}{n + \cfrac{(m+3n)^2}{n + \cdots}}}}.$$

EXAMPLE IV

Since in section 178 we found

$$\frac{\pi \cos \frac{m\pi}{n}}{n \sin \frac{m\pi}{n}} = \frac{1}{m} - \frac{1}{n-m} + \frac{1}{n+m} - \frac{1}{2n-m} + \frac{1}{2n+m} - \cdots,$$

we let $A = m$, $B = n - m$, $C = n + m$, $D = 2n - m$, $\cdots$ and write the continued fraction

$$\frac{\pi \cos \frac{m\pi}{n}}{n \sin \frac{m\pi}{n}} = \cfrac{1}{m + \cfrac{m^2}{n - 2m + \cfrac{(n-m)^2}{2m + \cfrac{(n+m)^2}{n - 2m + \cfrac{(2n-m)^2}{2m + \cdots}}}}}.$$

370. If a given series has the form

$$x = \frac{1}{A} - \frac{1}{AB} + \frac{1}{ABC} - \frac{1}{ABCD} + \frac{1}{ABCDE} - \cdots,$$ and we want to express it as a continued fraction, then we have

$$\alpha = \frac{b}{A}, \quad \beta = \frac{bc}{B-1}, \quad \gamma = \frac{Bcd}{(B-1)(C-1)},$$

$$\delta = \frac{Cde}{(C-1)(D-1)}, \quad \epsilon = \frac{Def}{(D-1)(E-1)}, \quad \ldots .$$

If we let:

$$b = A, \quad \text{then } \alpha = 1;$$

$$c = B - 1, \text{ then } \beta = A;$$

$$d = C - 1, \text{ then } \gamma = B;$$

$$e = D - 1, \text{ then } \delta = C;$$

$$f = E - 1, \text{ then } \epsilon = D;$$

etc. It follows that

$$x = \cfrac{1}{A + \cfrac{A}{B - 1 + \cfrac{B}{C - 1 + \cfrac{C}{D - 1 + \cfrac{D}{E - 1 + \cdots}}}}}.$$

EXAMPLE I

Since we have found that

$$\frac{1}{e} = 1 - \frac{1}{1} + \frac{1}{1\cdot 2} - \frac{1}{1\cdot 2\cdot 3} + \frac{1}{1\cdot 2\cdot 3\cdot 4} - \cdots \text{ or}$$

$$1 - \frac{1}{e} = \cfrac{1}{1 + \cfrac{1}{1 + \cfrac{2}{2 + \cfrac{3}{3 + \cfrac{4}{4 + \cfrac{5}{5 + \cdots}}}}}}.$$

In order to avoid asymmetry at the beginning, we write this as

$$\frac{1}{e - 1} = \cfrac{1}{1 + \cfrac{2}{2 + \cfrac{3}{3 + \cfrac{4}{4 + \cfrac{5}{5 + \cdots}}}}}.$$

EXAMPLE II

We have also found that the cosine of an arc which is equal to the radius can be written as

$x = 1 - \frac{1}{2} + \frac{1}{2}\cdot 12 - \frac{1}{2}\cdot 12\cdot 30 + \frac{1}{2}\cdot 12\cdot 30\cdot 56 - \cdots$. If we let

$A = 1,\ B = 2,\ C = 12,\ D = 30,\ E = 56, \cdots$, then we have the continued fraction

$$x = \cfrac{1}{1 + \cfrac{1}{1 + \cfrac{2}{11 + \cfrac{12}{29 + \cfrac{30}{55 + \cdots}}}}},$$

or

$$\frac{1}{x} - 1 = \cfrac{1}{1 + \cfrac{2}{11 + \cfrac{12}{29 + \cfrac{30}{55 + \cdots}}}}.$$

371. Let the series be of the form

$x = A - Bz + Cz^2 - Dz^3 + Ez^4 - Fz^5 + \cdots$, then

$$\alpha = Ab,\ \beta = \frac{Bbcz}{A - Bz},\ \gamma = \frac{ACcdz}{(A - Bz)(B - Cz)},$$

$$\delta = \frac{BDdez}{(B - Cz)(C - Dz)},\ \epsilon = \frac{CEefz}{(C - Dz)(D - Ez)}, \cdots$$. If we let

$b = 1$, then $\alpha = A$;

$c = A - Bz$, then $\beta = Bz$;

$d = B - Cz$, then $\gamma = ACz$;

$e = C - Dz$, then $\delta = BDz$;

etc. so that

$$x = \cfrac{A}{1 + \cfrac{Bz}{A - Bz + \cfrac{ACz}{B - Cz + \cfrac{BDz}{C - Dz + \cdots}}}}.$$

372. In order to obtain a more general result, we let

$$x = \frac{A}{L} - \frac{By}{Mz} + \frac{Cy^2}{Nz^2} + \frac{Dy^3}{Oz^3} - \frac{Ey^4}{Pz^4} + \cdots, \text{ then}$$

$$\alpha = \frac{Ab}{L}, \ \beta = \frac{BLbcy}{AMz - BLy}, \ \gamma = \frac{ACM^2cdyz}{(AMz - BLy)(BNz - CMy)},$$

$$\delta = \frac{BDN^2deyz}{(BNz - CMy)(COz - DNy)}, \ \cdots$$ When we choose the following values for $b, c, d, \cdots$, the values for $\alpha, \beta, \gamma, \cdots$ result:

$$b = L \qquad \alpha = A$$

$$c = AMz - BLy \qquad \beta = BL^2y$$

$$d = BNz - CMy \qquad \gamma = ACM^2yz$$

$$e = COz - DNy \qquad \delta = BDN^2yz$$

$$f = DPz - EOy \qquad \epsilon = CEO^2yz$$

etc. It follows that the given series is represented by the following continued fraction.

$$x = \cfrac{A}{L + \cfrac{BL^2y}{AMz - BLy + \cfrac{ACM^2yz}{BNz - CMy + \cdots}}}.$$

373. Let the given series have the form

$$x = \frac{A}{L} - \frac{ABy}{LMz} + \frac{ABCy^2}{LMNz^2} - \frac{ABCDy^3}{LMNOz^3} + \cdots, \text{ then}$$

$$\alpha = \frac{Ab}{L}, \ \beta = \frac{Bbcy}{Mz - By}, \ \gamma = \frac{CMcdyz}{(Mz - By)(Nz - Cy)},$$

$\delta = \dfrac{DNdeyz}{(Nz - Cy)(Oz - Dy)}$, $\epsilon = \dfrac{EOefyz}{(Oz - Dy)(Pz - Ey)}$, $\ldots$. In order to find integers we choose as follows:

$$b = Lz \qquad \alpha = Az$$

$$c = Mz - By \qquad \beta = BLyz$$

$$d = Nz - Cy \qquad \gamma = CMyz$$

$$e = Oz - Dy \qquad \delta = DNyz$$

$$f = Pz - Ey \qquad \epsilon = EOyz$$

etc. We then obtain the following continued fraction

$$x = \cfrac{Az}{Lz + \cfrac{BLyz}{Mz - By + \cfrac{CMyz}{Nz - Cy + \cfrac{DNyz}{Oz - Dy + \cdots}}}}.$$

374. In this way it is possible to find innumerable continued fractions which continue indefinitely and whose value is known. Since in previous chapters we have discussed infinite series whose sum is known, these series can be transformed into continued fractions with the same value. We have given enough examples to illustrate this point. It would be desirable to find a method by which any given continued fraction could be immediately evaluated. Although a continued fraction can be transformed into an infinite series whose sum can be investigated, still many of those series are so intricate that their sums, even though they may be simple, cannot , or can only with great difficulty, be found.

375. In order that this point may be seen more clearly, we note that there are continued fractions whose value we know from other sources, but from the infinite series into which they are transformed, we obtain no information. For

example, consider the continued fraction $x = \cfrac{1}{2 + \cfrac{1}{2 + \cfrac{1}{2 + \cfrac{1}{2 + \cdots}}}}$,

in which all of the denominators, are equal. If we use the method given above to form a series we have the sequence of fractions

$$\begin{array}{ccccccccc} 0, & 2, & 2, & 2, & 2, & 2, & 2, & \cdots & \\ \frac{1}{0}, & \frac{0}{1}, & \frac{1}{2}, & \frac{2}{5}, & \frac{5}{12}, & \frac{12}{29}, & \frac{29}{70}, & \cdots & . \end{array}$$

From this sequence we have the series

$x = 0 + \frac{1}{2} - \frac{1}{2 \cdot 5} + \frac{1}{5 \cdot 12} - \frac{1}{12 \cdot 29} + \frac{1}{29 \cdot 70} - \cdots$. If we combine the terms two at a time, we have $x = \frac{2}{1 \cdot 5} + \frac{2}{5 \cdot 29} + \frac{2}{29 \cdot 169} + \cdots$ or $x = \frac{1}{2} - \frac{2}{2 \cdot 12} - \frac{2}{12 \cdot 70} - \cdots$. We even have, since

$x = \frac{1}{4} - \frac{1}{2 \cdot 2 \cdot 5} + \frac{1}{2 \cdot 5 \cdot 12} - \frac{1}{2 \cdot 12 \cdot 29} + \cdots$, that

$x = \frac{1}{4} + \frac{1}{1 \cdot 5} - \frac{1}{2 \cdot 12} + \frac{1}{5 \cdot 29} - \frac{1}{12 \cdot 70} + \cdots$. Although this series is strongly convergent, we have no information about its sum.

376. We consider continued fractions of the kind in which the denominators are all equal or are periodically repeated. In this case, if some terms at the beginning are dropped, then the fraction keeps the same value. There is an easy method for finding the value. In the example, since

$$x = \cfrac{1}{2 + \cfrac{1}{2 + \cfrac{1}{2 + \cfrac{1}{2 + \cdots}}}},$$

$x = \dfrac{1}{2 + x}$ and $x^2 + 2x = 1$, so that $x + 1 = \sqrt{2}$. It follows that the value of the continued fraction is equal to $\sqrt{2} - 1$. The sequence of fractions we found above approximate this value more and more closely and very quickly. In fact there is hardly a quicker way of approximating this irrational number with rational numbers. Indeed $\sqrt{2} - 1$ is so close to $\dfrac{29}{70}$ that the error can hardly be detected. Notice that $\sqrt{2} - 1 = 0.41421356236$ while $\dfrac{29}{70} = 0.41428571428$, and that the error is in the hundred thousandth place.

377. Since continued fractions are such a convenient way of approximating the value of $\sqrt{2}$, I will now indicate a very easy way of approximating the square roots of some other numbers. For this purpose we let

$$x = \cfrac{1}{a + \cfrac{1}{a + \cfrac{1}{a + \cfrac{1}{a + \cfrac{1}{a + \cdots}}}}}.$$

Then $x = \dfrac{1}{a + x}$ and $x^2 + ax = 1$. It follows that

$x = -\dfrac{1}{2}a + \sqrt{1 + (1/4)a^2} = \dfrac{\sqrt{a^2 + 4} - a}{2}$. Now we can use continued fractions in order to find the square root of the number $a^2 + 4$. We illustrate this by choosing successively for a the values 1, 2, 3, 4, etc., in order to find $\sqrt{5}$, $\sqrt{2}$, $\sqrt{13}$, $\sqrt{5}$, $\sqrt{29}$, $\sqrt{10}$, $\sqrt{53}$, etc., where we have given the square root in its simplest form.

$$\begin{array}{cccccc} 1, & 1, & 1, & 1, & 1, & 1 \\ \frac{0}{1}, & \frac{1}{1}, & \frac{1}{2}, & \frac{2}{3}, & \frac{3}{5}, & \frac{5}{8} \end{array} \cdots = \frac{\sqrt{5}-1}{2}$$

$$\begin{array}{cccccc} 2, & 2, & 2, & 2, & 2, & 2 \\ \frac{0}{1}, & \frac{1}{2}, & \frac{2}{5}, & \frac{5}{12}, & \frac{12}{29}, & \frac{29}{70} \end{array} \cdots = \sqrt{2}-1$$

$$\begin{array}{cccccc} 3, & 3, & 3, & 3, & 3, & 3 \\ \frac{0}{1}, & \frac{1}{3}, & \frac{3}{10}, & \frac{10}{33}, & \frac{33}{109}, & \frac{109}{360} \end{array} \cdots = \frac{\sqrt{13}-3}{2}$$

$$\begin{array}{cccccc} 4, & 4, & 4, & 4, & 4, & 4 \\ \frac{0}{1}, & \frac{1}{4}, & \frac{4}{17}, & \frac{17}{72}, & \frac{72}{305}, & \frac{305}{1292} \end{array} \cdots = \sqrt{5}-2\,.$$

We should note that the approximation is more rapid the larger the value of a, for instance, in the last example $\sqrt{5} = 2 + \frac{305}{1292}$, where the error is less than $\frac{1}{1292 \cdot 5473}$, where 5473 is the denominator of the next fraction $\frac{1292}{5473}$.

378. This method does not give an approximation of the square root of all numbers, but only those which are the sum of two squares. In order to extend the method to include all numbers we let

$$x = \cfrac{1}{a + \cfrac{1}{b + \cfrac{1}{a + \cfrac{1}{b + \cfrac{1}{a + \cfrac{1}{b + \cdots}}}}}}.$$

Then $x = \cfrac{1}{a + \cfrac{1}{b + x}} = \dfrac{b + x}{ab + 1 + ax}$, and so $ax^2 + abx = b$.

It follows that

$$x = -\frac{1}{2}b \pm \sqrt{(1/4)b^2 + (b/a)} = \frac{-ab + \sqrt{a^2b^2 + 4ab}}{2a}$$. From this expression we can obtain the square roots of all numbers. For example, let $a = 2$ and $b = 7$, then $x = \frac{-14 + \sqrt{14 \cdot 18}}{4} = \frac{-7 + 3\sqrt{7}}{2}$. The approximate value of x is obtained from the following sequence of fractions.

$$\begin{array}{cccccc} 2, & 7, & 2, & 7, & 2, & 7 \\ \frac{0}{1}, & \frac{1}{2}, & \frac{7}{15}, & \frac{15}{32}, & \frac{112}{239}, & \frac{239}{512} \cdots . \end{array}$$

We have the approximate value of $\frac{-7 + 3\sqrt{7}}{2}$ is $\frac{239}{510}$, so that $\sqrt{7} = \frac{2024}{765} = 2.6457516$. The true approximate value of $\sqrt{7}$ is 2.64575131 and the error is less than $\frac{3}{10000000}$.

379. We extend this method even further by letting

$$x = \cfrac{1}{a + \cfrac{1}{b + \cfrac{1}{c + \cfrac{1}{a + \cfrac{1}{b + \cfrac{1}{c + \cfrac{1}{a + \cdots}}}}}}}.$$

It follows that

$$x = \cfrac{1}{a + \cfrac{1}{b + \cfrac{1}{c + x}}} = \cfrac{1}{a + \cfrac{c + x}{bx + bc + 1}}$$

$$= \frac{bx + bc + 1}{(ab + 1)x + abc + a + c} \text{ so that}$$

$$(ab + 1)x^2 + (abc + a - b + c)x = bc + 1 \text{, and}$$

$$x = \frac{-abc - a + b - c + \sqrt{(abc + a + b + c)^2 + 4}}{2(ab + 1)}.$$

We note that the quantity under the radical is again the sum of two squares, so that it is no more helpful than the first method. In a similar way, if four letters, a, b, c, d, are continually repeated in the denominators of continued fractions, then the result is like the second method, which contained only two letters, and so forth.

380. Just as continued fractions are very useful for the extraction of square roots, so also they can be used for solving quadratic equations. It is clear from the above calculation that x is the root of a quadratic equation, but conversely, if x is the root of any quadratic equation, we can use continued fractions to solve the equation. Let $x^2 = ax + b$ be a given quadratic equation, so that $x = a + \frac{b}{x}$. When we substitute in the last term for x the value already obtained we have $x = a + \cfrac{b}{a + \cfrac{b}{x}}$. In a similar way we proceed indefinitely to obtain the infinite continued fraction $x = a + \cfrac{b}{a + \cfrac{b}{a + \cfrac{b}{a + \cdots}}}$.

Since the numerators, b, are not equal to 1, this expression is not too convenient.

381. In order to use continued fractions in calculations we note that any ordinary fraction can be expressed as a continued fraction. Let the fraction be $x = \frac{A}{B}$ in which $A > B$. When A is divided by B we obtain a quotient equal to a and a remainder C. Then the previous divisor, B, is divided by the remainder C to give a quotient b with remainder D. Next we divide C by D.

This operation, which is the usual method for finding the greatest common divisor of A and B, is continued until it comes to an end, as follows. $A = aB + C$, so that $\frac{A}{b} = a + \frac{C}{B}$; $B = bC + D$, so that $\frac{B}{C} = b + \frac{D}{C}$, $\frac{C}{B} = \cfrac{1}{b + \frac{D}{C}}$;

$C = cD + E$, so that $\frac{C}{D} = c + \frac{E}{D}$, $\frac{D}{C} = \cfrac{1}{c + \frac{E}{D}}$; $D = dE + F$, so that

$\frac{D}{E} = d + \frac{F}{E}$, $\frac{E}{D} = \cfrac{1}{d + \frac{F}{E}}$; etc. It follows by substitution that

$$x = \frac{A}{B} = a + \frac{c}{B} = A + \cfrac{1}{b + \frac{D}{C}} = a + \cfrac{1}{b + \cfrac{1}{c + \frac{E}{D}}} = \cdots$$

, so that x can be expressed in terms of $a, b, c, d,$ etc. as follows.

$$x = a + \cfrac{1}{b + \cfrac{1}{c + \cfrac{1}{d + \cfrac{1}{e + \cfrac{1}{f + \cdots}}}}}.$$

EXAMPLE I

Let the given fraction be $\frac{1461}{59}$. We convert this fraction into a continued fraction with all numerators equal to 1. We begin the operation which leads to the greatest common divisor of 59 and 1461.

$$\frac{1461}{59} = 24 + \frac{45}{59}$$

$$\frac{59}{45} = 1 + \frac{14}{45}$$

$$\frac{45}{14} = 3 + \frac{3}{14}$$

$$\frac{14}{3} = 4 + \frac{2}{3}$$

$$\frac{3}{2} = 1 + \frac{1}{2}$$

$$\frac{2}{1} = 2 + 0$$

It follows that $\dfrac{1461}{59} = 24 + \cfrac{1}{1 + \cfrac{1}{3 + \cfrac{1}{4 + \cfrac{1}{1 + \cfrac{1}{2}}}}}$.

EXAMPLE II

Decimal fractions can also be transformed into continued fractions. Since $\sqrt{2} = 1.41421356 = \dfrac{141421356}{100000000,}$ we begin the operation with

$$\frac{141421356}{100000000} = 1 + \frac{41421356}{100000000} \qquad \frac{100000000}{41421356} = 2 + \frac{17157288}{41421356}$$

$$\frac{41421356}{17157288} = 1 + \frac{7106780}{17157288} \qquad \frac{17157288}{7106780} = 2 + \frac{2943728}{7106780}$$

$$\frac{7106780}{2943728} = 2 + \frac{1219324}{2943728} \qquad \frac{2943728}{1219324} = 2 + \frac{505080}{1219324}$$

$\dfrac{1219324}{505080} = 2 + \dfrac{209364}{505080}$ etc. From these calculations we see that the denominators are always 2, so that

$$\sqrt{2} = 1 + \cfrac{1}{2 + \cfrac{1}{2 + \cfrac{1}{2 + \cfrac{1}{2 + \cdots}}}},$$

but this is the expression already found earlier.

EXAMPLE III

Something especially deserving our attention is the number e, whose logarithm is 1, and whose value is 2.718281828459. It follows that

$\frac{e-1}{2} = 0.8591409142295$. If we express this decimal fraction as a continued fraction, using the method given, we obtain the following.

$$\frac{10000000000000}{8591409142295} = 1 + \frac{1408590857704}{8591409142295}$$

$$\frac{8591409142295}{1408590857704} = 6 + \frac{139863996071}{1408590857704}$$

$$\frac{1408590857704}{139863996071} = 10 + \frac{9950896994}{139863996071}$$

$$\frac{139863996071}{9950896994} = 14 + \frac{551438155}{9950896994}$$

$$\frac{9950896994}{551438155} = 18 + \frac{25010204}{551438155}$$

$\frac{551438155}{25010204} = 22 + \frac{1213667}{25010204}$ etc. If the value for e at the beginning had been more exact, then the sequence of quotients would have been 1, 6, 10, 14, 18, 22, 26, 30, 34, $\cdots$ which form the terms of a geometric progression. It follows that

$$\frac{e-1}{2} = \cfrac{1}{1+\cfrac{1}{6+\cfrac{1}{10+\cfrac{1}{14+\cfrac{1}{18+\cfrac{1}{22+\cdots}}}}}} .$$

This result can be confirmed by infinitesimal calculus.

382. Since fractions arise from this operation which very quickly approxi-

mate the value of the expression, this method can be used to express decimal fractions by ordinary fractions which approximate them. Indeed, if the given fraction has a very large numerator and denominator, then a fraction expressed by smaller numbers can be found which does not give the exact value, but is a very close approximation. This is the problem discussed by WALLIS and has an easy solution in that we find fractions, expressed by smaller numbers, which almost equal the given fraction expressed in large numbers. Our fractions, obtained by this method, have a value so close to the continued fraction from which they come, that there are no other numbers, unless they be larger, which give a closer approximation.

EXAMPLE I

We would like to find a fraction which expresses the ratio of the circumference of a circle to the diameter such that no more accurate fraction can be found unless larger numbers are used. If the decimal equivalent $3.1415926535 \cdots$ is expressed by our method of continued division, the sequence of quotients is $3, 7, 15, 1, 292, 1, 1, \cdots$. From this sequence we form the fractions $\frac{1}{0}, \frac{3}{1}, \frac{22}{7}, \frac{333}{106}, \frac{355}{113}, \frac{103993}{33102}, \cdots$. The second fraction already shows that the ratio of the diameter to circumference to be 1:3, and is certainly the most accurate approximation unless larger numbers are used. The third fraction gives the *Archemedian* ratio of 7:22, and the fourth fraction give the *Metian* ratio which is so close to the true value that the error is less than $\frac{1}{113 \cdot 33102}$. In addition, these fractions are alternately greater and less than the true value.

EXAMPLE II

We would like to express the approximate ratio of one day to one solar year in smallest possible numbers. This year is 365 days, 5 hours, 48 minutes, and 55 seconds. That means that one year is 365 $\frac{20935}{86400}$ days. We need be concerned only with the fraction, which gives the sequence of quotients 4, 71, 1, 6, 1, 2, 2, 4 and the sequence of fractions $\frac{0}{1}, \frac{1}{4}, \frac{7}{29}, \frac{8}{33}, \frac{55}{227}, \frac{63}{260}, \frac{181}{747}$. The hours, minutes, and seconds which exceed 365 days make about one day in four years, and this is the origin of the *Julian* calendar. More exact, however, is the eight days in 33 years, or 181 days in 747 years. For this reason, in 400 years there are 97 extra days, while the *Julian* calendar gives 100 extra days. This is the reason that the *Gregorian* calendar in 400 years converts three years, which would be leap years, into ordinary years.

END OF THE FIRST BOOK.